Oil Well Drilling Technology

OIL WELL DRILLING TECHNOLOGY

BY
ARTHUR W. McCRAY
AND
FRANK W. COLE

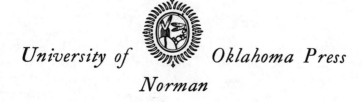

University of Oklahoma Press
Norman

LIBRARY OF CONGRESS CATALOG CARD NUMBER: 59-7486

Copyright © 1959 by the University of Oklahoma Press, Norman, Publishing Division of the University of Oklahoma. Manufactured in the U.S.A. First edition, 1959; second printing, 1960; third printing, 1967; fourth printing, 1973; fifth printing, 1976; sixth printing, 1979; seventh printing, 1981.

Acknowledgments

The technology of drilling has been developed through the efforts of many individuals, companies, and organizations. We freely acknowledge our indebtedness to all who have made these contributions and only regret that we cannot thank each and every individual by name.

We do, however, wish to say a special word of appreciation to Mr. Jack H. Abernathy, past president of the American Association of Oilwell Drilling Contractors, and to Dean W. H. Carson, of the College of Engineering, University of Oklahoma, for their encouragement in this work. We must also extend special thanks to the Education and Training Committee of the A.A.O.D.C. for assistance in making the writing of this book possible, and particularly to Mr. H. W. Davidson, chairman, and to Mr. J. J. Harrigan, who has served most willingly as special liaison member during the preparation of this volume. The general interest of all members of the organization is deeply appreciated.

We also wish to thank the University of Oklahoma for the many facilities it possesses which have greatly aided us in our research and writing.

Arthur W. McCray
Frank W. Cole

Norman, Oklahoma

Table of Contents

CHAPTER

Acknowledgments	v
1. Oil and Gas Wells	3
2. The Rocks and Fluids of Oil and Gas Provinces	5
Petroleum Reservoirs	7
Rocks	11
Rock Properties	13
Fluids Found in the Rocks	14
Flow of Fluids through Rocks	17
Units of Permeability	19
3. Discovery of Petroleum	22
Gravity Methods	23
Torsion Balance	27
Pendulum	28
Gravimeter	29
Interpretation of Gravity Data	31
Magnetic Methods of Geophysical Prospecting	32
Corrections for Reduction of Magnetic Data	34
Interpretation of Magnetic Data	35
Air-borne Magnetometer	36
Seismic Exploration Principles	37
Refraction Seismograph	38
Reflection Seismograph	38
Seismic Equipment	39
Seismograph Equipment	39
Interpretation of Data	39
Radioactive Prospecting Methods	40
Electrical Methods of Geophysical Prospecting	40
Geochemical Methods of Petroleum Prospecting	40
4. Acquisition of Leases	41
Oil and Gas Lease	43
Typical Oil and Gas Lease	49
Government and Indian Land	52
5. Earth Temperatures	53
Geothermal Gradients	53
Mud-Circulation Temperature Gradients	56
Temperature Logs Made after Cementing Casing	58
The Transfer of Heat	59

viii OIL WELL DRILLING TECHNOLOGY

6. Pressure Relations in the Earth and in Bore Holes....... 61
   The Hydrostatic Heads of Liquids..................... 61
   The Hydrostatic Heads of Mud and Cement Slurries..... 63
   Total Overburden Pressure........................... 63
   Formation Pressures................................. 67
   Pressure Relations in Bore Holes..................... 70
7. Drilling Fluids... 76
   The Functions of Drilling Fluids..................... 76
   Descriptions of Drilling Fluids...................... 77
   The Composition of Water-Base Mud.................... 80
   Mud Properties and Their Measurement................ 87
   Mud Mixing and Weighting Calculations............... 103
   The Yield of a Clay................................. 116
8. Planning the Well...................................... 119
9. Power Plants... 125
   Basic Units of Power-Plant Design.................... 126
   Drilling-Rig Power Plants............................ 133
      Steam Engine...................................... 133
      Internal Combustion Engines....................... 137
      Electric Power.................................... 141
      Turbines.. 151
      Free-Piston Engines............................... 152
      Auxiliary Equipment............................... 153
      Cable-Tool Power Plants........................... 153
      Power-Transmission Mechanisms..................... 155
10. Rotary Operations, Drill Strings, and Bits............. 162
    Swivels.. 162
    Kellys... 163
    Rotary Tables.. 164
    Drill Pipe and Tool Joints........................... 166
    Drill Collars.. 170
    Rotary Drill Bits.................................... 172
11. Hoisting Operations.................................... 189
    Hoists and Their Operation........................... 191
    Hoisting Horsepower.................................. 196
    Component Parts of the Hoisting System............... 198
       Derricks.. 198
       Masts... 204
    Construction of Derricks and Masts................... 205
    Derrick Substructures................................ 205
    Miscellaneous Hoisting Equipment..................... 208
    Weight Indicators.................................... 208

Buoyancy	213
Wire-Line Usage	216
Running Casing	226
Coring Operations	227
Evaluation of Wire-Line Service	228
Calculating Length of Line on a Drum	229
Draw Works	230
12. Drilling-Fluid Circulation	236
The Circulating System	236
Drilling-Fluid Circulation Requirements	241
Flow Pressure Losses—Bernoulli's Theorem	242
Types of Flow and Flow Patterns	243
The Reynolds Number	245
The Fanning Friction Factor	247
The Characteristics of Viscous and Plastic Flow	250
Turbulent-Flow Viscosity	260
Flow-Friction Pressure Losses—Turbulent Flow	263
Pressure Drop through Bit Nozzles	265
Hydraulic Horsepower	269
Lifting of Drill Cuttings and Cavings	270
Required Circulation Rates for Deep Air-Gas Drilling	271
13. Drilling Practices	280
Rotary Drilling	280
Choice of Bit	280
Drilling Weights and Rotary Speeds	281
Hydraulic Effects	292
Air-Gas Drilling	304
Straight-Hole Drilling	306
Directional Drilling	315
Diamond Coring	323
Cable-Tool Operations	324
14. Fishing Tools and Practices	331
Fishing for Tubular Products	332
Inside Fishing Tools	332
Outside Fishing Tools	334
Hydraulic and Impact Tools	335
Special Equipment	338
Miscellaneous Fishing Equipment	340
Cable-Tool Fishing Devices	342
15. Development of Drilling Systems	343
Purpose of Drilling	343
Effects of Rock Properties on Bit Performance	343

Theories of Rock Failure	346
Effects of Drilling Fluid on Bit Performance	348
Metallurgical Influence	351
Special Drilling Systems	351
Trends in Development	357
16. Coring	359
Rotary Coring	359
Conventional Coring	361
Wire-Line Coring	364
Diamond Core Drilling	366
Design of Equipment	366
Reverse-Circulation Coring	367
Side-Wall Coring	368
Cable-Tool Coring	368
Cable-Tool Core Barrel	368
Chip Coring	370
Preservation of Cores	370
17. Formation Evaluation	371
Drill-Stem Testing	371
Drill-Stem Test Procedure	372
Logging-Cable Formation Testing	376
Core Analysis	378
Cable-Tool Core Analysis	381
Logging	381
18. Pressure Surges and Anomalies	384
Causes and Effects of Pressure Anomalies	384
Blow-outs Caused by Pressure Reductions During Hoisting	384
Mud Losses Caused by Pipe Movement	385
Pressure—Gel Strength Relationships	389
Viscous-Flow Effects	393
Acceleration Effects	393
Plastic- and Turbulent-Flow Effects	395
Summary	405
19. Casing and Casing-String Design	406
Functions and Requirements of Casing	406
Properties of Casing	407
Testing of Properties	409
Design Safety Factors	410
Biaxial Stresses	414
Mathematical Relations Used in Casing-String Design	416
Graphical Solution of Collapse Change Points	418
Examples of Casing-String Design	420

Analytical Solution of Collapse Change Points	436
Casing Landing Practices	439
20. Cementing Operations	443
Composition of Oil Well Cements	444
Effects of High Pressures and Temperatures on Cement Properties	448
Special Oil Well Cements	449
Bentonitic Cements	450
Pozzolanic Cements	452
Perlite Cements	452
Diatomaceous Earth Cement	453
Gypsum Cements	454
Resin Cements	454
Diesel Oil Cement	454
Contamination of Cement	454
Squeeze Cementing	456
Cementing Methods	456
Stage Cementing	462
21. Drilling Economics	463
Equipment Costs	467
Slim-Hole Drilling	471
Air-Gas Drilling	472
Methods of Reducing Drilling Costs	475
Payment of Drilling Charges	478
General Trends in Drilling Costs	483
About the Figures	487
Index	489

Oil Well Drilling Technology

CHAPTER 1

Oil and Gas Wells

The first well in the United States for the purpose of producing oil was drilled near Titusville, Pennsylvania, in 1859 to a depth of 69 feet, under the supervision of Colonel Edwin L. Drake, and is known as the Drake Well. In this operation, beds of caving gravel were penetrated by drilling ahead and subsequently driving pipe downward a few inches at a time. Shortly afterwards, wells were drilled to a depth of 1,000 feet by rope-suspended iron bars which could be carried on the shoulders of two men. A few years later, strings of iron pipe were lowered into the wells to conduct the oil to the surface. Feed bags filled with dry grain were lashed around the outside of the pipe so that, when the grain became wet and swelled, troublesome water from overlying shallow strata would be sealed off. After the Roberts' torpedo was patented, subsurface explosions were used to create fractures in the rock so that oil could flow more plentifully into the wells. Modern drillers and equipment, capable of drilling to depths greater than 20,000 feet, may sometimes seem unrelated to such crude beginnings; nevertheless, the same qualities of ingenuity, the fascination of conducting unseen underground operations, and the lure of unknown geological formations are as evident today as they have been throughout the past.

The modern drilling industry engages the services of thousands of men and a complicated array of machinery and materials—steel pipe, wire rope, cement, drilling-fluid materials, and the like—and requires a vast network of transportation facilities. At the present time there are approximately 3,600 rotary rigs in the United States and Canada, drilling from 45,000 to 60,000 wells a year, and approximately half of the proceeds from the sale of crude oil is utilized for the drilling of additional wells.

In addition to being a fascinating industry, modern drilling involves great responsibility. On-the-spot decisions which must be made in the course of drilling a well invariably concern the many thousands of dollars invested in the hole already drilled and in the drilling equipment. Thus, responsible supervision and management are based on an understanding of fundamental technological principles as well as human relations. This involves keeping abreast of technological advancements, which have proceeded so rapidly that practices at any one time have seldom been economically competitive with practices a decade later.

Oil and gas wells are a matter of concern to the general public as well as to persons employed in the petroleum industry, for these wells are the conduits through which flow the raw materials that are converted into motor fuels, lubricants, and an increasing variety of petrochemicals. Indeed, the development of petroleum resources since the middle of the nineteenth century has greatly accelerated technological advancement.

These advances made in the drilling industry have been encouraged and in many cases sponsored by various interested organizations, among them the American Petroleum Institute, the American Association of Oilwell Drilling Contractors, the American Institute of Mining, Metallurgical, and Petroleum Engineers, the Petroleum branch of the American Society of Mechanical Engineers, and the Petroleum Industry Sub-Committee, Industry Division of the American Institute of Electrical Engineers. All such organizations have contributed to the growth and effectiveness of the drilling industry through their publications and technical research.

CHAPTER 2

The Rocks and Fluids of Oil and Gas Provinces

Petroleum occurs principally in marine sedimentary rocks. This is an important fact for the petroleum industry and will be discussed in some detail. Although geologists and other scientists are not in complete agreement concerning the precise origin of petroleum, sufficient data are available to enable certain conclusions to be drawn. Since petroleum is organic in nature, organic matter, or material which could be converted to organic matter, must have been present during the oil-accumulation period. Definite processes must have been underway to convert this organic matter into the petroleum deposits which are discovered by the drilling bit. These processes may have been (1) chemical, (2) bacterial, (3) radioactive, or (4) a combination of these three.

Present theories regarding the accumulation of petroleum favor the belief that the source bed is not necessarily the formation in which the deposit is located at the time of discovery. The consensus is that shale is probably a general "source rock" for petroleum, although much evidence can be presented to refute this opinion. It is postulated that petroleum originated in the shales, and then, as the overburden increased and the shales were compacted, it was forced into adjacent rocks that were more permeable than the shale.

The migration and accumulation of petroleum is a subject of primary importance to all persons interested in the petroleum-producing industry. The question, why is petroleum where it is today, if properly answered, could save millions of dollars each year in the search for oil. The "anticlinal" or "structural" theory is generally accepted as the primary explanation for the migration and accumulation of petroleum. Briefly, this theory holds that oil, as it leaves the source bed and begins its period of migration, will move into a more permeable bed, which is already filled with water. The oil will displace the water ahead of it until the oil accumulates, on account of differences in density, in structurally high places, or until some impermeable zone, such as a fault resulting in the juxtaposition of an impermeable formation in the flow path or a sandstone changing to shale, stops the flow of fluids. Therefore, structural feature is the principal consideration, aside from the oil itself, in the location of petroleum deposits, and it is on this assumption that most geophysical prospecting methods are based.

A thorough analysis of forces present during the migration and accumulation of petroleum has been made by M. King Hubbert.[1] He cites laboratory data and field examples which show that where the ground waters are in motion, this movement can have a significant effect upon oil accumulation, the direction of ground-water movement having an appreciable effect upon the location of the oil zone (see Fig. 2–1).

A petroleum reservoir is the permeable rock stratum which the petroleum occupies at the time it is discovered by drilling. Sedimentary rocks, especially sandstone, dolomites, and limestones, are the most common petroleum reservoirs, since most igneous and metamorphic rocks are impermeable. Exceptions to this rule are the

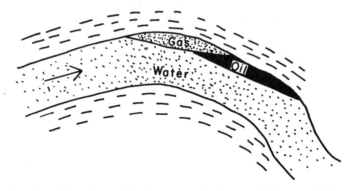

Fig. 2–1. Illustration of the hydrodynamic theory of oil accumulation.

permeable igneous or metamorphic rocks which are in communication with permeable sedimentary rocks from which the petroleum could migrate. This fact has been borne out in the search for petroleum, for several deposits have been found in igneous and metamorphic rocks. However, in all instances, a logical explanation exists for the migration of petroleum into these rocks, an example of which is the petroleum in serpentine rocks at Litton Springs Field in Caldwell County, Texas.

Geological time is an important element to be considered when searching for underground deposits of petroleum. If the structural theory of accumulation of petroleum is important, then the age of the structures in relation to the time of migration of the petroleum is important. If the structure was formed after the petroleum had migrated through the rock strata, there would be no chance for petroleum to

[1] M. King Hubbert, "Entrapment of Oil under Hydrodynamic Conditions," *A.A.P.G. Bulletin*, Vol. XXXVII, No. 8 (1953), 1954.

occur in the structure. This is a possible explanation for many structures which might logically be expected to contain petroleum but do not.

Petroleum has been found in sediments in every period from Precambrian through Pliocene.

Petroleum Reservoirs

In order for petroleum to occur in the isolated deposits in which it is now found, it must have been trapped by some means. As the petroleum migrated from its source bed through other permeable strata under forces of gravity and pressure, it would continue to move unless prevented by some barrier. According to the structural theory of oil accumulation, oil can be trapped as it moves through a permeable strata which may be higher in one spot, on account of folding or for other causes, than in the surrounding area. The oil, which is displacing the formation water ahead of it, moves into this high portion of the strata, and as its specific gravity is less than that of the water, it is isolated in this structurally high position and cannot escape.

Geological features requisite to the accumulation of petroleum are these:

1. A permeable, porous stratum which is the container or reservoir for the petroleum.
2. An overlying impermeable bed.
3. A structural feature, or permeability barrier, or a combination of the two which will "trap" the petroleum, preventing its further migration.

Many methods of classification of petroleum reservoirs have been proposed. However, one of the simplest and most useful to the petroleum engineer is based on the principal feature responsible for trapping the petroleum and will be used in this book. It encompasses three broad classifications: (1) structural trap reservoirs, (2) stratigraphic trap reservoirs, and (3) combination trap reservoirs.

A structural trap reservoir is one in which the principal confining element is the structural position of the rocks. Examples of structural traps are domes, folds, faults, and unconformities. In domes and folds the petroleum occupies a position at the top of the structure and is effectively trapped there. Petroleum is trapped by faults and unconformities when an impervious stratum is in juxtaposition with a permeable stratum containing petroleum. Examples of structural trap reservoirs are illustrated in Figs. 2–2 and 2–3.

Stratigraphic trap reservoirs are those in which the principal confining element is stratigraphic or lithologic. These lateral changes in

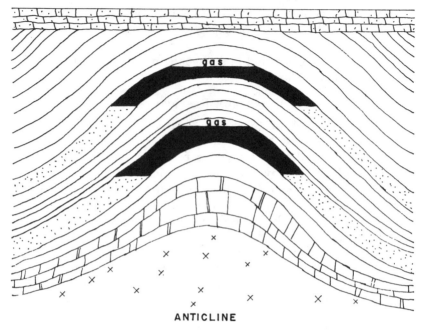

ANTICLINE

FIG. 2-2. Structural trap.

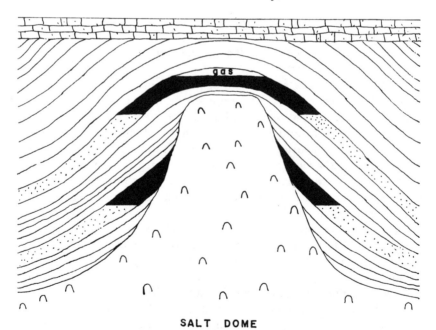

SALT DOME

FIG. 2-3. Structural trap.

lithology in the permeable rock stratum cause the rock to become impermeable, thus effectively preventing the oil from migrating, and so a trap is formed. Many stratigraphic traps will also have some structural influence. Typical stratigraphic traps are the so-called shoestring lenses, formed as sandbars or channel deposits, and reef deposits. Illustrations of typical stratigraphic traps are shown in Figs. 2–4 and 2–5.

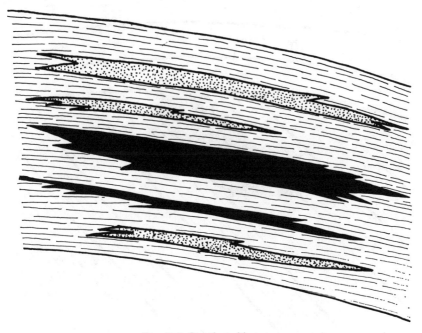

Fig. 2–4. Stratigraphic trap.

The combination trap reservoir is one in which structural features and lithology changes combine to form the trap. Figures 2–6 and 2–7 illustrate typical combination trap reservoirs.

Tilted rock strata present additional problems in drilling operations. Except in special circumstances a vertical or near vertical hole is required. When a much harder formation is encountered in penetrating highly tilted rock strata, the drilling bit will tend to follow the line of least resistance, which in this case will not be vertical. This problem will be discussed more fully in Chapter 13.

An additional factor requisite to the discovery of petroleum is that the deposit must be within reach of the drill bit. Therefore, the search for petroleum can be concentrated in areas which contain (1) marine sedimentary rocks, which are (2) within reach of the drill bit. The latter factor is quite important and presents one of the principal problems encountered in the oil-producing industry. An outstanding example of one solution of this problem is the development of drilling methods in the relatively deep waters of the continental shelves. For many years geologists have been aware of the large marine sedi-

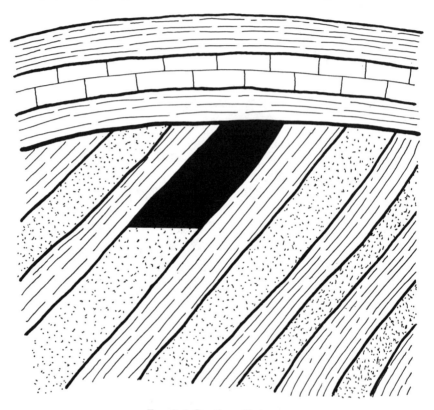

Fig. 2–5. Stratigraphic trap.

mentary deposits covered by waters ranging from a few feet to several thousand feet deep. Piece by piece, equipment has been designed to permit drilling in areas covered by water. Of course, experience was first gained by drilling in areas where the water was relatively shallow, such as lakes and bayous. Then with this experience as a guide, equipment was designed for drilling in deeper water.

As the easily located deposits of petroleum get scarcer and the search for petroleum intensifies, perhaps the most logical new horizons for petroleum are the deeper sediments. There are several obvious advantages in searching for petroleum in deeper horizons: (1) in many cases drilling can be concentrated in areas where shallower production already exists, which reduces the inherent risks; (2) if production already exists in the areas from shallower structures, the cost of developing and producing will be less than if the property were isolated; and (3) geophysical information may already be available to help plan the exploration program.

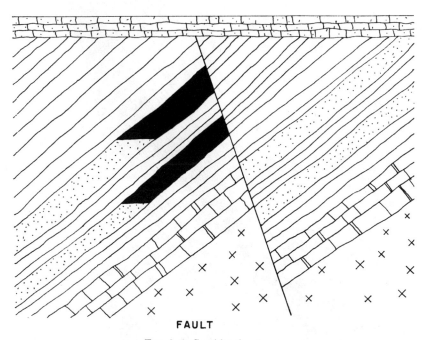

FIG. 2–6. Combination trap.

Serious mechanical problems are encountered in deep drilling operations which must be solved before this type of drilling can be pursued economically. High temperatures, high formation pressures, and large stresses on equipment are problems which are accentuated in deep drilling operations. These problems will be discussed in more detail in Chapters 5 and 6.

ROCKS

As has been said, nearly all rocks containing deposits of petroleum are of sedimentary origin; therefore, a study of petroleum-reservoir rocks can be principally restricted to sedimentary rocks. A. I. Levorsen has suggested a simple, broad classification of reservoir rocks

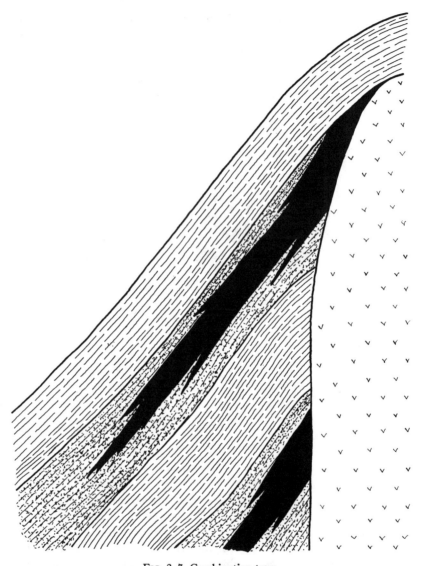

Fig. 2-7. Combination trap.

which is based essentially on the origin of the rock.[2] This classification divides all reservoir rocks into three groups: (1) fragmental (clastic), (2) chemical (precipitated), and (3) miscellaneous. Fragmental rocks are those rocks which have been formed by the bonding together of

[2] A. I. Levorsen, *Geology of Petroleum* (San Francisco, W. H. Freeman and Company, 1954), 51.

mineral and rock particles eroded from older rocks. The cementing or bonding agent is called the matrix. Carbonates, silicate, or clays comprise the principal cementing material in the fragmental rocks. Sandstones, conglomerates, graywackes, and arkoses are the most common fragmental sedimentary reservoir rocks, but fragmental limestones are often encountered. Technically, shale, although not a common reservoir rock, would be considered a fragmental or clastic rock.

The chemically precipitated reservoir rocks are those formed in a quiescent period of gelogical history, when transportation of fragmental material was at a minimum. The principal chemically precipitated reservoir rocks are limestones, dolomites, chalks, and cherty dolomites or limestones, the last of which may contain large proportions of silica.

In the miscellaneous group of reservoir rocks are placed all rocks not listed in the first two classifications. This group would include the permeable igneous and metamorphic rocks, such as basalt flows, serpentine, granite, volcanic tuff, pyroclastic, etc. The amount of petroleum which has been produced from this last group is infinitesimal when compared to the amount produced from the first two groups.

The drilling engineer is interested not only in those reservoir rocks containing petroleum, but also in all other rocks which must be penetrated in order to reach the producing horizon. A knowledge of the composition of these rocks will help him solve many of his most perplexing problems, such as rate of penetration, lost circulation, and abnormal pressures.

Rock Properties

Three reservoir-rock properties of primary importance are porosity, fluid saturation, and permeability. Porosity is defined as that fraction of the bulk volume of the material not occupied by solids. It is a measure of the storage capacity of rocks and is usually expressed on a percentage basis. Two different types of porosity have been defined, absolute porosity and effective porosity. Absolute porosity is the percentage of void spaces in the rock as compared with that of the total rock volume. Effective porosity is the percentage of the interconnected void spaces in the rock as compared with that of the total rock volume. Since the principal concern of the oilman is fluids that can flow through the rock, effective porosity is the porosity used in most work dealing with petroleum engineering. However, it is essential to gain a clear understanding of the differences in the two types of porosity and the methods used to obtain each value. Absolute porosity can be determined using the following formula:

$$\text{Absolute Porosity, per cent} = \frac{\text{Bulk Volume} - \text{Grain Volume}}{\text{Bulk Volume}} \times 100$$

Effective porosity can be determined by using the relationship shown below:

$$\text{Effective Porosity, per cent} = \frac{\text{Interconnected Pore Volume}}{\text{Bulk Volume}} \times 100$$

Several satisfactory methods for determining porosities of rocks have been developed, and it is not considered necessary to repeat these procedures here.

Fluid saturations are important in that they express the percentage of the rock void spaces filled with various fluids. For example, an oil saturation of 30 per cent means that 30 per cent of the pore volume is filled with oil.

Permeability is a measure of the ease of flow of a fluid through a porous medium; it will be discussed in more detail later in this chapter.

Fluids Found in the Rocks

As most petroleum reservoirs are marine in origin, it is logical to suppose that some marine water will be found in these reservoirs. According to the commonly accepted theory of oil migration and accumulation, as the oil moved from the source bed into more permeable strata, the oil displaced the water from the rock. This displacement process was not, however, 100 per cent effective in removing the water from the rock, and some water was retained. This principle is in agreement with current concepts concerning fluid-displacement processes. The distribution of water throughout the petroleum reservoir is governed principally by capillary forces. The distribution of water in a theoretical oil and gas reservoir is shown in Fig. 2–8. The water that was not displaced by the moving petroleum is called connate, or interstitial, water. It is immobile and will not be produced, since, if it could be produced, there is no reason why it should not have been displaced by the petroleum as it invaded the rock. If the water is mobile, then it must be assumed that a water zone is near by.

The water found in petroleum reservoirs varies from brackish to saline, although it is normally saline in character.

The word "petroleum" refers to those organic deposits, either liquid

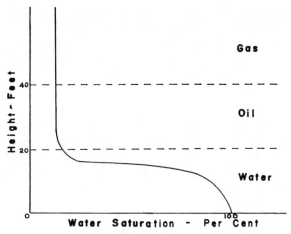

FIG. 2–8. Distribution of water saturation in an oil and gas reservoir.

or gaseous, composed principally of hydrocarbons and located at or below the surface of the earth. The word "hydrocarbon," as the name implies, is a compound consisting of carbon and hydrogen. The simplest hydrocarbons exist in the gaseous phase at normal conditions of temperature and pressure and are called natural gas. Petroleum may also exist in the liquid state, and when it is so found, it is normally called crude oil. Petroleum is a complex mixture of various types and amounts of hydrocarbon compounds, and any attempt to classify it in a simple manner is fraught with hazards. For instance, many hydrocarbon compounds can exist as either gas, liquid, or solid, depending upon the temperature and pressures to which the compounds are subjected.

The hydrocarbons existing in a gaseous phase, or natural gas, as they will be referred to hereafter, are composed principally of

TABLE 2-1
ANALYSIS OF A TYPICAL NATURAL GAS

Component	Volume Percentage
Methane, CH_4	90.12
Ethane, C_2H_6	4.03
Propane, C_3H_8	1.52
n-Butane, C_4H_{10}	0.73
iso-Butane, C_4H_{10}	0.36
Pentane, C_5H_{12}	0.50
Hexane, C_6H_{14}	0.16
Heptane, C_7H_{16} and heavier	0.25
Carbon Dioxide, CO_2	0.60
Nitrogen, N_2	1.73

methane, ethane, propane, and butane. Methane is the lightest component and is the principal component in most natural gases. Heavier hydrocarbons, such as pentane, hexane, and heptane may also be present in smaller amounts. Impurities found in natural gas are nitrogen, hydrogen, sulfur, and oxygen. The density of a typical natural gas is less than the density of air. Composition of a typical natural gas is shown in Table 2–1.

Hydrocarbons are classed according to their molecular structure. Each series is characterized by the ratio of hydrogen to carbon atoms and the method of bonding between carbon-carbon atoms and carbon-hydrogen atoms. The most important classes, or series, of hydrocarbons are shown in Table 2–2.

TABLE 2–2
CLASSES OF HYDROCARBONS*

Name	Formula
Paraffin	$C_n H_{2n+2}$
Naphthene	$C_n H_{2n}$
Benzene	$C_n H_{2n-6}$
Olefin	$C_n H_{2n}$
Diolefin	$C_n H_{2n-2}$
Acetylene	$C_n H_{2n-2}$

* Partial list embodying only the principal classes of hydrocarbons.

Crude oils are more complex mixtures of hydrocarbons, which are liquid in their natural state in the reservoir. Attempts have been made to classify crude oils on the basis of the principal types of hydrocarbons present in the crude. However, even the most complete classification would still be inadequate, as no crude oil has ever been completely separated into its individual components. A broad classification which has apparently been satisfactory for most general uses includes the paraffin-base, asphaltic-base, and mixed-base crudes. These terms indicate the composition of the principal hydrocarbon compounds in the crude. Paraffin-base crudes contain principally the paraffinic hydrocarbons; asphaltic-base crudes are composed predominantly of the naphthenic hydrocarbons, and mixed-base crudes are those crudes which are intermediate in composition. Paraffin-base crudes are usually lighter in color and have a lower specific gravity than the asphaltic-base crudes.

Crude oils may also contain, in small amounts, compounds other than hydrocarbons. The impurities found in crude oils are essentially the same as those found in natural gas, namely sulfur, nitrogen, and oxygen.

Certain physical properties of crude oils are normally used to identify the crude. The most commonly measured poperties are (1) specific gravity (or API gravity), (2) viscosity, (3) color, and (4) odor. The specific gravity is a dimensionless number which is a measure of the density of a substance compared to the density of pure water at an arbitrary temperature and pressure. In the United States this reference temperature and pressure is 60 degrees Fahrenheit and one atmosphere. An arbitrary gravity scale which has been developed and is in use throughout the United States and many other parts of the world is the API (American Petroleum Institute) gravity scale. The formula for converting API gravity to specific gravity is:

$$\text{Specific Gravity} = \frac{141.5}{131.5 + °\text{API}}$$

The European Baumé gravity scale is also abritrary and the conversion formula is shown below:

$$\text{Specific Gravity} = \frac{140}{130 + °\text{Baumé}}$$

As the API or Baumé gravity increases, the specific gravity decreases. Tables are available for correcting the observed readings when the temperature is not exactly 60 degrees Fahrenheit. The gravity of crude oils is the most commonly determined property of a crude oil principally because most crude-oil pricing structures are related to gravity.

Properties of drilling fluids may be radically changed by contamination with the reservoir fluids. In many cases it has been necessary to discard complete mud systems on account of contamination by the reservoir fluids. Therefore a knowledge of the nature of the reservoir fluids and the effects of these fluids on the drilling muds and the drilling activities is essential in order to cope with these problems as they arise. More complete treatment of these problems is given in Chapter 7.

FLOW OF FLUIDS THROUGH ROCKS

In order to produce oil or gas in quantities sufficient to pay the cost of drilling and completing a well, the oil or gas must flow from relatively long distances through the reservoir rock into the well bore through small pore openings, often of capillary size.

The French engineer Henri Dárcy is responsible for development of the basic theory concerning flow of fluids through porous media. Although he worked with flow through sand filters used in city water systems, his conclusions have been found to apply equally well to viscous flow through any porous medium. Dárcy found that the velocity of flow through any porous medium is proportional to the pressure drop per unit length, and inversely proportional to the viscosity of the flowing fluid. His finding, stated in differential form, is

$$v = -\frac{k}{\mu}\frac{dp}{dl}$$

where

$v =$ apparent velocity (found by dividing the flow rate by the cross-sectional area across which flow occurs)
$k =$ proportionality constant (permeability)
$\mu =$ viscosity of flowing fluid
$\dfrac{dp}{dl} =$ pressure drop per unit length

The minus sign indicates that dp and dl are measured in opposite directions.

Dárcy found that the proportionality constant, k, for a given system was independent of the type of fluid flowing, the flow rate, or pressures imposed. In other words, it was a physical property of the porous medium. Dárcy's original work has been verified by a large number of experiments by later research workers, the only modification being when L. J. Klinkenberg presented data showing that permeabilities measured when gas was the flowing fluid were higher than when liquid was the flowing fluid.[3] He also found that as the pressure increased, the gas permeability decreased. Klinkenberg advanced the theory that permeability is a function of the mean free path of the molecules in the flowing fluid. As the pressure increased, the mean free path decreased, and the gas behaved more like a liquid. Klinkenberg also found that when a liquid is flowing through a porous medium, the liquid has a zero velocity at the solid surface; however, when gas is flowing, the gas has some finite velocity at the solid surface. He found further that when the apparent permeability as determined with a gas is plotted versus the reciprocal of the mean flowing pressure, a straight-line relationship exists. When this line is extrapolated to the point where the reciprocal of the mean pressure is zero (i.e., infinite mean pressure), the extrapolated gas permeability will equal the liquid permeability.

[3] L. J. Klinkenberg, "The Permeability of Porous Media to Liquids and Gases," *API Drilling and Production Practice* (1941), 200.

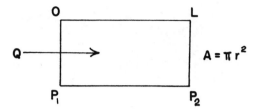

Fig. 2-9. Linear flow model.

The differential form of Dárcy's Law can be expanded to fit any particular geometric flow pattern, provided the limits of the system can be described and integration of the resulting equation is possible. For example, flow through linear systems, radial systems, hemispherical systems, and spherical systems can be described quite simply by expanding the basic Dárcy equation. The linear flow system is the type of flow commonly used in the laboratory. Consider a porous medium with the dimensions shown in Fig. 2-9. Beginning with the Dárcy equation:

$$\frac{Q}{A} = v = -\frac{k}{\mu}\frac{dp}{dl}$$

separating the differential terms and integrating between the limits in Fig. 2-9 results in the following expression:

$$\frac{Q}{A}\int_0^L dl = -\frac{k}{\mu}\int_{P_1}^{P_2} dp \qquad (2\text{-}1)$$

integrating

$$\frac{Q}{A}(L) = -\frac{k}{\mu}(P_2 - P_1)$$

eliminating the minus sign

$$\frac{Q}{A}(L) = \frac{k}{\mu}(P_1 - P_2)$$

and solving for k:

$$k = \frac{Q\mu L}{A(P_1 - P_2)} \qquad (2\text{-}2)$$

Units of Permeability

When a fluid with a visocisity of one centipoise is flowing at a rate of one cubic centimeter per second across a cross-sectional area of one square centimeter, under a pressure differential of one atmosphere per centimeter of length, the permeability of the porous medium is one Dárcy.

One Dárcy is a high permeability, considering the permeability of most petroleum reservoirs, and in order to avoid the use of fractions, most permeabilities are reported in millidarcies. A millidarcy is one-thousandth of a Dárcy.

Equation (2–2) is valid for flow of an incompressible fluid; however, when a compressible fluid is flowing, the mass rate of flow measured at the outflow end of the porous medium is not the same as the mass rate of flow within the porous medium. If the compressible fluid obeys Boyle's Law, the average mean flow rate through the porous medium can be determined in the following manner:

$$P_a Q_a = P_m Q_m = PQ \qquad (2\text{-}3)$$

where

Q_a = flow rate at outflow end.
P_a = pressure at outflow end.

P_m = mean flowing pressure = $\dfrac{P_1 + P_2}{2}$.

Q_m = mean flow rate at Pm

A point can be selected at some spot in the core where the flow rate is Q, and the pressure at that point is P. Integration of the basic Dárcy equation can be made on the basis of this point. To do this, both sides of equation 2–1 are multiplied by P, which results in

$$\frac{PQ}{A}\int_0^L dl = -\frac{k}{\mu}\int_{P_1}^{P_2} P\,dp$$

$P_m Q_m$ can be substituted for PQ:

$$\frac{P_m Q_m}{A}\int_0^L dl = -\frac{k}{\mu}\int_{P}^{P_2} P\,dp$$

Integrating:

$$\frac{P_m Q_m}{A}(L) = -\frac{k}{\mu}\frac{(P_2^2 - P_1^2)}{2}$$

simplifying and eliminating the minus sign:

$$k = \frac{Q_m P_m \mu L}{A \dfrac{(P_1 - P_2)(P_1 + P_2)}{2}}$$

However,

$$P_m = \frac{P_1 + P_2}{2}$$

then

$$k = \frac{Q_m \left(\dfrac{P_1 + P_2}{2}\right) \mu L}{A \dfrac{(P_1 - P_2)(P_1 + P_2)}{2}}$$

Again simplifying:

$$k = \frac{Q_m \mu L}{A(P_1 - P_2)} \qquad (2\text{-}4)$$

Equation (2-4) is the expression used when a compressible fluid is flowing. To determine the permeability, the flow rate used in the equation must be the mean rate of flow.

An important factor in drilling through the reservoir rocks is their permeability. When drilling through extremely permeable zones, loss of circulation can occur. In permeable zones there may also be abnormally high mud filtrate losses, which in turn cause thick mud cakes to form on the wall of the hole being drilled and present potential problems such as stuck drill pipe. When drilling extremely permeable zones with air, reservoir fluids may flow into the well bore at such large rates that the air cannot remove cuttings from the hole fast enough to permit drilling to continue with air.

CHAPTER 3

Discovery of Petroleum

The occurrence, at the surface, of petroleum in some form, either solid, liquid, or gas, is a direct indication of the presence of petroleum in an area. Surface indications of petroleum are common throughout many parts of the world, and in the early history of the petroleum industry, drilling was concentrated in these areas. As the areas near surface indications were more or less thoroughly explored and it became necessary to employ other methods for locating petroleum reservoirs, increasing use was made of surface and subsurface geology. For many years these were valuable oil-finding tools for the petroleum industry, and, in fact, they remain important today, although subsurface geological studies are perhaps more important than surface geological studies, principally because much surface geological work has already been done.

As the demand for petroleum increased and most of the easily located structures had been tested, it became obvious that new methods of locating petroleum reservoirs were needed. In many areas of the world, large deposists of sedimentary beds of marine origin were known to exist, but these sediments and the structural features in them were overlaid by relatively thick horizontal beds resting unconformably on the older surface, or else the surface outcrops were difficult to map, so that very little surface geological work could be accomplished.

Since the structural theory of petroleum accumulation had been well established and the elevation of sediments was in many cases due to a rise in the basement rocks, the petroleum industry adopted certain mining techniques used for locating ore deposits. These methods all involved measurement of certain basic physical phenomena of the earth. Dobrin defines geophysical prospecting as "the art of searching for concealed deposits of hydrocarbons or useful minerals by measurement, with instruments on the surface, of the physical properties of materials within the earth."[1] Geophysical prospecting for oil began in the early twentieth century and at the present time is a very important tool in petroleum prospecting.

A relatively few basic laws of physics, such as Newton's Law, Ohm's Law, and Snell's Law, form the basis for most geophysical

[1] Milton B. Dobrin, *Introduction to Geophysical Prospecting* (New York, McGraw-Hill Book Company, Inc., 1952), 1.

prospecting techniques. All of these methods involve measurement of some physical property of the underlying sediments, in which a change in a measured value indicates the possibility of some change in the underlying sediments. Thus, changes in measured values, or "anomalies" as they are often called, interest the petroleum prospector because they may represent structural changes within the earth. The principal effort in geophysical exploration is the location of structures in areas potentially favorable for oil accumulation. Geophysical prospecting is, then, an indirect method of oil finding.

Both surface and subsurface techniques are used in geophysical exploration. The principal methods which have been utilized are these:

Surface:
 1. Gravity
 2. Magnetic
 3. Seismic
 4. Radioactive
 5. Electrical
 6. Geochemical

Subsurface:
 1. Drilling time logs
 2. Caliper logs
 3. Density logs
 4. Magnetic logs
 5. Continuous velocity logs
 6. Fluorologs
 7. Electrical logs
 8. Radioactivity logs
 9. Geothermal correlations
 10. Gas analysis of drilling mud

Subsurface geophyiscal exploration tools will not be discussed here because they are all directed toward the interpretation of data obtained from well bores and the petroleum engineer will become familiar with their use and application in the couse of his work. However, since he may not come into direct contact with the surface geophysical methods, principles of the major techniques will be outlined briefly.

Gravity Methods

Gravity methods are designed to detect variations in the gravitational field of the earth caused by density variations of rocks and thus locate anomalies several miles below the surface. There are notable differences in densities of the various types of rocks which comprise the earth's crust. Table 3–1 shows typical densities of some of the rocks in the earth.

TABLE 3–1
DENSITY OF ROCKS IN THE EARTH

Rock	Average Density
Unconsolidated sediments	1.5–2.0
Shale	1.5–2.8
Sandstone	1.8–2.5
Salt	2.2–2.4
Limestone	1.8–2.5
Granite	2.5–2.8
Basalt	2.4–3.1

Sir Isaac Newton, in experiments dealing with gravitational acceleration, first expressed the gravitational attraction between two bodies in the form of a definite equation. Newton's Law of Gravitation states that the force of attraction between two masses, when placed some distance, s, apart from each other, is proportional to the product of the two masses and inversely proportional to the square of the distance between them. The mathematical expression of this postulate is shown below:

$$F = \frac{\gamma m_1 m_2}{s^2} \tag{3-1}$$

where

F = force, dynes
m_1 = mass, grams
m_2 = mass, grams
s = distance between centers of the masses, centimeters
γ = proportionality constant, commonly known as the universal gravitational constant

The first actual laboratory measurement of the universal gravitational constant was made in 1798 by Lord Henry Cavendish. The equipment developed by him, shown in Fig. 3–1, consisted of a small weight suspended at each end of the rod. Large external weights were suspended near each of the small weights in such a manner that their attraction would cause the beam to rotate. When the torsional constant of the suspending wire, the distance between weights, and the masses of the two weights were known, the universal gravitational constant could be calculated. Lord Cavendish found the universal gravitational constant to be 0.00000006658. His experiment has been duplicated recently, and the currently accepted value of 0.0000000667 is remarkably close to the original value determined by Cavendish.

FIG. 3-1. Cavendish torsion balance.

When m_1, m_2 and s, are each unity, then $F=\gamma$. The equipment used by Cavendish was probably the first successful gravity measuring device, and provided the basic design of the torsion balance, a standard instrument used in gravitational measurements.

Newton's Law of Gravitation can be combined with his Second Law of Motion to show the acceleration of a body.

From Newton's Second Law of Motion:

$$F = ma \qquad (3\text{-}2)$$

where

F = force, dynes
m = mass of object, gm.
a = acceleration, cm./sec.^{2}

or

$$a = \frac{F}{m} \qquad (3\text{-}3)$$

From Equation (3-1),

$$\frac{F}{m_1} = \frac{\gamma m_2}{s^2} \qquad (3\text{-}4)$$

therefore

$$a = \frac{\gamma m_2}{s^2} \qquad (3\text{-}5)$$

The acceleration determined in this manner is a measure of the gravitational force acting at any point. The gravitational acceleration at the earth's surface has been found to be about 980.6 cm./sec.² at 45 degrees latitude and sea level. It should be remembered that this number is not a constant, but depends on the latitude, and is a minimum at the equator. A common unit of gravitational acceleration is the "gal" (in honor of Galileo), which is equivalent to the acceleration in the c.g.s. (centimeter-gram-second) system. Therefore, the gravitational acceleration at the earth's surface is also approximately 980.6 gals at 45 degrees latitude and sea level. In geophysical work, interest lies in the change in gravitational acceleration at any point, and the gal becomes a rather large and cumbersome unit of measurement. The milligal (mg.) is the most commonly used gravitational measuring unit (1 mg. = 0.001 gal). The variation in the earth's gravitational acceleration is a principal point of interest to those searching for potentially oil-bearing structures. Variations in gravitational acceleration of as little as one-tenth of a milligal often indicate structures of economic importance; therefore, any gravity measuring instrument must have a minimum sensitivity of at least one-tenth of a milligal in order to be an effective geophysical prospecting tool.

Before gravity data can be properly interpreted, it must be corrected for (1) drift or tidal effect, (2) free-air effect (elevation differences), (3) Bouguer effect, (4) terrain effect, (5) normal or latitude effect, (6) regional effect (under certain conditions), and (7) isostatic effect (under certain conditions).

Tidal corrections. The effects of the sun and moon on the water masses of the earth are well known. The periodic rising and falling of large bodies of water are caused by the attraction of the sun and moon. The earth itself is also acted on by these two bodies in the same manner, although to a much smaller degree. These changes in the earth are, however, detectable by sensitive gravity-measuring devices, and readings must be corrected for earth tidal gravity variations. The tidal variation is normally less than 0.3 mg.

Free-air corrections. Gravity readings must be corrected for the elevation at the observed point, for as the distance from the center of the earth increases, the gravitational acceleration is reduced. This correction is often referred to as the free-air correction because it does not take into consideration the attraction of any material above the base level. The free-air correction is added to the observed reading if the station is above base level, and subtracted if the station is below base level.

Bouguer correction. The Bouguer correction is designed to correct for the gravitational attraction of the rocks above the base level, so that the corrected gravity data can be reduced to a common base level.

Terrain corrections. In hilly country it is necessary to correct for variations in the terrain. The material which is higher than the station will cause a force component in the upward direction, opposing the force of gravity. To correct for this, the correction must be added to the observed reading. A depression, on the other hand, will cause a lower observed reading, and to compensate for depressions, a correction must be added. It is interesting to note that the terrain correction is always *added* to the observed reading, regardless of whether the topographic feature is a depression or a high point.

Latitude corrections. The earth is not a perfect sphere, the equatorial radius being slightly larger than the polar radius. The acceleration of gravity is less at the equator than at the poles because the centrifugal effect of the earth's rotation is at a maximum at the equator and also because the distance to the center of the earth is greater at the equator. The centifugal force opposing the gravitational force results in a reduced gravitational acceleration, and the greater distance from the equator results in a reduced gravitational acceleration. Although both of these effects reduce the gravitational acceleration, the over-all reduction is modified somewhat by the additional mass that lies between the equator and the center of the earth. The actual difference in gravity between the equator and the poles is about five gals.

Regional corrections. In many areas, large deep-seated structures may exist which tend to obscure the gravitational changes caused by the smaller features which are of more immediate concern to the geophysicist. It is common practice to eliminate these regional changes in gravitational attraction in order to study in more detail the smaller structural features.

Isostatic corrections. Stated very simply, the principle of isostasy means that the material underlying mountains and other features having a high elevation must be less dense than the material underlying the low-lying sediments, such as those underneath the ocean. Proof of the principle is supplied by many gravity measurements, and in certain instances it may be necessary to correct observed gravity data for this phenomenon.

Three different types of gravity measuring instruments have been used in petroleum geophyiscal prospecting: (1) the torsion balance, (2) the pendulum, and (3) the gravity meter.

Torsion Balance

The torsion balance, which was discussed briefly earlier in this chapter, was the first successful gravity-measuring instrument. It measures a change in the gravitational field rather than the actual intensity of the field. The Cavendish Balance was the original torsion balance. The Eötvös Torsion Balance was a modification of the original Cavendish Balance, and is shown in Fig. 3–2. The balance consists essentially of two weights in different horizontal planes, attached to a horizontal rod which is suspended by a torsion wire. The standard Eötvös Balance measures the change in gravitational acceleration, or gravitational gradient, between the two ends of the

Fig. 3–2. Eötvös torsion balance.

instrument. When the earth's gravitational field is changing, a horizontal force causes the balance to rotate, the magnitude of rotation being dependent upon the resistance of the torsion wire to turning as well as upon the gravitational gradient.

Torsion balances are very sensitive instruments, and only a limited number of readings can be obtained in a day's time. In addition, correcting for terrain effects requires tedious and time-consuming computations. These two factors have limited the use of the torsion balance in petroleum geophysical prospecting, especially since the development of more rapid methods of gravity prospecting. The torsion balance was introduced on the Gulf Coast in 1922, and the Nash Salt Dome, located in 1924, was the first structure in the United States to be located by geophysical prospecting instruments. The torsion balance was quite successful in locating buried salt domes, and in the period 1924–32, many such salt domes were located by the torsion balance.

Pendulum

The pendulum, a rigid rod supported at one end in such a manner that it is free to swing in one plane about the point of support, has been used to a limited extent in petroleum prospecting. The period of a pendulum can be expressed by the equation

$$T = 2\pi \sqrt{\frac{I}{mgh}} \qquad (3\text{–}6)$$

where

T = period of the pendulum
I = moment of inertia about the pendulum support
m = mass of the pendulum
g = gravitational acceleration
h = distance from center of gravity to the point of support

If the period of the pendulum is known, the gravitational acceleration can be directly computed. In practice, rather than measuring the actual gravitational acceleration at any point, the difference in gravitational acceleration at two points is determined by noting the difference in periods at the two stations. The pendulum has never been used extensively in petroleum prospecting work because only a limited number of readings could be made each day and the sensitivity of the instrument was not great enough. In the United States the gravity pendulum developed by the Gulf Oil Corporation has been used in prospecting for oil.

Gravimeter

The gravimeter, or gravity meter, measures small variations in gravitational acceleration by recording the differences in weight of a constant mass when subjected to changes in the gravitational field. These changes in weight are a measure of the change in the gravitational attraction between two points. There are two fundamental types of gravimeters, the stable and the unstable. All gravimeters are basically alike in that they are a mass supported in some manner by a spring. This instrument has essentially replaced all other methods of measuring gravity in petroleum prospecting.

Stable Gravimeters

The Hartley gravimeter, shown in Fig. 3–3 was one of the earliest and simplest gravimeters. Changes in weight of the mass, W, caused deflection of beam, B. The beam was returned to its original position by means of a micrometer screw, M, and adjusting spring, A. The mechanics of this system are such that a relatively large adjustment of the micrometer screw is necessary to produce a small change in the beam's position. The sensitivity of this instrument is only about one milligal, which is not sensitive enough for most prospecting.

Gulf Gravimeter

The Gulf gravimeter uses a coiled flat spring for measuring gravitational changes. A weight is attached to the coiled spring, and as the gravitational field changes, the weight of the mass changes, causing the coiled spring to elongate or contract. The rotation of the spring is measured with an optical system which greatly magnifies the motion. The sensitivity of the Gulf instrument is approximately 0.025 milligal, which is entirely satisfactory for most petroleum geophysical prospecting.

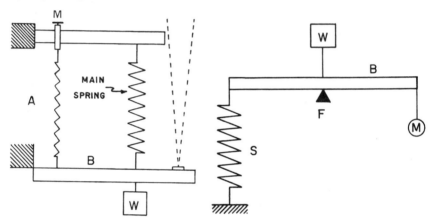

FIG. 3–3. Hartley gravimeter.   FIG. 3–4. Thyssen gravimeter.

Unstable Gravimeters

Thyssen gravimeter: The principle of unstable gravimeters is illustrated clearly in the Thyssen instrument, shown in Fig. 3–4. The mass, M, is suspended from a rigid bar, B, supported by a fulcrum, F, and a spring, S. An auxiliary weight, W, located directly above the fulcrum, is responsible for the unstable characteristics of this instrument. As the beam is tilted by changes in the weight of mass, M, the auxiliary weight moves into such a position that a moment of inertia is created, thus increasing the tilting of the beam. The sensitivity of the Thyssen gravimeter is about 0.3 milligal.

Humble gravimeter: The Humble gravimeter, shown in Fig. 3–5 is an unstable type of gravimeter. Instability is provided by the auxiliary spring, A, which, when the instrument is in equilibrium, acts through the support point, B. When the gravitational acceleration changes, the mass, M, causes movement of the beam, thus shifting the axis of the auxiliary spring away from the support point and creating a moment of inertia, amplifying the deflection of the beam. This instrument has a sensitivity of about 0.2 milligal.

In addition to the gravity corrections which must be made, gravity meters contain certain inherent errors. An important element in the design of gravimeters is temperature. Most of the instruments are extremely sensitive to changes in temperature, and some method of maintaining a constant temperature within the instrument is usually required. All gravity meters are subject to drift, which is the change

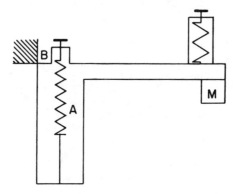

Fig. 3–5. Humble gravimeter.

in gravity readings on account of changes in the instrument itself, such as spring creep. The drift correction must be determined in order to obtain the true reading. Drift is usually determined by taking two readings at each station during the day, the time of the two readings being as far apart as practical, or the instrument can be returned several times each day to a reference station, with a drift curve being determined from these readings.

Gravity meters are usually calibrated at two points where the relative or absolute values of gravity are known precisely. These absolute values can be determined in many areas with a pendulum.

Interpretation of Gravity Data

When the gravity data have been obtained and the proper corrections made, the results are plotted on a map and contoured. However, this map cannot be interpreted as if it were a structure map. In an area where there is no independent information, it is impossible to translate gravity data into reliable estimates of the structure.

Since the density variation in the subsurface rocks is the principal factor governing the interpretation of gravity data, a knowledge of the density of the various rocks which might be encountered is essential. As shown in Table 3–1, the density of some sedimentary rocks may approach that of some of the basement igneous rocks which may be responsible for structural formations. When this is the case, there would be little or no indication in the gravity data that an upwarping of the basement rocks existed. Therefore, additional control is necessary before gravity data can be properly evaluated. This can take the form of other geophysical data or subsurface geological information from wells drilled in the area. Because of these problems gravity methods of prospecting are normally considered to be reconnaissance in nature.

Magnetic Methods of Geophysical Prospecting

The magnetic method of geophyiscal prospecting and the gravity method are basically similar. Both measure changes in physical properties of the subsurface rocks. In the gravity method, changes in rock density are determined, while in the magnetic method, changes in the magnetic character of the rocks are determined. The magnetic method is more complicated than the gravity method, on account of the following factors: (1) The magnetic properties of a rock have both magnitude and direction, while the gravitation force has magnitude only; (2) magnetic effects from rocks may be greatly influenced by relatively small amounts of certain minerals in the rocks, while gravity forces are the result of bulk properties of the rock.

A basic concept of magnetism is the occurrence of poles. A simple illustration of this concept of poles is obtained when a bar magnet is placed beneath a sheet of paper on which iron filings have been spread at random. As the bar magnet comes into position beneath the paper, the filings will tend to orient themselves along the "lines of force" of the magnet. These lines of force extend from a point near one end of the magnet to a point near the other end, these points being defined as the magnetic "poles." These lines of force are a measure of the magnetic-field strength of the system. If two poles of strength, P_1 and P_2, are separated by a distance, r, the magnetic force between the poles is proportional to the product of the pole strengths and inversely proportional to the square of the distance between the poles. This can be expressed mathematically as

$$F = \frac{CP_1P_2}{r^2} \tag{3-7}$$

where

F = force between the poles
P_1 = magnetic strength of pole 1
P_2 = magnetic strength of pole 2
r = distance between poles
C = constant

The magnetic-field strength is defined as the force per unit of pole strength exicted on a pole.
Therefore,

$$H = \frac{F}{P_1} = \frac{CP}{r^2} \tag{3-8}$$

where

H = magnetic field strength

A unit magnetic field is one in which a unit magnetic pole would experience a force of one dyne. This unit magnetic field is one dyne per unit pole, or, more commonly, one oersted.

The magnetic method of prospecting measures the variations in intensity of the magnetic field. However, the standard unit of measure, the oersted, is too large a unit for convenient use, and in practice, the gamma, which is 1×10^{-5} oersted, is the commonly used unit of magnetic-field intensity in geophysical prospecting.

The earth itself is a gigantic magnet. A bar magnet that is free to move will align itself with the earth's "lines of force." One end of the magnet will always point in the general direction of the north geographic pole, which is the negative pole, while the other end of the bar magnet will point in the general direction of the south geographic pole, which is the positive pole. The magnetic poles are near to, but do not coincide with, the geographic poles.

Poles always exist in pairs. It is impossible to separate the poles, as a very simple experiment will demonstrate: If a bar magnet having a north pole and a south pole is cut into several sections, each section will have a north pole and south pole.

Since all geophysical magnetic prospecting measurements are made in the magnetic field of the earth, something of the character of this magnetic field must be known.

The following changes are known to occur in the earth's magnetic field:

1. A relatively large, slow change caused by changes within the earth's interior. These are called "secular" changes.

2. Relatively small changes, which are somewhat erratic, but may occur at regular intervals. These are called "diurnal" variations.
3. Variations due to the magnetic inhomogeneities of the earth's crust.
4. Relatively large, erratic magnetic disturbances, which are unpredictable and are thought to be due to some type of solar activity. These are called "magnetic storms." Magnetic prospecting activities normally cease during these "storms" as there is no way to correct the data being taken.

The "dip needle" was one of the earliest instruments used in mining prospecting. It is a simple compass needle which is free to move in a vertical plane. An adjustable weight attached to one side of the needle is used to balance it. Then, when the dip needle is moved to another location, any change in the vertical component of the earth's magnetic field causes an unbalance of forces, resulting in a declination of the needle. The dip needle has a sensitivity of about 300 gammas, which makes it too insensitive for petroleum prospecting.

The Schmidt type of vertical magnetic-field balance is the most commonly used magnetic instrument for prospecting. This instrument, shown in Fig. 3–6, consists essentially of a bar magnet pivoted at a point away from the center of gravity, C. This balance is arranged so that the magnet is horizontal and measures only the vertical component of the magnetic field. The instrument is designed to measure changes in the magnetic field. The magnetic field creates a torque around the pivot, P, which is opposed by the torque of the gravitational pull caused by the pivot's not being placed at the center of gravity of the instrument. These forces are opposite in direction, and when they are equal, the balance is in a horizontal position. When the balance is moved to another position where the magnetic attraction has changed, the forces will be unbalanced and the magnet will move about its fulcrum. The magnitude of this movement is an indication of the change in magnetic field. This instrument is the standard type used and has a sensitivity of about one or two gammas.

Corrections for Reduction of Magnetic Data

Temperature corrections. A temperature correction must be made as temperature changes cause changes in the moment of the magnet as well as expansion or contraction of the individual parts of the magnet. Most magnetometers are well insulated, so that a very small temperature correction is necessary.

Terrain corrections. Terrain corrections have the same effect on both magnetic data and gravity data, provided the rocks are magnetic in character. The terrain correction is not usually applied, for in most areas the surface rocks have very little magnetic attraction.

Normal correction. The actual magnetic attraction varies from point to point on the earth, and in addition to this normal variation, the large, slow variations, or secular variations, in magnetic attraction must be known. Magnetic charts are available from the U. S. Coast and Geodetic Survey which show the regional magnetism of most areas in the United States.

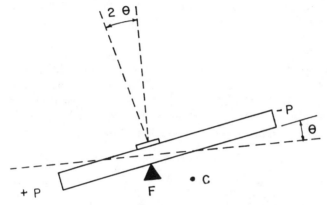

FIG. 3–6. Schmidt-type vertical magnetic-field balance.

Diurnal corrections. Magnetic data may be corrected for diurnal variations by returning the instrument to a base station as frequently as possible for check readings. However, it is generally considered to be more satisfactory to have two instruments, one to be kept at a base station for obtaining frequent readings which can be compared with the field-instrument readings. When using two instruments, it is necessary that they be calibrated together.

Interpretation of Magnetic Data

The magnetic method of prospecting is successful principally because of the difference in magnetic character of the basement rocks and the overlying sedimentary rocks. After the magnetic field data have been reduced (i.e., corrected), any variations, or anomalies, in data should be due entirely to changes in the magnetic character of the subsurface rocks. Measurements of the magnetic character of many rocks have shown that the magnetic attraction of igneous and metamorphic rocks is normally many times greater than that of sedimentary rocks. Magnetite (Fe_3O_4) appears to be the most common of the magnetic minerals, and the magnetic properties of most rocks can be traced almost entirely to it. Igneous and metamorphic rocks, which normally contain varying amounts of magnetite, will be more or less magnetic, depending on the percentage of magnetite present in them. Sedimentary rocks, on the other hand, usually contain very little magnetite, and therefore their magnetic quality is usually low. Thus, magnetic anomalies can normally be considered to be change in the magnetic attraction of the basement rocks. Increasing magnetic attraction can be attributed to uplifting of the basement rocks, thickening of the basement rocks, changes in the magnetic character of the basement rocks, or changes in the magnetic character of the overlying sedimentary rocks.

Since factors other than upwarping of the basement rocks can cause magnetic anomalies, it is necessary to have additional information before magnetic data can be used to select actual subsurface structures. Magnetic data have the same physical limitations as gravity data, and therefore the magnetic method is usually referred to as a reconnaissance tool in geophysical explorations.

Air-borne Magnetometer

Aerial magnetometer surveys were made as early as 1921, when a survey of an ore deposit in Sweden was made from a balloon. During World War II an air-borne magnetometer was developed for use in antisubmarine warfare. This instrument was modified and adapted for limited use in petroleum exploration work during the war years, and by 1946 several air-borne magnetometers were available for use. Most instruments of this kind measure the total earth magnetic field, rather than only the vertical or the horizontal component. The sensitivity of most air-borne magnetometers approaches one gamma.

Advantages of an Air-borne Magnetometer

1. Speed.
2. Reduction of effects of disturbing near-surface magnetic anomalies, such as ore bodies, railroads, and pipe lines.
3. Data obtained easily in very rugged terrain.
4. Unit cost for large surveys less than the unit cost if a ground magnetometer is used.

Disadvantages of Air-borne Magnetometer

1. High minimum cost.
2. Normally limited in detailing micromagnetic anomalies.

Seismic Exploration Principles

Seismic prospecting depends upon the introduction of artificial energy into the earth. This method was a natural development from the study of earthquakes. In principle, the seismic exploration technique is based on the propagation of sound waves through the earth, these waves being initiated by an explosion at or near the surface of the earth. The principles governing the behavior of these waves are similar to those governing the behavior of light waves. The two basic types of sound waves are longitudinal, or primary, waves, and transverse, or secondary, waves. Longitudinal waves are those in which the motion of the particles through which a wave is traveling is parallel to the direction of wave propagation. Transverse waves are

TABLE 3–2

Longitudinal Wave Velocities in Rocks*

Rock	Longitudinal Wave Velocity (feet per second)
Alluvium, Surface	1,640– 6,600
Shale	4,600–14,100
Sandstone	7,200–13,400
Salt	14,400–21,400
Limestone	11,000–20,000
Granite	13,100–18,700
Basalt	18,300

* From Birch's *Handbook of Physical Constants*.

sound waves which are traveling in a direction perpendicular to the motion of the particles of the medium. Longitudinal and transverse waves have also been called body waves because they travel through the body of the medium. Longitudinal waves always travel at a higher velocity than transverse waves. Longitudinal and transverse wave velocities are dependent upon the elastic properties of the medium through which they are traveling. Table 3–2 shows longitudinal wave velocities in various rocks.

When a wave traveling through a medium encounters a physical change in the wave-transmitting character of the medium, according to optical principles, part of the energy of the wave is "reflected" and part is "refracted." Thus transverse and longitudinal waves can be both reflected and refracted. Refraction is governed by Snell's Law, which states that the ratio of the sine of the angle of incidence to the sine of the angle of refraction in a substance is a constant. This can be expressed as

$$\frac{\sin i}{\sin r} = C = \frac{V_1}{V_2} \qquad (3\text{--}9)$$

or

$$\frac{\sin i}{V_1} = C \qquad (3\text{--}10)$$

where

i = angle of incidence
r = angle of refraction
V_1 = velocity of medium 1
V_2 = velocity of medium 2
C = constant

The laws governing the reflection of sound waves are based on the Dutch physicist Huygens' work with light waves. Huygens' principle states that the progress of a wave can be predicted as every point in a wave front acts like a new source of waves. It also explains the behavior of a reflected wave. A wave passing through a medium whose physical properties change will have part of its energy reflected in such a manner that the angle of incidence is equal to the angle of reflection. Both longitudinal and transverse waves can be reflected.

Seismic principles are based on the laws governing refraction and reflection. Thus there are two different types of seismic methods of geophysical prospecting, refraction seismic prospecting and reflection seismic prospecting. Normally, both of these methods are concerned with measurement of longitudinal waves, since these waves have a greater velocity than transverse waves.

Refraction Seismograph

The refraction seismograph is based on further observations from Snell's Law that at some critical angle of incidence where the angle of refraction is 90 degrees, the refracted wave will not enter the second medium but will travel along the surface of the two mediums. According to Huygens' principle, the refracted wave traveling along the surface of the second medium will also be source for new waves. These new waves move out in all directions, and those that move into the upper medium and back to the surface of the ground are recorded by seismic instruments.

Reflection Seismograph

The reflection seismograph, which is the most extensively used of all geophysical prospecting methods, simply measures the reflected wave produced when the sound wave encounters a medium discontinuity.

Seismic data must be corrected for elevation differences between shot points and detector points and a weathering correction which compensates for the low velocity of the sound waves as they pass through the weathered zone at the surface. Almost all seismic shots are detonated beneath this surface-weathered zone, which varies from a few feet to a few hundred feet in depth.

Seismic Equipment

The equipment for refraction and reflection seismic prospecting is essentially the same. The principal difference is that, in order to record the refracted longitudinal waves first, the refraction recording instruments are spaced a relatively great distance from the shot point, while in reflection seismic work, the recording, or pick-up, instruments are spaced relatively close to the shot point.

Seismograph Equipment

Field reflection seismograph operations are essentially these: the shot hole drilled some distance below the surface, in which the explosive is placed; detectors, or geophones, spaced to record the sound waves reflected after detonation of the explosive; and the amplifier-recording system, usually located in a panel truck, for recording the waves picked up by the geophones. An accurate time system is required to measure the time lapse between detonation of the shot and response from the geophones.

In addition to this equipment, a core drilling rig is an essential feature of a seismograph crew. The core drilling rig works ahead of the remainder of the crew, drilling core holes at previously determined locations.

Interpretation of Data

When a shot is detonated, the geophones will receive the sound-wave impulses from all sources where there are physical changes in the character of the underlying rocks. The reflection records are masses of wavy lines which are of value only to an experienced geophysicist who has a thorough knowledge of the area in which he is working. Each line on the record is an indication of the existence at the surface of some type of sound wave. However, with a knowledge of the velocity of sound waves through the underlying sediments—the time required for the sound wave to travel to the reflecting horizon and return to the surface—an indication is obtained of the depth to a particular marker horizon. In this manner some knowledge of the structure of the underlying rocks can be determined. Seismic methods of prospecting give more subsurface detail than any other geophysical method in use at the present time. The reflection seismic method is used to a much greater extent than the refraction seismic method because of the difficulty of interpreting refraction shooting data in many areas.

Radioactive Prospecting Methods

Surface radioactivity measurements have been used in an effort to learn something about the subsurface formations and fluids. This method had not been very successful because of the many variables present which hampered good correlations, but new developments have recently revived an interest in this tool.

Electrical Methods of Geophysical Prospecting

Many attempts have been made to develop satisfactory methods of electrical prospecting. To date these methods have not been satisfactory for one or more of the following reasons:
1. The depth of penetration is shallow.
2. The resistivities of oil-bearing formations, although somewhat greater than resistivities of salt water-bearing formations, are considerably higher than the resistivities of ore bodies.
3. Many ore deposits are located near the surface and will obscure the results of any electrical determinations.

Geochemical Methods of Petroleum Prospecting

A method of petroleum prospecting which has been used for many years is the geochemical method, which involves a chemical analysis of surface soil samples in an attempt to determine whether petroleum deposits are probable. This is a direct method of prospecting, in that the principal effort is directed toward the location of petroleum, not structures.

Geochemical prospecting is based on the theory that if underlying permeable sediments contain petroleum, some of the petroleum constitutents, principally the less dense, gaseous material, will finally escape through the overlying cap rock and reach the surface. Therefore, an attempt is made to locate minute traces of petroleum in the surface soil. Geochemical methods of locating petroleum have not been used extensively, probably because of the lack of sufficient basic scientific data. A reliable method of this nature would have obvious advantages in that deposits of petroleum could be located directly rather than indirectly, as is the current practice.

CHAPTER 4

Acquisition of Leases

It has been generally accepted by the courts that the *unqualified owner* of a property not only owns the surface area but also has title to all the strata underlying his property and everything above it.

Therefore, for purposes of ownership, the earth is divided vertically from the center of the earth to the outer reaches of the atmosphere. There are certain exceptions, such as the free use of the space above property by aircraft, but these will not concern us here since the content of this book is directed principally to the surface and subsurface portions of the land.

The unqualified owner of a tract of land, therefore, has the right to extract such materials, including petroleum, as he desires from the subsurface portion of his property. However, the business of drilling for and producing petroleum is highly specialized, normally requiring large sums of money in addition to detailed technical knowledge. For these reasons, very few landowners, as such, attempt to produce their own petroleum; instead, they usually permit some oil company or individual, whose principal business is the production of petroleum, to search for and produce any petroleum which may lie under their property.

It has been generally conceded that the earth can be divided horizontally as well as vertically for purposes of ownership. The unqualified owner of property can sell his subsurface rights to another person, and since the value of subsurface mineral deposits may be much greater than the value of the surface of the land, there quite often is a heavy traffic in the sale of subsurface rights. However, rather than completely surrender his interest in any minerals which may be present beneath the surface of his land, the owner usually *leases* the subsurface to a company or individual for the purpose of drilling for and producing petroleum. The oil and gas lease is distinguished from an outright grant of the minerals in that it is actually a permit to explore and develop the property.

Various oil-producing states in the United States have developed different concepts of the ownership of oil and gas. There are basically two principal concepts regarding the ownership of oil and gas in place, that of absolute ownership in place, and that of nonownership in place. According to the ownership-in-place theory, the oil and gas are regarded as a part of the land, and therefore the owners of the land are also the owners of the minerals in place beneath the surface. As the migratory nature of petroleum is well known, those states subscribing to the ownership-in-place concept recognize that when the oil or gas has moved out from under a particular piece of property, title to it is lost. Arkansas, Colorado, Kansas, Michigan, Mississippi, Montana, Ohio, Pennsylvania, Tennessee, Texas, and West Virginia all subscribe generally to ownership in place.

Under the non-ownership-in-place concept, oil and gas are considered to be migratory in nature and ownership of the land does not guarantee ownership of the subsurface oil and gas. The owner of the land owns only the petroleum which can be reduced to possession by production from his property. California, Illinois, Indiana, Kentucky, Louisiana, New Mexico, New York, Oklahoma, and Wyoming normally adhere to the nonownership-in-place concept.

The theory of ownership of oil and gas in most other countries is markedly different from that of the United States. The government of each of these is the sole owner of all minerals lying under the surface of the land, and therefore all negotiations for drilling and producing petroleum are made directly with the government concerned. Canada is something of an exception, in that the subsurface rights may be owned by individuals or large corporations as well as by the government.

A company which is interested in conducting preliminary geophysical work in an area must first obtain permission from the owners of the subsurface rights. There are basically three methods of doing this: (1) outright purchase of the subsurface rights either with or without the surface rights; (2) obtaining an oil and gas lease on the property beforehand; or (3) obtaining permission from the surface-property owner to enter on the property and perform the necessary work, either with or without an option to lease the land for drilling and producing oil and gas.

The first method is seldom used because of the large investment in property that would be required to conduct an adequate exploration program. The oil operator is normally primarily interested only in the production of oil and gas from the property, and since the surface acreage may have significant value for crops, grazing, or other uses, the surface and subsurface rights are usually separated. Furthermore some states limit the purchase of land by corporations to that amount absolutely necessary to conduct their business operations.

Obtaining an oil and gas lease prior to performing the initial exploratory work has both advantages and disadvantages. Since the oil and gas lease, which will be discussed in more detail later, permits use of the necessary surface acreage to conduct drilling and producing operations, there is no actual need for the oil operator to purchase surface acreage in order to conduct his operations. A lease can normally be obtained much more cheaply before exploratory work has been done, because the owner of the mineral rights knows that if a company has obtained geophysical information and is still interested, the area must have potentialities for oil and gas production, and the price will naturally increase. On the other hand, the oil operator may not wish to expend large sums of his capital for acreage which may have little value after a geophysical survey has been made. The procedure very often employed, and generally preferred, therefore, is to secure an option to lease the acreage at the time arrangements are made to perform the necessary geophysical work. This procedure has the obvious advantage of permitting the operator to take advantage of the exploration work, which he alone has paid for, in selecting the acreage he desires.

Oil and Gas Lease

The oil and gas lease is the legal instrument which has been responsible for the rapid development of the oil industry. It grants persons or organizations with specialized knowledge the authority to enter upon the surface of land they do not own and proceed in the search for underground deposits of petroleum. The oil and gas lease has evolved from a very crude instrument to a complex document. Although there is no standard lease, the basic requirements are such that most leases are similar. The eight principal elements of an oil and gas lease include: (1) principal parties, (2) date, (3) granting clause, (4) habendum clause, (5) royalty clause, (6) drilling and delay rental clause, (7) description of the property leased, and (8) special considerations.

Principal parties. The lease agreement must specifically name the individuals involved in the transaction. The owner of the subsurface rights is called the *lessor*, and the person obtaining the right to search for the oil and gas is called the *lessee*. It is interesting to note that although all the lessors must sign the agreement, the lessee is not required to sign it.

Date. Although an undated lease may not be void and usually takes effect upon its execution and delivery, the date on the lease agreement is very important, since it may become necessary to determine precedence of documents, in which case the date would be of utmost importance. Where the date of execution of the lease and the date in the body of the lease are at variance, the date in the body of the lease is usually the controlling one.

Granting clause. The granting clause specifies pecisely what the lessor is granting to the lessee. It is a highly important feature of the lease agreement in that it specifies the nature of the lessee's interest. In the granting clause the lessee secures the right to search for and produce oil and gas. The lessee has the right to use the surface of the land when and where required in normal drilling and producing operations even though the agreement does not state this fact specifically. However, so much controversy has arisen over the use of the surface in drilling and producing operations that most granting clauses do state the extent to which the surface can be used by the lessee. Oil and gas are classified as minerals, and if the lessor grants to the lessee the right to search for and produce *minerals*, then oil and gas are included in this term; however, if the term "oil and gas" is used, then it does not include other minerals, such as coal, silver, etc. In some of the older lease agreements, the term "petroleum" was used in the granting clause, and for many years the issue of the exact meaning of the word "petroleum," has been brought before the courts to determine whether it includes both oil and gas. Most leases written at the present time are for oil and gas only.

The granting clause also specifies the consideration paid the lessor by the lessee. A valid contract requires a *consideration*, that is, a *gain*, to the person granting the lease. The consideration is usually in the form of cash, but may be in the form of an oil payment or other type of payment which has some monetary value. A very nominal consideration, of as little as one dollar, has been deemed sufficient consideration in the past in most states, but recent trends in court decisions indicate that a larger consideration may be necessary. Practically speaking, in order to obtain a lease, especially a desirable one, a substantial cash consideration may be required.

Habendum clause. The typical habendum clause provides that "this lease shall remain in force for a term of———years and as long thereafter as oil, gas, casinghead gas, casinghead gasoline, or any of the products covered by this lease is or can be produced." Thus this clause fixes the duration of the lessee's interest. If production is not obtained within the time specified, often called the *primary term*, the lease will expire by its own terms.

Royalty clause. The royalty clause is the principal inducement, aside from the cash consideration, for the property owner to sign the agreement. Royalty is the payment of a part of the oil and gas produced under a lease. The landowner's royalty is free and clear of any drilling or operating expenses which the lessee may incur in producing the oil, and usually varies from one-eighth to one-sixth of the gross production, although there is no fixed royalty and variations have been recorded from less than one-eighth to more than seven-eighths. This latter royalty was paid, of course, in a special case involving a small tract of land in the center of known oil-producing acreage, and the oil company agreed to pay the unusually large royalty in order to control the entire reservoir.

Drilling and delay rental clause. The courts have ruled that one of the primary considerations in the oil and gas lease is the early development of the property, whether or not there is any statement about it in the lease itself. The drilling and delay rental clause specifies the manner in which early drilling can be deferred. Drilling can normally be deferred, under the terms of this clause, for a specified period by the payment of a sum of money, called *delay rental*, to the lessor. However, in no event can drilling be deferred past the primary term without voiding the lease.

Description of the property. An accurate description of the property involved in the lease agreement is necessary. In order to describe adequately a specific tract of land, some system of description is required. As early as 1784, the Continental Congress of the United States appointed a committee to devise a system of land measurement and description. In 1805, the present system of land measurement, known as the "Rectangular System" was adopted and is used today in most of the oil-producing states.

In this system, the states are surveyed by the General Land Office of the United States government by first locating, as a starting point, a north-south line in the approximate center of the state or area being surveyed. This north-south line is known as the "prime or principal meridian." Next an east-west base line, running at right angles to the principal meridian, is established. The basic unit of measurement is six miles. Therefore, north-south lines are spaced six miles apart, beginning on each side of the principal meridian. East-west lines are also spaced six miles apart, beginning on both sides of the base line. Each of the north-south lines is known as a range line, and is further designated as the east or west range line depending on its location with respect to the principal meridian.

These north-south and east-west lines form a unit area which is six miles square. This unit area is called a *township*. Each township is numbered both in a north and a south direction, commencing at the base line. Townships are further subdivided into 36 *sections*, these sections being numbered 1 through 36, beginning at the northeast corner of the township. Each of these sections contains approximately 640 acres, and each 640-acre section is subdivided into four tracts of approximately 160 acres each. These tracts are identified with respect to their geographical location within the section; for example, the northeast quarter. The 160-acre tracts can be further subdivided into four equal tracts of 40 acres each, which can also be described by their geographical location within the tract. Subdivision can continue as far as necessary in this same fashion.

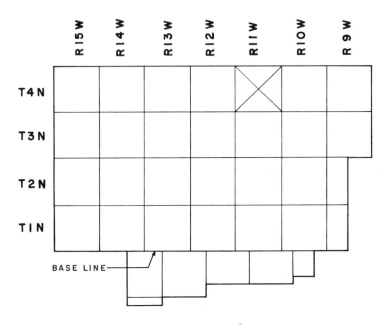

Comanche County

Oklahoma

FIG. 4–1. Survey of Comanche County, Oklahoma.

Figure 4–1 shows the survey of Comanche County, Oklahoma. The principal meridian of Oklahoma is called the Indian Meridian, and is located in the approximate center of the state. The base line of the Oklahoma survey is near the southern part of the state. Township 4 North, Range 11 West has been marked on Fig. 4–1. This township is shown in Fig. 4–2 with all of the sections numbered. A 40-acre tract in Section 6 has been marked on the figure. This 40-acre tract is properly described as the SW $\frac{1}{4}$, SE $\frac{1}{4}$ Section 6, Township 4 North,

ACQUISITION OF LEASES

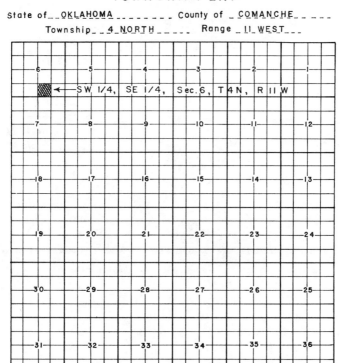

Fig. 4–2. Township plat.

Range 11 West. The south 20 acres of this 40-acre tract would properly be described as the S $\frac{1}{2}$, SW $\frac{1}{4}$, SE $\frac{1}{4}$, Section 6, Township 4 North, Range 11 West.

Areas in some oil-producing states were not laid out in this systematic manner, but were surveyed in parcels, as in parts of Texas. Figure 4–3 shows a typical survey of a county in Texas. Proper identification of a property in this part of Texas is made by specifying the county and survey. If the property does not constitute all of the survey, additional markers, such as trees, streams, or rocks, must be used in the description since most surveys have irregular geometric patterns. As shown in Fig. 4–3, small surveys may be difficult to locate.

Special considerations. Several additional clauses may be inserted in a lease agreement to protect or more fully describe rights and duties of either the lessee or the lessor. Typical examples of clauses describing the rights of the lessor are:

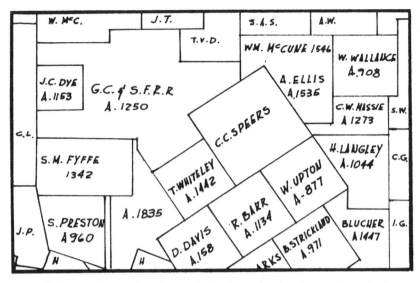

Fig. 4–3. Survey of a portion of a county in Texas showing the very irregular tracts.

1. Free use of gas for household purposes.
2. Drilling restrictions near buildings
3. Damage to crops.
4. Burying of pipe lines.
5. Additional development.

Typical examples of rights of the lessee which may be inserted in the lease agreement are:

1. Erection of buildings and equipment.
2. Removal of buildings and equipment, including casing.
3. Right to unitize or pool lands.
4. Use of oil, gas, and water.
5. Length of time for beginning of remedial operations on a one-well lease after expiration of primary term.

Over a period of years as the oil and gas lease has evolved from its original crude state to its present form, several covenants have been implied and are actually a part of the lease agreement, even though they are not set down in writing. These include:

1. Right of either party to assign all or part of his interest.
2. Right of the lessee to use such surface facilities as may be necessary to properly exploit his mineral lease.
3. Duty of the lessee to protect the lessor from drainage of petroleum to other leases.
4. Duty of the lessee to adequately develop the property.

ACQUISITION OF LEASES

TYPICAL OIL AND GAS LEASE

Form 88 (Producers Revised)

THIS AGREEMENT, Entered into this the _____ day of _____, 195__, between _____

_____ hereinafter called lessor, and _____
hereinafter called lessee, does witness:

1. That lessor, for and in consideration of the sum of _____ Dollars in hand paid and of the covenants and agreements hereinafter contained to be performed by the lessee, has this day granted, leased, and let and by these presents does hereby grant, lease, and let exclusively unto the lessee the hereinafter described land, and with the right to unitize this lease or any part thereof with other oil and gas leases as to all or any part of the lands covered thereby as hereinafter provided, for the purpose of carrying on geological, geophysical, and other exploratory work, including core drilling, and drilling, mining, and operating for, producing, and saving all of the oil, gas, casinghead gas, casinghead gasoline, and all other gases and their respective constituent vapors, and for constructing roads, laying pipe lines, building tanks, storing oil, building powers, stations, telephone lines, and other structures thereon necessary or convenient for the economical operation of said land alone or conjointly with neighboring lands, to produce, save, take care of, and manufacture of all such substances, and for housing and boarding employees, said tract of land with any reversionary rights therein being situated in the County of _____, State of _____, and described as follows:
in Section _____, Township _____, Range _____, and containing _____ acres, more or less.

2. This lease shall remain in force for a term of _____ years and as long thereafter as oil, gas, casinghead gas, casinghead gasoline, or any of the products covered by this lease is or can be produced.

3. The lessee shall deliver to lessor as royalty, free of cost, on the lease, or into the pipe line to which lessee may connect its wells the equal one-eighth part of all oil produced and saved from the leased premises, or at the lessee's option may pay to the lessor for such one-eighth royalty the market price for oil of like grade and gravity prevailing on the day such oil is run into the pipe line or into storage tanks.

4. The lessee shall pay to lessor for gas produced from any oil well and used by the lessee for the manufacture of gasoline or any other product as royalty $\frac{1}{8}$ of the market value of such gas at the mouth of the well; if said gas is sold by the lessee, then as royalty $\frac{1}{8}$ of the proceeds of the sale thereof at the mouth of the well. The lessee shall pay lessor as royalty $\frac{1}{8}$ of the proceeds from the sale of gas as such at the mouth of the well where gas only is found and where such gas is not sold or used, lessee shall pay or tender annually at the end of each yearly period during which such gas is not sold or used, as royalty, an amount equal to the delay rental provided in paragraph 5 hereof, and while said royalty is so paid or tendered this lease shall be held as a producing lease under paragraph 2 hereof; the lessor to have gas free of charge from any gas well on the leased premises for stoves and inside lights in the principal dwelling house of said land by making his own connections with the well, the use of such gas to be at the lessor's sole risk and expense.

5. If operations for the drilling of a well for oil or gas are not commenced on said land on or before the _____ day of _____, 19__, this lease shall terminate as to both parties, unless the lessee shall on or before said date pay or tender to the lessor or for the lessor's credit in the _____ Bank at _____
_____, or its successors, which Bank and its successors are the lessor's agent and shall continue as the depository of any and all sums payable under this lease regardless of changes of ownership in said land or in the oil and gas or in the rentals to accrue hereunder, the sum of _____ Dollars, which shall operate as

a rental and cover the privilege of deferring the commencement of operations for drilling for a period of one year. In like manner and upon like payments or tenders the commencement of operations for drilling may further be deferred for like periods successively. All payments or tenders may be made by check or draft of lessee or any asignee thereof, mailed or delivered on or before the rental paying date, either direct to lessor or assigns or to said depository bank, and it is understood and agreed that the consideration first recited herein, the down payment, covers not only the privilege granted to the date when said first rental is payable as aforesaid, but also the lessee's option of extending that period as aforesaid and any and all other rights conferred. Notwithstanding the death of the lessor or his successors in interest, the payment or tender of rentals in the manner above shall be binding on the heirs, devisees, executors, and administrators of such persons.

6. If at any time prior to the discovery of oil or gas on this land and during the term of this lease, the lessee shall drill a dry hole. or holes on this land, this lease shall not terminate, provided operations for the drilling of a well shall be commenced by the next ensuing rental paying date, or provided the lessee begins or resumes the payment of rentals in the manner and amount hereinabove provided, and in this event the preceding paragraphs hereof governing the payment of rentals and the manner and effect thereof shall continue in force.

7. In case said lessor owns a less interest in the above-described land than the entire and undivided fee simple estate therein, then the royalties and rentals herein provided for shall be paid the said lessor only in the proportion which his interest bears to the whole and undivided fee. However, such rental shall be increased at the next succeeding rental anniversary after any reversion occurs to cover the interest so acquired.

8. The lessee shall have the right to use, free of cost, gas, oil, and water found on said land for its operations thereon, except water from the wells of the lessor. When required by lessor, the lessee shall bury its pipe lines below plow depth and shall pay for damage caused by its operations to growing crops on said land. No well shall be drilled nearer than 200 feet to the house or barn now on said premises without written consent of the lessor. Lessee shall have the right at any time during, or after the expiration of, this lease to remove all machinery, fixtures, houses, buildings, and other structures placed on said premises, including the right to draw and remove all casing, but lessee shall be under no obligation to do so, nor shall lessee be under any obligation to restore the surface to its original condition, where any alterations or changes were due to operations reasonably necessary under this lease.

9. If the estate of either party hereto is assigned (and the privilege of assigning in whole or in part is expressly allowed), the covenants hereof shall extend to the heirs, devisees, executors, administrators, successors, and assignees, but no change of ownership in the land or in the rentals or royalties or any sum due under this lease shall be binding on the lessee until it has been furnished with either the original recorded instrument of conveyance or a duly certified copy thereof or a certified copy of the will of any deceased owner and of the probate thereof, or certified copy of the proceedings showing appointment of an administrator for the estate of any deceased owner, whichever is appropriate, together with all original recorded instruments of conveyance or duly certified copies thereof necessary in showing a complete chain of title back of the lessor to the full interest claimed, and all advance payments of rentals made hereunder before receipt of said documents shall be binding on any direct or, indirect assignee, grantee, devisee, administrator, executor, or heir of lessor.

10. If the leased premises are now or shall hereafter be owned in severalty or in separate tracts, the premises nevertheless shall be developed and operated as one lease, and all royalties accruing hereunder shall be treated as an entirety and shall be divided among and paid to such separate owners in the proportion that the acreage owned by each separate owner bears to the entire leased acreage. There shall be no obligation on the part of the lessee to offset wells on separate tracts into which the land covered by this lease may be hereafter divided by sale, devise, descent, or other-

wise or to furnish separate measuring or receiving tanks. It is hereby agreed that in the event this lease shall be assigned as to a part or as to parts of the above-described land and the holder or owner of such part or parts shall make default in the payment of the proportionate part of the rent due from him or them, such default shall not operate to defeat or affect this lease insofar as it covers a part of said land upon which the lessee or any assignee hereof shall make due payment of said rentals.

11. Lessor hereby warrants and agrees to defend the title to the land herein described and agrees that the lessee, at its option, may pay and discharge in whole or part any taxes, mortgages, or other liens existing, levied, or assessed on or against the above described lands and, in event it exercises such option, it shall be subrogated to the rights of any holder or holders thereof and may reimburse itself by applying to the discharge of any such mortgage, tax, or other lien, any royalty or rentals accruing hereunder.

12. Notwithstanding anything in the lease contained to the contrary, it is expressly agreed that if lessee shall commence operations for drilling at any time while this lease is in force, this lease shall remain in force and its terms shall continue so long as such operations are prosecuted, and if production results therefrom, then as long as production continues.

13. If within the primary term of this lease, production on the leased premises shall cease from any cause, this lease shall not terminate provided operations for the drilling of a well shall be commenced before or on the next ensuring rental paying date; or, provided lessee begins or resumes the payment of rentals in the manner and amount hereinbefore provided. If, after the expiration of the primary term of this lease, production on the leased premises shall cease from any cause, this lease shall not terminate provided lessee resumes operations for drilling a well within sixty (60) days from such cessation and this lease shall remain in force during the prosecution of such operations and, if production results therefrom, then as long as production continues.

14. Lessee may at any time surrender or cancel this lease in whole or in part by delivering or mailing such release to the lessor, or by placing same of record in the proper county. In case said lease is surrendered and canceled as to only a portion of the acreage covered thereby, then all payments and liabilities thereafter accruing under the terms of said lease as to the portion canceled shall cease and any rentals thereafter paid may be apportioned on an acreage basis, but as to the portion of the acreage not released the terms and provisions of this lease shall continue and remain in full force and effect for all purposes.

15. All provisions hereof, express or implied, shall be subject to all federal and state laws and the orders, rules, or regulations (and interpretations thereof) of all governmental agencies administering the same, and this lease shall not be in any way terminated wholly or partially nor shall the lessee be liable in damages for failure to comply with any of the express or implied provisions hereof if such failure accords with any such laws, orders, rules, or regulations (or interpretations thereof). If lessee should be prevented during the last six months of the primary term hereof from drilling a well hereunder by the order of any constituted authority having jurisdiction thereover, or if lessee should be unable during said period to drill a well hereunder due to equipment necessary in the drilling thereof not being available on account of any cause, the primary term of this lease shall continue until six months after said order is suspended and/or said equipment is available, but the lessee shall pay delay rentals herein provided during such extended time.

16. The unitization of this lease or any portion thereof with any other lease or leases or portions thereof shall be accomplished by the execution and filing by lessee in the recording office of said county of an instrument declaring its purpose to unitize and describing the lease and land unitized, which unitization shall cover the gas rights only and comprise an area not exceeding approximately 640 acres. The royalty provided for herein with respect to gas from gas wells shall be apportioned among the owners of such royalty on minerals produced in the unitized area in the proportion

that their interests in the minerals under the lands within such unitized area bear to the minerals under all of the lands in the unitized area. Any well drilled on such unit shall be for all purposes a well under this lease and shall satisfy the rental provision of the lease as to all of the land covered thereby. Provided, however, lessee shall be under no obligation, express or implied, to drill more than one gas well on said Unit.

17. This lease and all its terms, conditions and stipulations shall extend to and be binding on all successors of said lessor or lessee.

IN WITNESS WHEREOF, we sign the day and year first above written.

_____ (SEAL) _____ (SEAL)
_____ (SEAL) _____ (SEAL)
_____ (SEAL) _____ (SEAL)
_____ (SEAL) _____ (SEAL)

Before actual drilling operations begin, the lessee must determine that title to the property on which he obtained the lease actually belongs to the lessor or lessors. Many lawsuits have developed as a result of inadequate checking of titles before drilling operations commenced.

The lessee, as owner of the oil and gas lease, may decide for one reason or another not to drill a well on the property covered by the lease. If he knows someone who is interested, however, he may *farm out* the lease to a third party, usually receiving an overrriding royalty. The lessee may help finance the drilling of the first well with *bottom-hole* money, a cash sum paid when a specific depth has been reached, or with *dry-hole* money, a cash sum paid only if the well is a dry hole. Dry-hole money is a much more popular form of contribution, the theory being that if the well is completed as a producer, financial assistance will not be required nearly as much as if the well had been a dry hole.

Government and Indian Land

Where government or Indian land is concerned, leasing regulations are considerably different from those for privately owned land. Much government and Indian lands are leased on open bid, which results, in many cases, in the payment of large bonuses.

CHAPTER 5

Earth Temperatures

The temperatures at the various depth levels to which wells are drilled have great influence on the properties of both the reservoir fluids and the materials used in drilling and completing oil wells. Temperatures within the rocks increase from the surface downward. The effects of the higher temperatures upon the reservoir fluids are generally beneficial, chiefly because the viscosity of crude oil is greatly reduced at higher temperatures, and lower viscosities permit the oil to flow more freely through the small openings in the reservoir rocks. Conversely, the higher temperatures encountered at the greater depths usually have adverse effects upon the materials used in drilling and completing wells. Both the treating chemicals and the clays used in drilling muds tend to become ineffective or unstable at high temperatures. The thickening and setting of portland cement slurries are accelerated by high temperatures as well as by high pressures, and special slow-setting cements must be used in deep wells. The properties of such materials as thermal setting plastics must be controlled according to the temperatures which may be encountered in the well. The strength and toughness of the steel used in drill pipe and bits decrease at high temperatures. New alloys will be required for the deeper wells of the future, where temperatures above 400 degrees Fahrenheit will be encountered. Accordingly, the temperatures occurring in rocks and in bore holes are of interest and importance.

GEOTHERMAL GRADIENTS

The temperature in the rocks at any place and depth, particularly where no disturbing events have occurred and where the rocks are in termal equilibrium, is known as the *formation temperature*. The geothermal gradient is the relation between depth and natural formation temperatures. There is normally a rise in temperature with increasing depth, and the average rise in temperature in various areas commonly varies from about one degree Fahrenheit per 60 feet to about one degree Fahrenheit per 100 feet. The theory has been advanced that the earth has an extremely hot core, but it is not known whether the heat radiated by the earth's surface is coming from the heat contained in the core or is partially produced by radioactive disintegration. Heat flows through the rocks by conduction. Since some minerals

54   OIL WELL DRILLING TECHNOLOGY

and some rocks are better conductors of heat than others, it follows that the average temperature gradient in the rocks will vary from one region to another. The average temperature of the rocks at the surface varies with the seasons, although seasonal variations probably do not extend as far as one hundred feet deep. The average tempera-

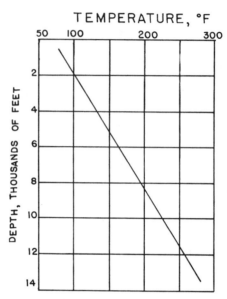

FIG. 5-1. Gulf Coast geothermal gradient. (After Guyod.)

ture of the rocks at the earth's surface varies with latitude. For the Gulf Coast region of the United States, Guyod[1] gives the relation

$$T = 70 + \frac{h}{64}$$

where T is the temperature in degrees Fahrenheit prevailing below depths of about one hundred feet and h is the depth in feet. This relation, illustrated in Fig. 5-1, gives the geothermal temperature except in the vicinity of salt domes. According to data presented by White,[2] the geothermal temperatures at Point Barrow, Alaska, between depth of 100 and 1,200 feet may be expressed approximately as

[1] H. Guyod, "Temperature Well Logging," a series of seven articles published in *The Oil Weekly*, October 21, 28, November 4, 11, December 2, 9, and 16, 1946, reprinted by the Halliburton Oil Well Cementing Company, Duncan, Okla.

[2] F. L. White, "Investigation of Properties of Oil Well Cements for Use in Low Temperature Formations," Master's thesis, School of Petroleum Engineering, University of Oklahoma, 1952.

$$T = 10 + \frac{h}{41}$$

The latter relation is in the permafrost, which at that location is about 930 feet in thickness. These equations indicate an average surface rock temperature of 70 degrees Fahrenheit in the Gulf Coast area and 10 degrees Fahrenheit at Point Barrow, Alaska. The lower limit to which such linear relations may be used has not been determined.

The geothermal gradients as used are average gradients which apply over hundreds, and usually thousands, of feet of depth. It is presumed that the same total amount of heat flows from the interior of the earth outward to the surface through each square mile of the earth's surface. Under such conditions of equal total heat flow, the geothermal gradient at a particular location should actually be a series of joined line segments corresponding to the thickness and thermal conductivity of each rock stratum under that location. Inspection of the thermal conductivities of the common rocks as given in Table 5–1 shows that such line segments would have approximately the same slope.

TABLE 5–1

Heat Conductivities of Rocks ($\times 10^3$ c.g.s. units)

(after Guyod)

Coal, lignite	0.5 to 1.0
Clay	2.0 to 3.0
Chalk	2.0 to 3.0
Shale	2.0 to 4.0
Sandstone	3.0 to 5.0
Porous limestone	3.0 to 5.0
Dense limestone	5.0 to 8.0
Rock salt	8.0 to 15.0
Basalt	5.0 to 7.0
Granite	5.0 to 8.0
Slate	6.0

Attempts have been made to obtain temperature logs of wells which would permit the identificaton or differentiation of strata penetrated by the well bore. However, the contrast of the thermal properties of rocks is usually much less than the contrasts of the electrical and radioactivity properties, and logs based on the latter properties are in common use. Also, most logs are urgently needed just after a well has been drilled, or worked over, and during both of these operations drilling mud is customarily circulated in the well bore. Such circulation of liquids in the well modifies and smooths out temperature gradients in the area of the well bore so that thermal distinctions between the strata are masked. Consequently, temperature logs are not commonly used for distinguishing rock strata.

Mud-Circulation Temperature Gradients

During the drilling of a well by the rotary method, drilling mud is customarily circulated down through the interior of hollow drill pipe. It then returns from the bottom of the hole to the surface through the annular space between the outside of the drill pipe and the wall of the bore hole and inside of any casing placed in the well. If the well is several thousand feet deep, the rocks at the bottom of the hole will be hot, according to the geothermal temperature, while the rocks at the surface will be relatively cool. The drilling fluid picks up heat from the rocks in the bottom part of the hole and the drilled rock fragments, and it loses heat to the rocks in the top part of the hole and in the surface mud pits. Under mud-circulating conditions, the cooler mud pumped into the top of the drill pipe picks up heat in its downward travel from the warmer mud returning from the bottom. In this sense, the drill pipe acts as a somewhat inefficient heat exchanger which tends to maintain a temperature difference between the top and bottom of the hole. Inspection of the conditions shows that if the rate of drilling-mud circulation were infinitely slow, the temperature at any depth in the hole would be about equal to the temperature of the geothermal gradient. Conversely, if the rate of drilling-mud circulation were infinitely rapid, the drilling mud in the hole would have the same temperature from top to bottom. The latter condition is approached by the more powerful modern drilling rigs which circulate mud at a rate in excess of twenty barrels per minute. The mud tends to assume a temperature about midway between the ground temperature at the top of the hole and the formation temperature at the bottom of the hole. The mud temperature increases as the depth of the well increases.

When mud circulation is stopped in a drilling well, the temperature of the mud begins to shift back to the geothermal gradient. The mud in the top part of the hole becomes cooler and the mud in the bottom part of the hole becomes warmer. The temperature in the mud column of a well drilled to a depth of several thousand feet may not return completely to the equilibrium geothermal gradient for a period of two or three months, particularly in the sections of the hole where drilling fluid was circulated for a considerable period of time. In general, however, there is a comparatively rapid change in temperature in the mud column for six to twelve hours, except in depth regions approximately half the total depth of the well, where little change in temperature occurs. The amounts of the temperature shifts which occur within such short time intervals depend partially upon the length of time that mud has been circulated in various portions of the hole, for time governs the extent of temperature adjustments in the rock surrounding the bore hole. Usually, mud will have been circulated in the bottom portion of the hole for only a short period of time. Consequently, on temperature logs run about twenty-four hours after the cementing of casing, it is common to find a curved section at the bottom of the log where temperatures have changed (increased) much more rapidly than in other sections of the hole. Another manifestation of the comparatively rapid return to formation temperature in the bottom sections of a drilling well has occurred in wells using drilling muds which unfortunately solidified at the high formation temperatures encountered below about 12,000 feet. Mud circulation must be halted in order to pull the drill pipe and replace a worn bit. In some instances it has been necessary to drill solidified mud out the bottom one hundred feet of hole, where the mud became too hot and solidified during the time required to make a round trip with the drill pipe.

The geothermal gradient and drilling-mud circulation gradient relationships are shown in Fig. 5–2. An illustration of a transient tem-

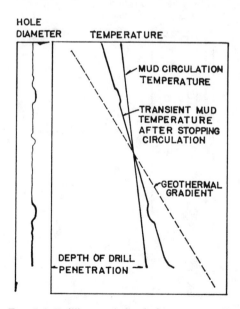

FIG. 5–2. Drilling-mud circulating temperature.

perature gradient, such as might be obtained about five hours after mud circulation was stopped in a well ten inches in diameter, is also included. In this case the mud temperature is adjusting back toward the geothermal gradient. A caliper log at the side of the figure shows hole enlargements such as occur in some shale sections. Note that a lag in temperature change occurs at the places of hole enlargement. This happens because there is comparatively more drilling mud at those places. The specific heat of the water-base mud is greater than the specific heat of the rock minerals, and more heat must be transferred into or out of the mud to obtain comparable temperature changes. The additional heat flow requires more time for adjustment to equilibrium temperature, thus producing a temporary temperature anomaly on the log at such places of hole enlargement.

Temperature Logs Made after Cementing Casing

Temperature surveys are often made about twenty-four hours after cement has been pumped behind the string of casing run into a well. Such a log is illustrated in Fig. 5-3. The setting of portland cement and posmix cement are reactions which produce heat. The resulting temperature rise makes it possible to detect the presence of the cement behind the casing. This figure illustrates the correlation which

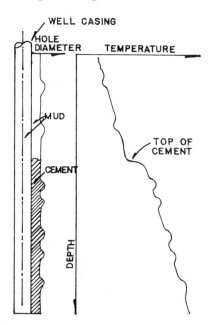

Fig. 5-3. Temperature log made after cementing casing.

is usually obtained between the caliper log, which measures hole diameter, and the temperature log. More cement is required between the pipe and the rock at the places of hole enlargement, with the result that the temperature rise from the setting of the cement tends to be proportional to the diameter of the hole.

The Transfer of Heat

All of the temperature effects which are observed in wells depend upon or are influenced by the transfer of heat. Heat is transferred by (1) conduction, (2) convection, and (3) radiation.

Heat flows through the rocks of the earth almost entirely by conduction. The fundamental law of heat transfer by condution is known as Fourier's Law, which may be expressed as follows:

$$\frac{dq}{dt} = KA\frac{dT}{dx}$$

where

$\frac{dq}{dt}$ = the time rate of heat flow, as Btu per hour

$\frac{dT}{dx}$ = the temperature gradient in the directions of heat flow, as degrees Fahrenheit per foot

A = the cross-sectional area through which heat flow occurs, square feet

K = the thermal conductivity constant for a particular material, Btu per hour per square foot per degree Fahrenheit per foot.

The thermal conductivities of some rock minerals are given in Table 5–1. There are no large differences in the thermal conductivities of the common sedimentary rock minerals, which largely eliminates this property as a basis for well logging. However, the difference in the conductivities of sediments has occasionally shown up in temperature logs.

The complete solution of heat-transfer problems usually involves heat flow through a series of substances of different heat-conducting abilities. Often such calculations must take into account the stagnant liquid or gaseous films which tend to adhere to a solid rock or metal wall. For such complete solutions, reference should be made to texts in mechanical and chemical engineering which give extensive treatment to such problems.

Heat transfer by convection consists of transport of the material containing the heat. It is usual to consider only liquids and gases in this sense. In drilling operations, a great amount of heat is transferred by convection in the circulation of the drilling fluid. In this manner the bottom of the hole is cooled and the top of the hole is warmed. The amount of heat transferred by convection is sensibly the heat above some base temperature of heat transfer. Thus, the heat transferred by convection depends upon the mass of the substance which is transferred, the difference in temperature, and heat capacitance of the substance expressed in Btu per pound per degree Fahrenheit.

Heat transfer by radiation is of little importance within the strata of the earth or in the bore hole. It may be important, however, in cooling surface equipment. According to Stefan's Law, all bodies radiate heat in proportion to the fourth power of their absolute temperature, which is degrees Fahrenheit plus 460. The total heat radiated depends upon the area of radiation and upon the efficiency of the surface radiating heat (or absorbing radiation). It is well known, for example, that an oil tank painted with aluminum paint will absorb less heat from the sun and consequently suffer less evaporation of crude oil than a tank of dark color. If it is desirable to have a piece of equipment lose the maximum amount of heat by radiation, that piece of equipment should not have a smooth and bright surface.[3]

[3] For additional information, two good sources are E. A. Nichols, "Geothermal Gradients in Mid-Continent and Gulf Coast Oil Fields," *Trans. A.I.M.E. Petroleum Division*, Vol. 170 (1947), 44–50; and R. E. Keen, "Drilling the World's Hottest Well," *The Petroleum Engineer*, Vol. XXIX, No. 11 (October, 1957), B-21-29.

CHAPTER 6

Pressure Relations in the Earth and in Bore Holes

Pressures play an important part in the life of an oil field from the time when drilling on the first well is started until the field is abandoned. Geological processes and forces have been in operation for thousands of years prior to the development of a field, and pressure relations have been established in the rocks in accordance with the geological conditions. The drilling of a well into the strata is an event which disturbs the pressure equilibrium. A blowout, i.e., the destructive and wasteful blowing of oil and gas out of a wild well, results from losing control of the pressures within the rocks. Other types of fluid movement, both desirable and undesirable, are caused by unbalanced pressure forces in the rocks and bore hole. Accordingly, the proper control of pressures is of great importance to all who are concerned with the drillling and completing of a well. The following sections describe the various types of hydrostatic pressures which are encountered in drilling and completion operations. These pressures are sometimes modified by flowing conditions, where flow friction imposes an additional pressure in the system. Such cases are discussed in Chapter 12. As a general rule, the additional pressures caused by flowing conditions are but a small fraction of the hydrostatic pressures described in the following sections.

THE HYDROSTATIC HEADS OF LIQUIDS

The pressure exerted by, or within, a column of liquid, which is caused by the weight of the liquid, depends upon the height of the liquid above a particular point at which the pressure is being considered. Such pressure is referred to as the hydrostatic head. These pressures are commonly expressed in pounds per square inch (psi) or in feet of a liquid of known constant density. Fundamentally, such pressures are equal to the product of the density of the liquid, the height of the column of liquid, and the force of gravity operating on the liquid. However, in general engineering calculations, the force of gravity is assumed constant and included as such in the calculations. By this convention, the hydrostatic head of a column of liquid may be expressed as

$$p = h\rho \qquad (6\text{--}1)$$

where

p = hydrostatic pressure, lb./ft.²
h = height of the liquid column, ft.
ρ = density of the liquid, lb./ft.³

The above equation is in fundamental engineering units. However, pressure is most conveniently expressed in pounds per square inch and density is most conveniently expressed as a specific gravity with reference to the density of water. Accordingly, the following substitutions are made:

$$p = 144\ P$$
$$\rho = 62.4\ S$$

where

P = pressure, psi
S = specific gravity, water = 1
144 = in.²/ft.²
62.4 = pounds per cubic foot of water

These substitutions give the following equation:

$$144\ P = 62.4\ Sh$$
$$P = 0.433\ Sh \tag{6-2}$$

Example: Calculate the hydrostatic head exerted by a column of 40-degree API oil at the depth of 5,000 feet.

In this example, the density of the oil is given in degrees API, and this must first be convered to specific gravity by the following relation:

$$S = \frac{141.5}{131.5 + °API}$$
$$= \frac{141.5}{131.5 + 40}$$
$$= 0.825$$

Equation (6-2) may then be used to calculate the pressure.

$$P = 0.433\ Sh$$
$$= (0.443)(0.825)(5{,}000\ \text{ft.})$$
$$= 1{,}790\ \text{psi}$$

The Hydrostatic Heads of Mud and Cement Slurries

Drilling mud is fundamentally a fluid mixture of clay and water which contains suitable additive materials and chemicals. Cement slurries which are pumped down into wells, generally for the purpose of sealing off water or gas after the cement hardens, are fluid mixtures of water and portland cement. The densities of such mud and cement slurries are usually expressed in terms of pounds per gallon (ppg), based upon the U. S. gallon, which contains 231 cubic inches.

In order to develop an equation for hydrostatic head where the density of the fluid is expressed in pounds per gallon, the following substitutions are made in Equation (6–1).

$$p = 144\ P$$

$$\rho = \left(\frac{1,728\ \text{in.}^3/\text{ft.}^3}{231\ \text{in.}^3/\text{gal.}}\right)(G\ \text{lb./gal.})$$

where G = density, lb./gal., or ppg.

When these subsitutions are carried out, the results are as follows:

$$144\ P = \frac{1,728}{231}\ Gh$$

$$P = \frac{12}{231}\ Gh$$

$$P = 0.052\ Gh \qquad (6\text{–}3)$$

Example: Calculate the hydrostatic head exerted by a column of 10.5 ppg mud at the depth of 5,000 feet.

Equation (6–3) may be used directly for the solution.

$$P = 0.052\ Gh$$
$$= (0.052)(10.5\ \text{ppg})(5,000\ \text{ft.})$$
$$= 2,725\ \text{psi}$$

Total Overburden Pressure

The total overburden pressure is derived from the weight of the materials which lie above any particular depth point in the earth. The oil and gas provinces are composed of sedimentary rocks which were generally deposited in water, and the materials which lie above any particular depth point consist of the rock mineral grains and the saline water contained within the pore spaces between the mineral grains.

In the following paragraphs, the total overburden pressure is derived in terms of an average specific gravity of the water contained in the pores and an average specific gravity of the mineral grains. These conditions are approximate only. The first few hundred feet of rocks may contain fresh water with a specific gravity of 1, while the rocks at greater depths contain salt water whose specific gravity is greater than 1 but usually does not exceed 1.10. Quartz, a very common mineral, has a specific gravity of 2.65. The clays and calcite have specific gravities near 2.7, so that the range of specific gravities of the more common minerals which make up the bulk of the rocks is not too great. An average value of porosity in the rocks is also assumed. The fractional porosity is the ratio of the pore space to the total bulk volume of the rock. Sandstones have porosities which often vary between 10 and 20 per cent, although both lower and higher values are common. The porosity of shales depends largely upon the depth to which the shale has been buried and its subsequent compaction. Representative values may range between 2 and about 10 per cent.

An expression for the total overburden pressure may be arrived at by adding the weights of the mineral grains and the water and dividing by the total area which supports this weight.

$$\text{Total Overburden Pressure} = \frac{\text{Weight of Minerals} + \text{Weight of Water}}{\text{Area}}$$

Also, since weight equals volumes times density,

$$\text{Weight of Minerals} = (1-f)hA\rho_m$$

and,

$$\text{Weight of Water} = fhA\rho_{sw}$$

so that the equation may be expressed as

$$p = \frac{(1-f)hA\rho_m + fhA\rho_{sw}}{A}$$

$$p = (1-f)h\rho_m + fh\rho_{sw} \qquad (6\text{--}4)$$

where

p = total overburden pressure, lb./ft.^{2}
f = fractional porosity
ρ_m = average density of minerals, lb./ft.^{3}
ρ_{sw} = average density of salt water, lb./ft.^{3}
h = depth, ft.

With pressures expressed in psi and density given as specific gravities, Equation (6-4) may be written as

$$P = 0.433(1 - f)hS_m + 0.433fhS_{sw} \qquad (6\text{-}5)$$

where

P = total overburden pressure, psi
S_m = average specific gravity of minerals
S_{sw} = average specific gravity of salt water in pores of the rock.

The total overburden pressure gradient, in terms of psi per foot of depth, would be

$$\frac{P}{h} = 0.433(1 - f)S_m + 0.433fS_{sw} \qquad (6\text{-}6)$$

If a porosity of 10 per cent ($f = 0.10$) and an average specific gravity of minerals of 2.7 and an average specific gravity of 1.07 for salt water are assumed, the pressure gradient is calculated at 1.099 psi per foot of depth. Other representative values result in calculations of slightly less or slightly greater than 1 psi/ft. Since this is a convenient figure, and unless more precise information is available, the theoretical overburden pressure is commonly assumed to be 1 psi/ft.

The overburden pressure is that pressure which the rocks exert by virtue of the weight of the overlying strata. For example, if a hole drilled into the rocks of the earth is filled with a drilling fluid which is under pressure, the bore hole will tend to contain that pressure up to a limit of about 1 psi per foot of depth because of the weight pressure exerted by the overlying strata. The pressure which a bore hole can contain without fracturing of the surroundng rock will be affected to some extent by the very strength of the rock itself and by the residual stresses present in the rock as a result of bending and other tectonic forces. The former seldom amount to more than a few hudred psi, and it is found in practice that the bore hole generally ruptures at pressures less than the theoretical overburden pressure. This is illustrated in Fig. 6-1, which is based on data taken on 385 wells by Brown and Neil.[1]

The fracturing or rupturing of the rock surrounding the bore hole is commonly referred to as breaking down the formation. In this process, a fracture crevice is formed, and whole fluid, such as mud or

---

[1] R. W. Brown and G. H. Neill, "Hydraulic Horsepower Requirements for Well Treatments," *The Petroleum Engineer*, Vol. XXX, No. 3 (March, 1957), B-53-62.

cement slurry, escapes into the rock. This occurs in the hydrafrac process used for well stimulation in zones of lower permeability and sand grains are pumped into the well with oil or water so that the crevices formed are propped open by the sand grains. Similar escape of whole fluid from the bore hole may occur when a well drills into a natural crevice or large solution cavity in the rocks. However, such escape of whole fluid from the bore hole should be clearly differentiated from the loss of water or other liquid from the mud by the process of filtration, which occurs commonly in the bore hole opposite

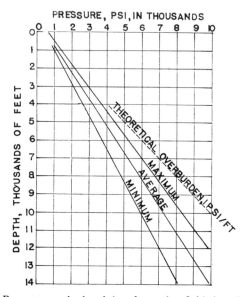

FIG. 6-1. Pressure required to inject fracturing fluids into formations.

all permeable formations for the reason that the pressure in the mud exceeds the pressure in the liquid within the pores of the rock.

Example: During the drilling of a well, a protective string of $10\frac{3}{4}$-inch casing was set and cemented at the depth of 3,000 feet. A blowout preventer, which provides for sealing the annular space between the drill pipe and the protective casing, was mounted on the top of the protective casing. The drilling mud weighs 9.2 ppg. Assuming that the well is full of mud, and that the formations will hold 70 per cent of the theoretical overburden pressure, how much pressure can be held against the well by the blowout preventer?

Solution: Since the casing may be assumed strong enough to contain internal pressures above the 3,000-foot setting depth, the shallowest depth subject to analysis is 3,000 feet.

Assumed bottom hole break-down pressure at 3,000 feet

$$= (0.70)(1 \text{ psi/ft.})(3,000 \text{ ft.})$$
$$= 2,100 \text{ psi}$$

Hydrostatic mud head $= 0.052\ Gh$

$$= (0.052)(9.2 \text{ ppg})(3,000 \text{ ft.})$$
$$= 1,435 \text{ psi}$$

Pressure held by blowout preventer

$$= 2,100 \text{ psi} - 1,435 \text{ psi}$$
$$= 665 \text{ psi}$$

Example: A formation is to be hydraulically fractured at the depth of 9,000 feet. The fracturing fluid has a specific gravity of 0.85. If the formation breaks down at 80 per cent of the theoretical overburden pressure, what pump pressure will be required for the break down?

Expected bottom hole break-down pressure

$$= (0.80)(1 \text{ psi/ft.})(9,000 \text{ ft.})$$
$$= 7,200 \text{ psi}$$

Hydrostatic head of fluid $= 0.433\ Sh$

$$= (0.433)(0.85)(9,000 \text{ ft.})$$
$$= 3,310 \text{ psi}$$

Required pump pressure $= 7,200 \text{ psi} - 3,310 \text{ psi}$
$$= 3,890 \text{ psi}$$

Formation Pressures

The formation pressure is the pressue of the fluid, either oil or gas or water, which exists within the pores of a rock formation. This pressure of the fluid within the pores has also been referred to as the rock pressure. Within oil or gas reservoirs, it is called the reservoir pressure. The formation pressure is distinct from the total overburden pressure and from the pressures which may exist within the rock crystals. By way of contrast, the formation pressure refers to the pressure within the mobile fluids contained in the pore spaces between the mineral grains. Fundamentally, the formation pressure must be less than the total overburden pressure, otherwise the fluid would fracture through the overlying formations and thus escape until the excess pressure would be relieved.

Normal formation pressures are attributable to the hydrostatic head of a column of water extending from the formation to the surface of the earth. The top of the column of water occurs at the water table, the depth at which water will stand in an open well, and this is usually less than two hundred feet below the surface. The top few hundred feet of this water may be fresh, but most of it is salt water. Equation (6–2) indicates that fresh water gives a pressure gradient of 0.433 psi per foot of depth. Within certain regions, it has been established that pressure gradients such as 0.45 or 0.46 psi per ft.

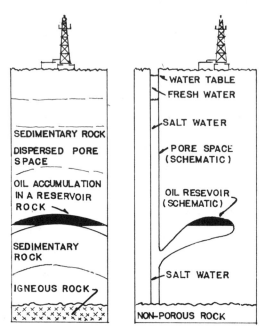

Fig. 6–2. Schematic diagram illustrating the origin of normal pressures in oil reservoirs.

apply. In the Gulf Coast region, the normal pressure gradient is 0.465 psi per foot of depth. This would correspond to salt water of 1.074 specific gravity, or a solution containing 10 per cent sodium chloride. In most cases, the salt water was present in the rocks at the time of deposition, and large quantities of it have been forced out of the shales particularly during the subsequent compaction process. The water occupies the pore spaces between the mineral grains and is disseminated throughout the rocks. Locally, accumulations of oil and gas occur within the pores of the rocks, but the oil and gas contact the water and their pressure is derived from the pressure of the water. Figure 6–2 is a schematic diagram in which all the small pores are

PRESSURE RELATIONS IN THE EARTH AND IN BORE HOLES 69

replaced by a large opening in order to illustrate the origin of normal formation pressures.

Abnormal formation pressures also occur. Unless otherwise modified, the term *abnormal formation pressures* refers to abnormally high formation pressures. Such pressures are greater than the pressure of a column of salt water reaching to the surface. Under these conditions, it is evident that the fluids within the pore spaces are partially sup-

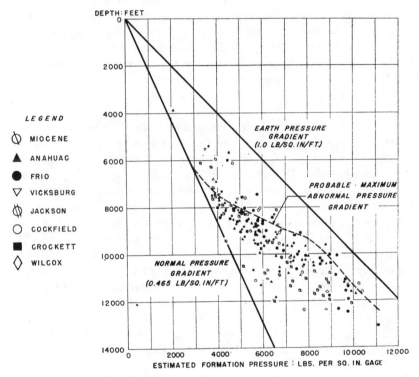

FIG. 6-3. Abnormal pressures in the Texas and Louisiana Gulf Coast region.

porting the weight of the overburden. The upper limit of abnormal pressures is the overburden pressure, about 1 psi per foot of depth, for the rocks cannot contain greater pressures than this. There are no other theoretical limitations concerning the exact amount or the depth limits where abnormal pressures may occur. Abnormal pressures are usually attributed to the compaction of rocks by the overlying sediments under such conditions that the water squeezed out of the shales cannot readily find its way to the surface through continuous layers of permeable rocks such as sandstones. The temperature rise which occurs in sediments when they become deeply buried, with the attendant expansion of the volume of the water in the rocks, could also be a contributing cause together with the compaction of the rocks. Abnormally high pressures are most common in areas containing younger sediments, such as the Gulf Coast region of the United States.

Abnormally high formation pressures may also occur from other causes. One class of such pressures occurs where a formation, a shallow sandstone for example, outcrops in near-by mountains at an elevation appreciably higher than the location of the well. The water entering the sandstone at its outcrop would influence the pressure encountered in the well. Another class of high pressures has occurred where shallow sandstones became charged with high-pressure gas which came from lower formations and entered the shallow sandstone through leaks in the casing.

Abnormally low formation pressures also occur. In such cases, the pressure in the formations is less than the pressure which would be exerted by a column of water extending to the surface. Such conditions cause little difficulty in drilling and completion operations. However, the regions of lower formation pressures sometimes coincide with those where light-weight mud, perhaps less than 10 ppg, must be maintained in order to avoid breaking down the formations and consequently losing mud circulation. No explanation for abnormally low pressures has been widely accepted. Possible explanations might be found in considering the escape of fluids such as gas through faults or other passageways in the rocks or in considering the cooling and shrinkage of fluids in cases where once deeply buried sediments have later become elevated nearer to the surface.

Example: Calculate the normal formation pressure which is to be expected at a depth of 8,000 feet in the Gulf Coast region.

Solution:

$$P = (0.465 \text{ psi/ft.})(8,000 \text{ ft.})$$
$$= 3,720 \text{ psi}$$

Pressure Relations in Bore Holes

It is fundamental that fluids, oil or gas or water, will flow into the well bore unless the pressures within the well bore are greater than, or equal to, the formation pressures. Such flows may be disastrous, as when oil or gas or salt water flows into the well bore from highly permeable formations and causes a blowout. In other cases such flows may be insignificant, where permeabilities and formation pressures are lower, as is the case in those regions and depth intervals where air or gas is circulated as the drilling fluid for the removal of cuttings during drilling. The following discussion is concerned primarily with those cases, which include most of the wells drilled, where formation pressures must be controlled by maintaining a pressure in the bore hole which is higher than the formation pressure.

In general, the density of the drilling mud is maintained at a sufficiently high value that the hydrostatic head of the drilling mud exceeds the formation pressures in the rocks exposed in the drilled hole. The difference between the hydrostatic mud pressure and the formation pressure normally increases with depth. For example, if the formation pressure gradient is 0.45 psi/ft. and the well is being drilled

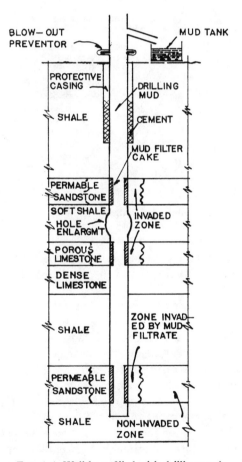

FIG. 6-4. Well bore filled with drilling mud.

with 10 ppg mud, the pressure difference would be $(0.052)(10)h - 0.45 h$, or $0.07 h$. The pressure difference would be 700 psi at 10,000 feet but only 350 psi at 5,000 feet. Such pressure differences, or safety allowances, are required in normal drilling operations. For example, gas may become mixed with the mud as a result of drilling through a rock containing high pressure gas, and such gas zones should not be drilled through too rapidly for that reason. The gas will expand as the mud circulates to the top of the hole and some lightening of the mud column will result. Again, when the drill pipe is raised, a pressure reduction occurs in the bottom sections of the hole, and the magnitude of

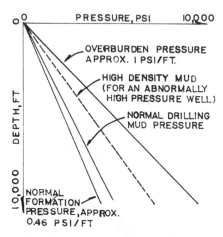

FIG. 6–5. Relation of drilling mud hydrostatic pressure to formation pressure and overburden pressure.

such pressure reductions depends upon the speed of raising the drill string, the clearances involved, and the condition of the drilling fluid, as discussed in Chapter 18.

Mechanical devices are also used at the top of the well for controlling pressures. These devices are fastened to the top of the deepest string of pipe which has been set and cemented in the well, for the deeper rocks exposed just below the bottom of the string of pipe (casing) can contain more pressure than the shallower rocks. These mechanical devices include valves, blowout preventers, and lubricators. Blowout preventers are equipped with rams or equivalent elements which can be operated to seal the annular space between the drill pipe and the string of casing to which the blowout preventer is attached. The lubricators sometimes used in cable-tool operations consist of pieces of pipe which are long enough to accommodate the piece of drilling equipment which is to be introduced into the well. The top of the lubricator is closed but is equipped with a stuffing box through which a cable passes to the piece of drilling equipment. The lubricator may be fastened to the top of the well casing, which has a valve shutting in the well pressure, and after the lubricator is so fastened, the valve may be opened and the piece of equipment lowered into the well.

In the usual case, where formation pressures are controlled by the density of the drilling mud, the lower limit of the mud density is established by the formation pressure plus a safety allowance. An upper limit of mud density is the overburden pressure gradient, which is about 1 psi per foot. Equation (6–3) may be solved for the equivalent mud density, which is about 19.2 ppg. Muds of densities just slightly greater than this theoretical figure have been circulated in drilling wells in a very few cases where extremely high pressures were encountered. On the other hand, in many areas the formations exposed in the well bore will break down at mud densities of 12, 10, or even as low as 9.5 ppg, so that the mud pumped into the well flows back into the formations and cannot be successfully circulated as a drilling fluid. In a large number of cases it has happened that the density had to be held within 0.2 ppg, with the well showing signs of beginning to blow out at the lower limit and with mud losses occurring in the formations at the higher limit.

Filtration from the mud occurs at all places where permeable formations are exposed in the bore hole. In the filtration process, only the liquid portion of the drilling fluid enters the rock formations. This is in contrast to the loss of whole mud which was described in the foregoing paragraph. Where the pressure of the drilling mud exceeds the pressure of the fluids present within the pores of a formation, the pressure difference will cause fluid flow. In the usual case where the clay crystals and other solid materials suspended in the drilling mud cannot enter the small pores of the rock formation, a filter cake composed of the mud solids forms on the exposed face of the formation, and the water associated with those solid particles flows on into the rock. It is evident that the drilling fluid must be of such quality that excessive filtration will not occur, otherwise the filter cake could build up to such thickness that the hole would become filled or bridged opposite a permeable formation with the possibility of sticking the drill pipe or interfering with mud circulation. Also, some oil sands are damaged by excesssive filtration, since they contain clay minerals which swell as a result of being in contact with the freshwater filtrate from the mud and partially close the pores of the sandstone.[2]

Example: A formation has a pressure of 3,720 psi at 8,000 feet. The operator desires to have a safety allowance of 600 psi opposite the formation. What is the required density of the drilling mud?

Solution: Equation (6–3) is rearranged and used as follows:

$$G = \frac{P}{0.052\,h}$$

$$= \frac{3{,}720 \text{ psi formation pressure} + 600 \text{ psi safety allowance}}{(0.052)(8{,}000 \text{ ft.})}$$

$$= 10.4 \text{ ppg}$$

Problems

1. An oil pool has a formation pressure of 3,780 psi at 8,400 feet. What mud weight (mud density) will be required to drill through this formation with a safety allowance of 700 psi? Is this a normal pressure formation? *Ans.:* 10.25.
2. If salt water could be used as a drilling fluid for drilling the formation described in Problem 1, the safety allowance might be reduced to 500 psi. How much salt would be required to mix up 750 barrels of salt water? (Note: One barrel = 42 gal. = 5.615 ft.^{3}.) *Ans.:* 37.4 tons.
3. A well may encounter a high pressure formation with 6,000 psi rock pressure at 8,000 foot depth. If 12.5 ppg mud is in the hole at that time, and if residual bending stresses in the rocks may reduce the pressure that the rocks can contain within the well bore to 80 per cent of the theoretical overburden pressure, what is the minimum depth at which the protective string of casing must be set? *Ans.:* 5,330 ft.
4. If the conditions anticipated in Problem 3 are encountered, how much pressure will the blowout preventer (which is fastened to the top of the protective casing and which can be closed tightly around the drill pipe) have to hold prior to increasing the mud weight? *Ans.:* 800 psi.

---

[2] Additional references of value for this chapter include: G. E. Cannon and R. S. Sullins, "Problems Encountered in Drilling Abnormal Pressure Formations," *API Drilling and Production Practice* (1946), 29–33; T. Gains, "Blowout Prevention," *The Petroleum Engineer*, Vol. XXIX, No. 2 (February, 1956), B-28–30; and J. M. Thorp and J. E. Bailey, "Problems in Completing High Pressure Gas Wells," *The Petroleum Engineer*, Vol. XXVIII, No. 9 (August, 1956), B-21.

5. A protective string of 10¾ inch 40.5 lb./ft. API seamless casing was set at 900 feet. At depth of 1,500 feet the well passed through a water sand containing a large amount of sodium sulphate in solution. The drill string consists of 180 feet of drill collars weighing 15,800 pounds, and the balance of the string is made up of 16.6 lb./ft. drill pipe. The drilling mud weighs 10.5 ppg. If the hole is not filled with mud when coming out of the hole with the drill pipe, at what depth will the mud begin to show sulphate contamination? Assume the pressure in the water sand corresponds to a gradient of 0.46 psi/ft. Note: The ID, or equivalent data, for the casing must be found from a table. Steel weights 488 lb./ft.³
Ans.: 3,080 ft. TD.

CHAPTER 7

Drilling Fluids

THE FUNCTIONS OF DRILLING FLUIDS

The rotary system of drilling requires the circulation of a drilling fluid in order to remove the drilled cuttings from the bottom of the hole and thus keep the bit and the bottom of the hole clean. Drilling fluids are usually pumped from the surface down through a hollow drill pipe to the bit and the bottom of the hole and returned to the surface through the annular space outside the drill pipe. Any cavings from the formations already drilled and exposed in the bore hole must be raised to the surface, together with the drill cuttings, by mud circulation. The cavings and larger drill cuttings are separated from the mud at the surface by flowing the mud through the moving screen of a shale shaker and by settling in the mud pits. The flowing drilling fluid cools the bit and the bottom of the hole. The mud usually offers some degree of lubrication between the drill pipe and the wall of the hole. Flows of oil, gas, and salt water into the well bore are commonly prevented by overbalancing or exceeding formation pressures with the hydrostatic pressure of the mud column.

One of the primary functions of a drilling mud is the maintenance and preservation of the hole already drilled. While shales are usually more easily drilled than other geological formations, they are less stable when exposed in the bore hole. Where the disintegration of a shale is confined to the exposed surface, hole enlargement commonly occurs gradually without endangering the drilling operations. However, where the disintegration results in any appreciable amount of shale sloughing into the hole, there is danger of stuck drill pipe, expensive fishing jobs, and even loss of the well. Such caving of large quantities of shale apparently is caused by disintegration of the caving section of shale or by an undermined condition brought about by disintegration of an underlying shale section. Muds of low filter-loss quality are commonly used to prevent the caving of shales. For example, muds of less than 10 c.c. API water loss are found to be satisfactory for drilling some shales, while muds of even lower filter loss are required for others. Such practices are common in the Gulf Coast, area, and the mud viscosities and gel strengths are maintained low to minimize pressure reductions caused by raising the drill pipe and to facilitate the settling of sand and the escape of entrained gas. On the other hand, muds of higher viscosities, over 100 seconds Marsh Funnel viscosity, with measurable gel strengths of the order of 20 to 30 grams API 10-minute gel strength, are required in some Mid-Continent areas to minimize caving and to sweep cavings continuously from the hole. The caving of unconsolidated formations, such as gravel beds found near the surface, is commonly controlled by using a thick mud which will develop a high gel strength after it flows into such loose materials.

The drilling fluid used must permit identification of drill cuttings and identification of any shows of oil or gas in the cuttings. It must permit the use of the desired logging methods and other well-completion practices. Finally, the drilling fluid should not impair the permeability of any oil- or gas-bearing formations penetrated by the well.

Descriptions of Drilling Fluids

All fluids used in a well bore during drilling operations may be classified as *drilling fluids*. The term is generally restricted to those fluids which are circulated in the bore hole in rotary drilling. Fluids employed for this purpose include gases, liquids, and solids suspended in liquids. Emulsions of oil in water or water in oil are also used for the suspension of solids. Combinations of two fluid streams, each with separate pumps or compressors, are used in some drilling operations.

The gases which have been used as drilling fluids include air, natural gas, exhaust gases from internal combustion engines, and commercial nitrogen. An objectional feature in the use of air is that explosive mixtures may be formed with natural gas produced from the formations penetrated by the drill. Mixtures of air and gas containing approximately 5 to 15 per cent natural gas are potentially explosive in cases where the natural gas is composed principally of methane. Furthermore, the oxygen in air and the carbon dioxide and other combustion products in exhaust gases are agents which commonly cause corrosion. In addition to explosion and corrosion control, the problems connected with the use of gases as drilling fluids include the control of pressures, as discussed in Chapter 6, and circulation-rate requirements and other special problems discussed under air and gas drilling in Chapters 12 and 13. Drill cuttings of the small dust-particle size are not as easily identified by the geologist as are the larger cuttings obtained in circulating liquids. On the other hand, the advantages of using a gas as a circulating fluid include the elimination of lost-circulation problems and the achievement of much faster drilling rates. Producing zones drilled in while using gas circulation suffer no damage from drilling fluids.

The liquids used as drilling fluids include fresh water, salt water, and crude oil. Next to air or natural gas, fresh water is generally conceded to be the best drilling fluid for hole-making purposes, presumably because of its combination of lower hydrostatic pressure and lower viscosity in comparison with other liquid systems. The use of fresh water is limited to those areas and depth intervals where its density is sufficient for the control of formation pressures and where it will not cause excessive caving of shales exposed in the bore hole. Salt water has the advantage of a slightly greater hydrostatic pressure. Crude oil exerts less hydrostatic pressure than water, according to the specific gravity of the particlar oil used.

Some oil sands are damaged by fresh water, as well as by the filtrate from fresh-water muds. One type of damage results from the swelling of clay particles in the pores of the sandstone so that oil or other fluids cannot readily flow through the sand and into the well bore. A second type of damage results from the large amount of water which may flow into an oil or gas zone and invade the formation for a considerable distance from the well bore. In rocks of low permeability, the presence of both oil and water in the small pores may drastically reduce the fluid conductivity of the rock. Salt water will produce less swelling of any clays present, although swelling is controlled by differences in both the kinds and the amounts of ions naturally present with the clays in comparison with the composition of the salt water. Inhibited muds are those which contain a sufficient amount of dissolved salt substances, usually including some calcium, that the water phase of the mud will not tend to leach the naturally occurring salts away from the clay particles in the sandstones and shales. Such inhibited muds reduce mud damage to producing formations and reduce hole enlargement in shale sections exposed in the bore hole. They also reduce the dispersion of shale drill cuttings into the mud and thereby reduce the tendency of the mud to increase in viscosity during the drilling of certain shale sections. It is generally believed that crude or refined oils will not damage oil sands either by producing swelling in any clays present or by unfavorably affecting the oil permeability of the formation.

The vast majority of drilling muds are suspensions of solids in liquids or in liquid emulsions. The densities of such systems can be adjusted between about 7 and 21 pounds per gallon, or 0.85 to 2.5 specific gravity. Where water is used as the liquid phase, the lower limit of the density is about 8.6 to 9 pounds per gallon. In addition to density, other important properties of such suspensions may be adjusted within suitable limits. The filtration quality may be controlled by having a portion of the solids consist of particles of such small size and nature that very little of the liquid phase will escape through the filter cake of solids formed around the bore hole. In addition to preventing the caving and hydration of most shales, this quality also restricts and reduces damage to oil- or gas-producing formations. Control over the viscosity and gel-forming character of such suspensions is achieved within limits by the amount and kind of solids in the suspension and by the use of chemicals which reduce the internal resistance of such suspensions so that they will flow easily and smoothly. The vast majority of drilling muds are suspensions of clays and other solids in water. They are referred to as *water-base muds*.

Oil-base muds are suspensions of solids in oil. High flash-point diesel oils are commonly used as the liquid phase, and the necessary finely dispersed solid is obtained by adding oxidized (air-blown) asphalt. Common weighting agents are used to increase the density. However, the viscosity and thixotropic properties are controlled by special soaps and other special chemicals. Several types of oil-base muds are commercially available and the control of such muds is commonly directed by a representative of the company whose products are being used in the mud system. Oil-base muds are used for such special purposes as preventing the caving of certain shales and particularly as completion muds for drilling into sensitive sands which are damaged by water.

Oil-emulsion muds are commonly of the oil-in-water type of emulsion in which small droplets of oil are dispersed in a continuous water phase. The amount of oil used varies up to 50 per cent of the volume of the mud, although commonly only 10 to 15 per cent is oil. The clays and other minerals and common mud-treating chemicals act as emulsifying agents. Additional agents such as soaps, some patented, are used as emulsifying agents. These emulsion muds are fundamentally water-base muds, and chemical control of them depends upon the type of water-base mud used in making the emulsion mud. Both fresh- and salt-water-emulsion muds are used. The presence of the oil commonly decreases the filtration loss of a mud, as indicated by surface testing, and emulsion muds are often used for their superior well-completion properties. The oil in the emulsion wets the surface of the steel bit, drill collar, and drill pipe, and so helps to reduce any tendency of drilling cuttings to stick to the teeth of the bit or the drill string. The drill pipe usually rotates more easily in the hole, and there is less sticking of the drill string. Faster drilling rates are often reported with oil-emulsion muds. Although the reasons for this are not well understood, they may include the manner in which the emulsion wets the rocks, particularly shales, or the cleaner teeth on the bit, or the reduced hydrostatic mud head resulting from the presence of the oil and additional entrained air or gas in the mud.

Water-in-oil-emulsion muds have been developed principally for well-completion purposes. In these muds, oil is the continuous phase and water is in the form of small droplets. They are referred to as *inverted emulsions*. Special soaps and surfactants are required in their preparation. These muds are not affected by salt, anhydrite, or cement contamination. The filtrate from such muds is oil.

Combinations of fluid streams have been used successfully where compressed air is pumped into the drill string along with the regular drilling fluid. The presence of the air lightens the mud column. The lower hydrostatic pressure has been used to reduce mud losses into formations and to obtain faster drilling rates.

The Composition of Water-Base Mud

Water-base mud consists basically of (1) a liquid phase, water or emulsion; (2) a colloidal phase, principally clays; (3) an inert phase, principally barite weight material and fine sand; and (4) a chemical phase, consisting of ions and substances in solution which influence and control the behavior of the colloidal materials such as clays.

Colloids. Colloidal material is necessary in a mud in order to produce suitable higher viscosities for removing cuttings and cavings from the hole and for suspending the inert material, such as finely ground barite. Colloids are substances composed of individual particles whose dimensions range between about 5 and 500 millimicrons. (One millimicron = 10^{-7} centimeter.) The shales which are drilled up by the bit often disperse in the mud and furnish all or part of the needed colloids from clays present in the shales. It is common for the younger shales of the Gulf Coast and other regions of comparatively recent deposition to furnish an excess of colloidal material, so that in drilling through such beds the mud must be continuously thinned with a stream of water. In other areas the colloids must be purchased and added to the mud, the principal material used for this purpose being finely ground bentonite. Bentonite is a rock deposit, and the desirable clay mineral in the rock is montmorillonite; however, the two terms are commonly used synonymously in the petroleum industry. When bentonite is added to water or to a mud and sufficiently agitated, as by pumping or jetting, the clay divides into crystals of such small size that they cannot be seen by any optical microscope. The small crystals are suspended in water by the thermal vibrations of the water molecules. The clay particles in turn support the larger (visible) particles, such as the inert finely ground barite weight material. In addition to yielding viscosity and suspending weight material, the bentonite type of clays produce a mud that has a low filtration loss. The flat, platelike clay crystals are presumed to overlap each other so tightly that they produce a filter cake of very low permeability.

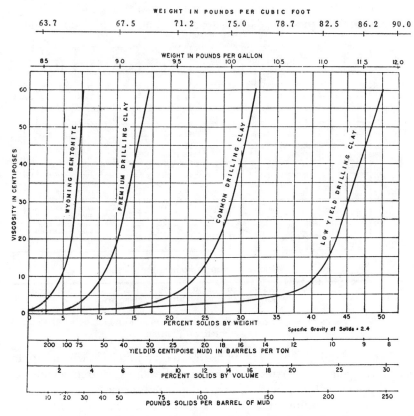

Fig. 7-1. Viscosity resulting from mixing various clays in water.

Other clays of similar properties but of lower quality than the bentonite clays have always been used to some extent in the drilling industry. In addition to being cheaper, some of them are less sensitive to contamination.

Special clays referred to as salt-water clays are required in muds made with saturated salt water. The principal mineral in these clays is attapulgite, which develops in needle-like crystals. The viscosity-producing ability of this clay in either fresh or salt water is about equal to that of the better-grade bentonites in fresh water. However, this clay is often slow to disperse and impart higher viscosity to the mud, particularly if the clay is very dry when added to the mud. For example, in the laboratory, high-speed mixers are required for mixing this clay into a suitable mud. The attapulgite clay is not adversely affected by salt contamination, as bentonite is. However, the salt-water clays do not have low filtration-loss qualities, and it is common to use starch in these muds in order to attain a low filter-loss mud.

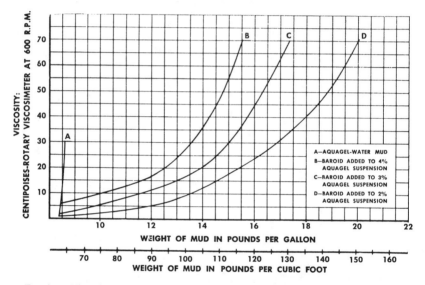

FIG. 7–2. Viscosity-weight relationship of minimum solid muds; aquagel-water muds weighted with baroid.

Two materials are in common use as auxiliary colloids which supplement the mud properties produced by the clays. Starch which has been pre-gelatinized with caustic is usually used in salt-water muds and also in lime-treated muds in order to obtain low filtration-loss quality. Gelatinized starch is an organic colloid which will reduce the filtration loss of all water-base muds. It is not affected by salt or calcium contamination. However, it is more expensive than clays and is used only where necessary. Fermentation or spoiling of the starch, which ends its usefulness, is prevented by using it in saturated salt water or in muds of high pH (pH of 12 or higher) or by adding about 0.5 pound of a formaldehyde-type germicide per barrel of the mud. Sodium carboxymethylcellulose (CMC) is used to reduce the filtration loss of muds. It is less effective in saturated salt water and consequently is little used in salt-water muds. While this material, which is a synthetic organic compound, is more expensive than starch, it requires less precautions against fermentation.

Inert solids. The inert solids in a drilling mud include the fine silica, quartz, and other inert mineral grains of small size which are present in rocks drilled by the bit and which become suspended in the mud. The inert solids which are added to a mud consist principally of finely ground weight material and lost circulation materials. The only commonly used weight material is barite, which has a specific gravity of about 4.3 as commercially available. The higher specific

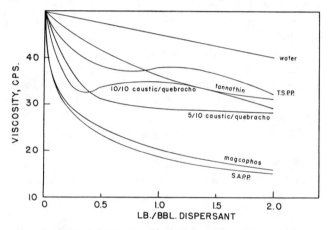

FIG. 7-3. Effect of chemical treatment on Magcogel.

gravity of barite, in comparison with the 2.5 to 2.7 specific gravity range of common clays, shales, sands and limestone, makes it desirable because a mud weighted with barite contains relatively less volume of solids and is more fluid. Barite is a comparatively soft mineral, softer than steel, a quality desirable from the standpoint of keeping down abrasion in mud-pump valves and cylinders. It is insoluble, and it is relatively inexpensive.

Lost circulation materials are added to a mud when losses of whole mud occur in crevices or cracks in the rocks exposed in the well bore. When such materials are used in the mud, it is necessary to bypass the shale shaker, which would otherwise screen the large individual pieces out of the mud. Many materials have been used in attempts to restore lost circulation. The commonly used materials include shredded cellophane flakes, mica flakes, cane fibers, wood fibers, ground walnut shells, and expanded minerals such as perlite. It is common to use two such materials in combination, one in smaller and the other in larger individual pieces. It is not common to use concentrations greater than four pounds per barrel, and usually less is used in the mud

Chemical phase. The chemical phase of water-base mud is important because it largely controls the colloidal phase, particularly the bentonitic type of clays. The chemical phase includes the soluble salts which enter the mud from the drill cuttings and distintegrated portions of the hole and those present in the make-up water added to the mud. The chemical phase also includes the soluble treating chemicals which are used for such purposes as reducing the viscosity and gel strength of the mud.

The bentonitic type of clays have a crystalline structure of such a nature that the mineral develops in thin, fragile sheets which break up into very small individual pieces. Water molecules and the ions and polar compounds present in the water are strongly attracted to the surfaces and broken edges which represent a discontinuity in the space lattice structure of the mineral. The positions to which ions are attracted result in each clay's having a definite capacity for adsorbing, or exchanging, ions. This property of a clay is referred to as its base exchange capacity, in which hydrogen is replaced by sodium and sodium is normally partially replaced by calcium, and so on, according to the degree of attraction between the clay and the particular ion and according to the concentrations of the various ions present in solution in the water phase. Figure 7–4 illustrates the titration of a bentonite in which hydrogen is being displaced by sodium. Sodium ions are usually associated with natural bentonites, and they are referred to as *sodium bentonites.* Figure 7–5 illustrates sodium being displaced by calcium. The degree of influence on the physical properties of a substance which can result from the presence of attracted ions and polar compounds depends upon the size of the individual particles of the substance. The individual particles of colloidal clays are sufficiently small that the chemical phase of the mud exerts a strong influence on the physical properties such as viscosity, gel strength, and filtration-loss quality of the mud.

The soluble salts which enter the mud from the drill cuttings and from the water used in mixing up the mud or added to it are usually undesirable and therefore represent contamination of the mud. Unless otherwise modified, mud contamination refers to salt (chloride) or calcium (from gypsum, anhydrite, or unset portland cement). Most soluble salts, when acting as contaminants, first cause an increase in the viscosity and gel strength of a mud, but these properties eventually decrease at higher concentrations of the contaminant as the mud becomes flocculated, with the individual clay particles gathering together into relatively inactive groups or flocs. The filter loss generally increases throughout the addition of a contaminant. Different types of ions are effective contaminants in different concentration ranges. For example, soda ash is used in treating anhydrite contamination in low-weight muds. The calcium in the anhydrite is precipitated out of solution as calcium carbonate while the sulfate from the anhydrite remains in solution. The influence of the sulfate

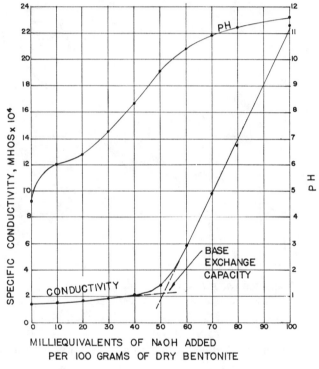

FIG. 7-4. Titration of hydrogen bentonite.

on the clay is much less than that of the calcium, and its effects are counteracted by adding organic thinners such as quebracho.

The chemicals used in treating muds include simple chemical compounds, complex (usually organic) compounds, and surfactants. Caustic soda, NaOH, is used in almost all muds except those treated with phosphates. Both the sodium and hydroxyl ions are beneficial to clays, provided that an organic thinner such as quebracho is also used, and tend to counteract the influence of such contaminants as the chloride ion. Bicarbonate of soda and soda ash are both used to precipitate calcium as calcium carbonate, with the choice between the two depending somewhat upon the desired effect on the pH of the mud, since soda ash will produce or maintain a higher pH value. Barium carbonate is used in muds which have received only light chemical treatments, such as caustic and quebracho, for precipitating both the calcium and sulfate contamination from gypsum or anhydrite. Insoluble calcium carbonate and barium sulfate are formed in this reaction. However, where considerable amounts of gypsum

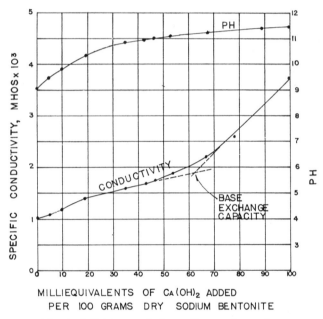

FIG. 7–5. Calcium replacing sodium on bentonite.

or anhydrite must be drilled, it is common to convert to a calcium-treated mud.

Calcium-treated muds are those to which calcium has been added. Lime, calcium hydroxide, is most commonly used for this purpose. The amount of lime added, usually from four to eight pounds per barrel, is sufficient to produce and pass beyond the viscosity increase commonly encountered when adding such chemicals to a mud. Caustic, organic thinners and starch are commonly used in such treatments to maintain good mud qualities. Such muds are insensitive to calcium contamination and have a high resistance to salt contamination, up to about 50,000 ppm NaCl. One of their desirable properties is the ability to retard the dispersion of shale into the mud. Another type of calcium treatment of muds is the addition of plaster of Paris, or gypsum, to produce "gypsum" muds. Starch and starch preservative are added to these muds for water-loss control.

The complex chemical compounds which have proved beneficial in reducing mud viscosities and gel strengths include sodium tetraphosphate. Sodium hexametaphosphate was formerly widely used, but since complex phosphates tend to become unstable and to break down at temperatures encountered below about 6,000 feet, calcium has largely replaced this chemical in treating muds. Quebracho, which is crude tannic acid, is widely used, usually in combination with caustic. Powdered lignite, a mineral of the coal series which contains complex organic acids, has proved widely successful. Calcium lignosulfonate, a wood by-product of the paper industry, is quite successful for calcium-treated muds.

The high temperatures encountered in deep wells have caused difficulties with ordinary lime-treated muds containing caustic. Such lime-base muds tend to solidify at temperatures above 250 degrees Fahrenheit, and this tendency severely interferes with drilling and well-completion operations. In this process, the caustic in the mud combines with the clay to form products similar to those in portland cement, and the calcium present speeds the reaction. Calcium surfactant drilling fluids were developed to overcome these difficulties in high-temperature wells. In addition to clays and barite, the mud system contains calcium sulfate, a water-loss reducing agent such as sodium carboxymethyl cellulose (CMC), and suitable surfactants. The surfactants include a primary surfactant which controls the rheological properties (viscosity and gelation) of the mud, a defoamant, and an emulsifier. The surfactant has been found by X-ray studies to be adsorbed in a mono-layer and to lie flat on the clay particles, thus reducing the viscosity effects of the particles and also, with the aid of the calcium ion, the tendency for shale drill cuttings and exposed formations to disperse into the mud. The mud is neutral, pH 7.0 to 8.0, so that high temperature reactions between clay and caustic do not occur. High temperature degradation of CMC was observed, and resulted in increased water loss, but this condition was handled with temperatures as high as 400 degrees Fahrenheit by further additions of CMC.

MUD PROPERTIES AND THEIR MEASUREMENT

The methods for measuring the properties of drilling fluids are described in detail in the bulletin *API RP 29*, published by the American Petroleum Institute (Dallas, Texas), which is the "Recommended

TABLE 7-1
Conversion Table for Mud Gradient

Gradient, psi per 1,000 ft. of Depth	Density lb. per gal.	Density lb. per cu. ft.	Specific Gravity	Gradient, psi per 1,000 ft. of Depth	Density lb. per gal.	Density lb. per cu. ft.	Specific Gravity
433	8.3	62.4	1.00	800	15.4	115.2	1.85
440	8.5	63.4	1.02	810	15.6	116.6	1.87
450	8.7	64.8	1.04	820	15.8	118.1	1.90
460	8.9	66.2	1.07	830	16.0	119.5	1.92
470	9.1	67.7	1.09	840	16.2	121.0	1.94
480	9.2	69.1	1.11	850	16.4	122.4	1.97
490	9.4	70.6	1.13	860	16.6	123.8	1.99
500	9.6	72.0	1.15	870	16.8	125.3	2.01
510	9.8	73.4	1.18	880	16.9	126.7	2.03
520	10.0	74.9	1.20	890	17.1	128.2	2.05
530	10.2	76.3	1.22	900	17.3	129.6	2.08
540	10.4	77.8	1.25	910	17.5	131.0	2.10
550	10.6	79.2	1.27	920	17.7	132.5	2.12
560	10.8	80.6	1.29	930	17.9	133.9	2.15
570	11.0	82.1	1.32	940	18.1	135.4	2.17
580	11.2	83.5	1.34	950	18.3	136.8	2.19
590	11.4	85.0	1.37	960	18.5	138.2	2.22
600	11.6	86.4	1.39	970	18.7	139.7	2.24
610	11.7	87.8	1.40	980	18.9	141.1	2.27
620	11.9	89.3	1.43	990	19.1	142.6	2.29
630	12.1	90.7	1.45	1,000	19.2	144.1	2.30
640	12.3	92.2	1.48	1,010	19.4	145.5	2.33
650	12.5	93.6	1.50	1,020	19.6	147.0	2.35
660	12.7	95.0	1.52	1,030	19.8	148.4	2.38
670	12.9	96.5	1.55	1,040	20.0	149.9	2.40
680	13.1	97.9	1.57	1,050	20.2	151.3	2.42
690	13.3	99.4	1.60	1,060	20.4	152.7	2.45
700	13.5	100.8	1.62	1,070	20.6	154.2	2.47
710	13.7	102.2	1.64	1,080	20.8	155.6	2.50
720	13.9	103.7	1.67	1,090	21.0	157.1	2.52
730	14.1	105.1	1.69	1,100	21.2	158.5	2.54
740	14.3	106.6	1.71	1,110	21.3	160.0	2.57
750	14.4	108.0	1.73	1,120	21.5	161.4	2.58
760	14.6	109.4	1.75	1,130	21.7	162.8	2.60
770	14.8	110.9	1.78	1,140	21.9	164.3	2.63
780	15.0	112.3	1.80	1,150	22.1	165.7	2.65
790	15.2	113.8	1.82				

Practice for Standard Field Procedure for Testing Drilling Fluids." The procedures given in *API RP 29* should be followed exactly, so that reproducible and comparable results may be assured. The significance of some of the common mud tests are described in the following paragraphs.

Density. The desnity of a drilling fluid is of great importance because it determines the hydrostatic pressure which the mud will exert at any particular depth. In the petroleum industry, the term *mud weight* is synonymous with mud density. Mud weight or density is usually expressed in pounds per U. S. gallon of 231 cubic inches, although it is also expressed in pounds per cubic foot in some areas. A newly adopted system consists of expressing the mud density in terms of its apparent hydrostatic head, and the units used are psi per thousand feet of depth. Any known volume can be weighed to determine the density; however, it is convenient to use a mud balance such as the one shown in Fig. 7-6.

Fig. 7-6. A mud balance.

Viscosity. The viscosity of a fluid is its internal resistance to flow. Newton's concept of viscosity may be explained by analogy to a deck of perfect playing cards. When the deck of cards is lying on a table, if the top card is moved along in a horizontal direction, each of the other cards will slide slightly, so that the total sliding motion will be divided equally among all of the cards. To arrive at the common unit of viscosity, consider the cards to be one centimeter square and consider the deck to be one centimeter high. Then, if a horizontal force of one dyne were sufficient to produce a relative velocity of one centimeter per second between the top and the bottom of the deck, the

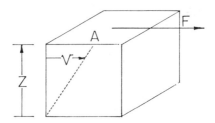

Fig. 7-7. Force and velocity relations in flowing liquid.

viscosity (internal resistance to relative motion) would equal one poise. This may be expressed mathematically by the following equation in differential form:

$$\frac{F}{A} = n\frac{dv}{dz}$$

where

F = force causing flow, dynes
A = lateral surface area over which the force is applied, sq. cm.
n = coefficient of viscosity, poises
dv/dz = velocity gradient perpendicular to the direction of flow, (cm./sec.)/(cm.), or sec.$^{-1}$

Water at room temperature has a viscosity of about one one-hundredth of a pose, or one centipoise. The centipoise is the preferred unit for expressing viscosity, both because it is a convenient size and because it furnishes an immediate comparison to a familiar substance, water.

Liquids may be divided into two general classes in regard to their flow characteristics. The first class includes those liquids such as water, glycerine, motor oils, kerosene, and similar liquids whose viscosity, at any given temperature and pressure, is constant. The liquids of this group are called *Newtonian fluids*, because they obey Newton's concept of viscosity, and they are sometimes referred to as true liquids. The smallest force applied to these fluids will cause them to start flowing slowly. There is no evidence of any permanent internal structure within the fluid.

The second class of liquids includes those whose viscosity is not constant at the particular temperature and pressure involved, but depends upon the additional factor of flow itself. These liquids are called *non-Newtonian fluids*. Slurries of portland cement in water and colloidal suspensions as in drilling muds are examples of this class. In particular, the thixotropic drilling muds—those which develop a gel (or jelly-like) structure upon quiescent standing—furnish an excellent example. When they start to flow, their measured or apparent viscosity is high, but it decreases as flow is continued and the gel structure is broken up by flow energy. This may be spoken of as the flow history of the fluid. However, even after prolonged flowing (sometimes referred to as shearing to equilibrium), the measured or apparent viscosity depends upon the rate of flow involved in the measurement. For the drilling fluids of interest, the viscosity generally decreases as the rate of flow increases and appears to approach some definite lower value at very high rates of flow. Within the non-

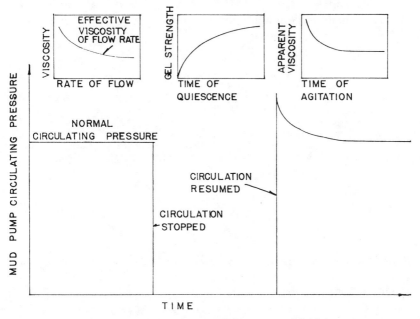

FIG. 7-8. Correlations between laboratory and field data.

Newtonian fluids, there is evidence of internal structure; and the extent of this structure and consequently the apparent viscosity depend upon the previous flow history and flowing conditions of the fluid. Figure 7-8 illustrates typical drilling-mud data and correlates it with field data.

One of the early workers in the field of plastic viscosity was E. C. Bingham, who developed a viscosity relationship for such fluids. Figure 7-9 presents graphically both Newton's and Bingham's relationships for shearing stress *vs.* the rate of shear (or flow), and both apply only at the slower rates of flow where turbulent flow does not exist. Bingham found that for flow through capillaries and small tubes, if the rate of flow were plotted *vs.* the pressure causing the flow, the data would fall on a fairly straight line, which did not pass through the origin but which had an intercept on the pressure axis. He gave the name of *Yield Point* (the Bingham Yield Point) to this intercept. Later experiments with rotating-cup viscometers, as well as with pipes up to two inches or more in diameter, have produced data which can be plotted in a similar manner. Two facts should be mentioned in connection with such plots. The line of data points is not always exactly straight; sometimes the curvature is in one direction, sometimes in the opposite direction. Also, deviations from the straight line quite commonly exist at very slow rates of flow, so that

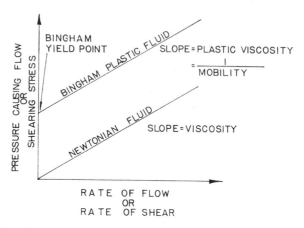

Fig. 7–9. Viscosity characteristics of Newtonian fluids and Bingham plastic fluids.

the intercept found by extrapolation to the pressure or shearing stress axis is only an apparent intercept. Nevertheless, the plot has considerable practical value, for it gives a fairly accurate description of the viscous nature of the fluid in the region of practical interest.

One of the standard instruments used in determining the viscosity of drilling fluids is the Stormer viscometer, illustrated in Fig. 7–10. This is a rotating-cup type of viscometer of standard dimensions, in which the inner cup rotates and the liquid is sheared between the rotating and the stationary cup. For cups, or cylinders, of nearly equal diameters, the rate of shear within the mud is nearly constant throughout the region between the two cylinders. The torque transmitted from the rotating cylinder to the stationary cylinder is, of course, constant throughout the region occupied by the fluid between the two cylinders, and the shear rate at any value of the radius between the two cylinders adjusts to this condition. The amount of torque so transmitted through the fluid depends chiefly upon the viscosity of the fluid between the two cups. (Density of the fluid has a minor effect in such instruments as the Stormer, where no cap is provided to prevent centrifugal force from causing the fluid level to rise near the outer edge and lower in the center portions.) On the other hand, capillary tubes, which are usually preferred for measuring the viscosities of true liquids, are less suitable for measuring the

Fig. 7-10. Stormer viscometer.

viscosities of such plastic fluids as drilling muds because the rate of fluid shear varies from zero at the center to a maximum at the wall of the capillary. It is inevitable that the determined viscosity of a plastic fluid depends upon the instrument used and the conditions of the measurement. Stormer viscometers are usually calibrated with mixtures of water and glycerine of known viscosity. Such mixtures are Newtonian fluids, and their viscosity changes less with slight changes in temperature than that of petroleum oils. The standard rotational speed of the cup in measuring viscosity with the Stormer instrument is 600 rpm. The results were formerly reported in grams weight required to drive the rotating cup at a speed of 600 rpm, but at present calibration data is used to convert the driving-weight data into centipoises.

The Fann viscometer has been widely adopted for measuring the viscosity of drilling fluids. This instrument, which is illustrated in Fig. 7–11, is a rotating-cup type of viscometer which is driven by an electric motor. The laboratory type of instruments are run from standard 110-volt, 60-cycle electric current and field models are available which run from an automobile battery. In this instrument the outer cup is rotated, and the torque produced on the inner cup

FIG. 7–11. Fann V-G meters—laboratory models.

gives a measure of the viscosity. Since the inner cup is held stationary, the dimensions of the vessel containing the mud are not critical, and the common mud cup is used for this purpose. The resisting torque and torque scale attached to the inner cup are adjusted so that the scale reading gives the viscosity (of a Newtonian fluid) directly in centipoises when the outer cup is rotated at 300 rpm. The general equation of such rotating cup viscometers is

$$\eta_A = (\text{const.}) \frac{R}{(\text{rpm})}$$

where η_A is the apparent Newtonian viscosity, R is the scale reading, and rpm designates the rotational speed in revolutions per minute.

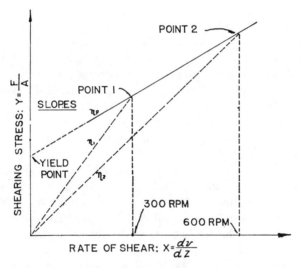

FIG. 7–12. Analysis of viscosity relations.

If the rotating speed were doubled, the scale reading would be doubled. Accordingly, the apparent viscosity of a drilling fluid at the standard speed of 600 rpm is one-half of the scale reading. The plastic viscosity, according to Bingham's concept, would be given by the slope of the line from Point 1 to Point 2 on Fig. 7–12. Where η_P is the plastic viscosity, this may be expressed mathematically as

$$\eta_P = \frac{y_2 - y_1}{x_2 - x_1}$$

The following geometric relations also apply:

$$y_2 = \eta_2 x_2$$
$$= 2\eta_2 x_1$$
$$y_1 = \eta_1 x_1$$

Substituting the relations gives the following:

$$\eta_P = \frac{2\eta_2 x_1 - \eta_1 x_1}{2x_1 - x_1}$$
$$= 2\eta_2 - \eta_1$$

However, the instrument is adjusted so that $R_{600} = 2\eta_2$ and $R_{300} = \eta_1$. Therefore,

$$\eta_P = R_{600} - R_{300}$$

where R_{600} and R_{300} are the scale readings at 600 and 300 rpm respectively. From the geometric conditions shown in Fig. 7–12 it is evident that the Yield Point is proportional to $y_1 - (y_2 - y_1)$, which is proportional to the reading at 300 rpm minus the plastic viscosity. The cups of the Fann instrument are proportioned so that the reading at 300 rpm minus the plastic viscosity is numerically equal to the Yield Point expressed in pounds per 100 square feet.

The first standard instrument used for measuring the viscosity of drilling fluids was the Marsh Funnel, illustrated in Fig. 7–13. The measurement is made by filling the instrument (1,500 c.c. capacity) and determining the time in seconds required for one U. S. quart to flow out through the tube at the bottom of the funnel. It is obvious that the density of the fluid influences the exit time. More important in most instances, the tendency of the fluid to build up gel structure during the time of the test and thereby develop a higher apparent viscosity also influences the time of efflux. This combination of viscosity and thixotropic-gel effects which are measured by the Marsh Funnel, while not subject to exact scientific interpretation, has made the instrument a favorite since it was first introduced in the drilling industry. In the Marsh Funnel fresh water has a viscosity of 26 seconds. Many good muds have viscosities between 34 and about 50 seconds. Highly thixotropic muds used to combat some caving shales may have viscosities of 90, 100, and occasionally 150 or more seconds.

Gel strength. Most drilling fluids exhibit some degree of thixotropy. A thixotropic fluid is one which develops a gel or jelly-like structure during quiescence but which reverts to a sol or liquid suspension upon agitation. The sol-gel reaction is reversible and essentially isothermal. Measurements of gel strength are measurements of shear

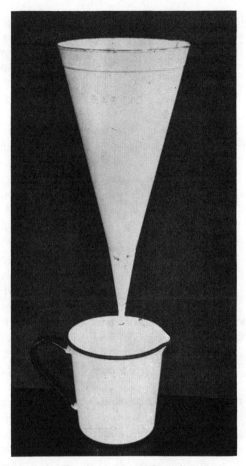

Fig. 7-13. Marsh Funnel.

strength. A definite area of rupture or sliding within the material is involved, and the force necessary to produce such sliding is determined in the measurement. Accordingly, gel strength could be reported in such units as pounds per square foot. The Stormer viscometer is commonly used for measuring gel strength by hanging gram weights on the cord which normally drives the rotating cup. The initial gel strength is measured as the grams weight required to turn the spindle between one-fourth and one-half turn when the measurement is made immediately after the mud has been thoroughly agitated. The ten-minute gel strength is measured as the number of grams weight required to cause the first steady motion of the spindle after a ten-minute quiescent period. The Stormer gel strength in grams multiplied by 0.00326 gives the gel strength in pounds per square foot. The Fann viscometer is also commonly used for measuring gel strength, and the dimensions of the instrument are such that the dial reading gives the gel strength directly in pounds per 100 square feet.

Drilling in most areas is facilitated by using muds of low gel strength. Such muds minimize pressure reductions at the bottom of the hole caused by withdrawing the drill pipe, and they permit small bubbles of gas to escape from the mud, thereby lessening the danger of blowouts. Muds having zero initial and zero ten-minute gel strengths are considered very desirable in the United States Gulf Coast area. Calcium contamination in fresh-water muds characteristically causes "flat" gels, such as 20-gram initial and 20-gram ten-minute gel strength, which would be referred to as a *high flat gel*. Muds having gel strengths greater than about 8-gram initial and 40-gram ten-minute gel strength are rarely used in deeper drilling. The treating chemicals used for reducing mud viscosity also reduce the gel strength.

A phenomenon which is the reverse of thixotropy has been noted in some muds in southern Louisiana. Under flowing conditions, a gel strength developed which was sufficient to stop rotation of the drill pipe. After it was left standing for twenty minutes, drilling could be resumed for about fifty minutes, with the process then being repeated. This phenomenon is apparently rare and is little understood at present.

Filtration-loss quality. The filtration-loss quality of a mud is commonly referred to as the water loss of a mud. Low water-loss muds are desirable so that thick filter cakes which might stick the drill pipe will not form in the hole on exposed faces of permeable zones. They also protect possible producing zones by limiting the amount of water which can seep into them from water-base muds. This quality of the mud is measured at the surface in a standard low pressure filter cell. A pressure of 100 psi ±5 per cent is applied to the cell, and the mud filter cake forms on a piece of filter paper which covers the bottom of the cell. The filter paper is supported by a 60–80 mesh screen, and water which passes through the filter cake and filter paper finds its way to the drain tube between the interlaced wires of the screen. The liquid filtrate is collected in a glass graduate, and the results of the test are reported as cubic centimeters of filtrate collected in thirty minutes time. The cell is then opened, and the filter paper with the mud filter cake is removed and lightly washed with water. The thickness of the cake is measured to the nearest thirty-second of an inch, usually by forcing a thin plastic rule down into the filter cake.

The rate of filtration in such a filter cell may be described mathematically as

$$\frac{dV}{dt} = \frac{CP}{\eta_L R}$$

where C is a proportionality constant involving the area of filtration, P is the driving pressure, η_L is the viscosity of the liquid filtrate, and R is the resistance to flow. Where the resistance to flow depends entirely on the mud filter cake already deposited, which is a fair approximation for the standard filter cell, the resistance may be approximated as

$$R = rwV$$

where r is the resistance per unit weight of mud solids, w is weight of solids per unit volume of filtrate collected, and V is the volume of liquid filtrate already collected. It is generally found, however, that the resistance of a drilling-mud filter cake is dependent on the applied pressure. The filter cake is compressible, and the higher the pressure is, the closer the individual particles in the filter cake are pressed together, so that higher pressures do not result in proportionately higher rates of filtration. This may be approximated by the power-type relation

$$r = r_0 P^b$$

where r_0 and b are experimentally determined constants. These relations may be substituted in the rate equation and integrated for a constant pressure filtration as follows:

$$\frac{dV}{dt} = \frac{CP}{\eta_L r_0 P^b w V}$$

$$\int_0^V V dV = \frac{CP^{1-b}}{\eta_L r_0 w} \int_0^t dt$$

$$V^2 = \frac{2CP^{1-b} t}{\eta_L r_0 w}$$

$$V = \sqrt{\frac{2CP^{1-b} t}{\eta_L r_0 w}}$$

$$V = \text{const.} \sqrt{t}$$

The above equation states that the volume of filtrate will be proportional to the square root of the time of filtration. Figure 7–14 illustrates the results of a water-loss test in which the volume of filtrate is plotted *vs.* time on a square-root scale. The line drawn through the data points commonly does not pass through the origin because of the "hold up" volume of water required to fill the voids in the wire screen under the filter paper. Where exact values of the water-loss quality of a mud are not required, the volume of filtrate collected in

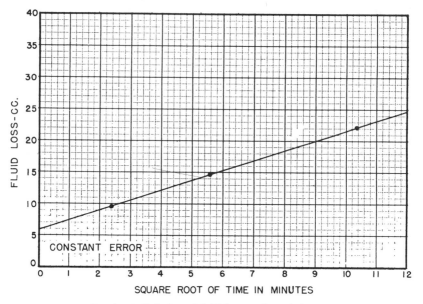

FIG. 7–14. Relation of fluid loss to elapsed time.

seven and one-half minutes may be multiplied by two to approximate the water-loss quality, since

$$\sqrt{30/7.5} = 2$$

This equation also states that the volume of filtrate collected should be inversely proportional to the square root of the viscosity of the liquid phase of the mud. This offers a means of comparing filtration rates at different temperatures; for example, where the viscosities of the liquid phase are known for the different temperatures considered, provided, however, that the temperature changes in no way affect the colloids in the mud. The effects of pressure are variable and depend upon the individual mud. If a mud should have an incompressible filter cake, the exponent b would have a value of zero and the amount of filtrate collected after equal times of filtration would vary as the square root of the pressure. If doubling the pressure doubles the resistance to flow through the filter cake, the exponent b would have a value of unity and the amount of filtrate collected after equal times of filtration would be independent of the pressure. Figure 7–15 illustrates data showing the effect of pressure on the water loss of drilling muds. Muds composed chiefly of bentonite have about the same water loss at 1,000 psi as they have at 100 psi. Water-loss tests are seldom made at pressures other than 100 psi, except where studies being made of the effects on the mud of both high pressures and high temperatures.

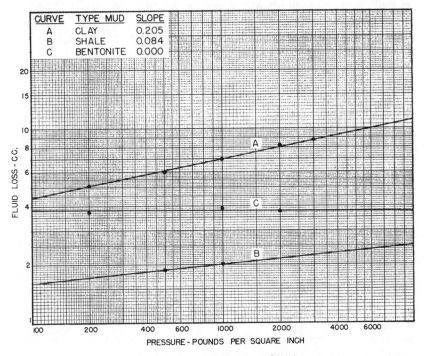

FIG. 7–15. Effect of pressure on fluid loss.

Sand content. The sand content of a mud is determined by diluting a known volume of mud with water and flowing it through a 200-mesh screen. The screen is washed with additional water until only particles too large to pass through the screen remain on it. Then the screen is inverted and the large particles are washed into a tapered graduated tube. The height to which the particles fill up the tube is taken as the volume of the sand grains, and the amount of sand so measured is converted (by the graduations on the tube) to a percentage of the volume of the original mud sample. Large amounts of sand are undesirable because of possible settling in the hole and because the sand is abrasive to mud-pump cylinders, liners, and valves. Sand contents of less than 1 per cent are desirable for the above reasons, although as much as 4 or more per cent must be tolerated in some of the rapid shallow-drilling operations.

Chemical tests. Most of the chemical tests used in controlling muds are made on the mud filtrate. However, the pH is usually measured by placing pHydrion paper on the surface of whole mud. The Chaney test for excess lime present in a mud also uses a sample of whole mud, as well as a sample of filtrate, and both are titrated with 0.02 normal sulfuric acid to the phenolphthalein end point in order to determine the amount of undissolved lime present in the mud. The salt content is measured with respect to the amount of chloride ion present in the filtrate. The chloride ion is precipitated with silver nitrate in a dilute neutral solution with potassium chromate indicator. The results are reported as ppm chloride or as ppm sodium chloride (ppm chloride $\times 1.65 =$ ppm sodium chloride), and it should be made clear which basis of reporting is used. The total alkalinity of muds whose pH is above approximately eleven is best measured by titrating the filtrate with acid. The *P alkalinity* is the number of c.c. of 0.02 normal sulfuric acid required to titrate one c.c. of filtrate to the phenolphthalein end point (pH $= 8.3$). The *M alkalinity* is the total number of c.c. of 0.02 normal sulfuric acid (including the P alkalinity) required to titrate one c.c. of filtrate to the methyl orange end point (pH $= 4.3$) The presence of calcium in a fresh-water mud is detected by adding drops of dilute ammonium oxalate to the filtrate to obtain a milky precipitate of calcium oxalate. The standard soap-solution test is used to obtain a measurement of the amount of calcium present in the filtrate. The basis of this test is that calcium (also any magnesium present) will precipitate the soap until used up, and no stable lather can be produced by shaking the container until some excess of soap has been added. The Versenate method is also used for determining the total calcium and magnesium present in the filtrate.

Pilot testing. Pilot testing of drilling muds consists of determining the physical properties of a mud, adding treating chemicals and/or mud materials to a relatively small sample of the mud, and subsequently determining the improvement in such physical properties as viscosity, gel strength, and filtration loss. Drilling muds are complex colloidal suspensions, and as no two are exactly alike, their responses to any treatment will vary at least slightly from one mud to another. The response of a mud to a treatment, or addition, of chemicals or materials can be predicted from the results obtained during pilot testing on small samples. Costly or even disastrous results of improper treatment can be avoided or minimized by noting the response in a small sample of the mud.

It is desirable to choose a sample of mud of such size that the weight in grams of chemicals or materials added to the sample will be numerically equal to the number of pounds added per barrel of mud. Since the weight ratio of the added material to the original material is the same in both cases, and since one barrel of water weighs 350 pounds, the weight ratio equation becomes

$$\frac{1 \text{ gr. material}}{(X \text{ c.c.})(\text{sp. gr.})} = \frac{1 \text{ lb. material}}{(350 \text{ lb.})(\text{sp. gr.})}$$

whence $X = 350$ c.c., the proper size of mud sample to use. For example, if suitable mud viscosity and gels were obtained by adding one-fourth gram of quebracho and one-fourth gram of caustic soda to a 350 c.c. sample of mud, the indicated treatment for a mud system containing a total of 800 barrels of mud in the hole and pits would be 200 pounds of caustic and 200 pounds of quebracho. The addition of these materials should be slow and spread out over one or two mud-circulation cycles so that all parts of the mud are treated evenly.

Mud Mixing and Weighting Calculations

Muds must be composed of materials which will impart suitable density, water-loss quality, viscosity, and thixotropic gel properties to the mud. The density of the mud is controlled by the liquids and solids contained in the mud. The chemicals used in treating the mud, if they are effective, are used in comparatively small amounts, so that they need not be considered in mud-mixing calculations. A comparatively few simple relations will handle the necessary mud-mixing calculations required for controlling the density of a drilling mud.

There are two general types of mud-mixing problems which must be considered. First, there is the problem of adding materials to a mud which has already been mixed. This arises when the density of the mud in a drilling well must be increased by adding weight material or decreased by adding water or oil. In such cases the present (original) volume of the mud is known, at least approximately, and its present density and the required final density are also known. The density of the material to be added is known, but the amount of material to be added must be calculated. Second, there is the problem of mixing up a pit or tank of mud. This occurs when there has been a great loss of mud at a drilling well, or when a well is to be worked over and the mud must be mixed up prior to circulating it into the well for pressure-control purposes. In such cases, the required (final) volume of the mud is known, as well as the required (final) density. The densities of the materials which will be mixed together to form the mud are also known quanities. The amounts of the colloidal materials to be used will be governed by the properties which they impart to the mud, but the amount of weighting material necessary to achieve a required final density may be calculated. It is desirable to calculate the amount of water which should be run into the pit or tank before

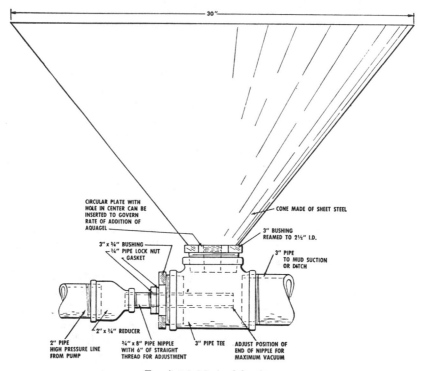

FIG. 7-16. Mud-mixing jet.

starting mud-mixing operations, so that the desired volume of mud will result after the solid materials are added.

Solid materials are commonly added to a mud through a hopper equipped with a jet. In this operation, the liquid mud or water is drawn out of the mud pit through the pump suction and pumped under pressure to the jet at the bottom of the hopper. The kinetic energy of the jet supplies the energy necessary to disperse the solid into the mud. Bentonitic clays in particular require a great amount of energy for dispersion, since they divide into invisibly small crystals with an enormous surface area. Such materials must be added slowly, usually at rates between two and five minutes per sack of bentonite, for it is very difficult to disperse any lumps which form in the mud. Both clays and weighting material have been added slowly by hand at the mud-pump suction. Liquids are usually added to the mud in a small stream, which is a small fraction of the total mud flowing, either at the mud-pump suction or at the shale shaker. Make-up water, which is required in drilling to compensate for filtration losses into the formations and for thinning the mud when drilled shales are dispersing into it, is added as a stream of water at the vibrating screen of the shale shaker, which separates drill cuttings out of the mud returning from the well. Most mud systems are equipped with jets so arranged that an auxiliary mud pump can be used to circulate one or more high-velocity mud streams down onto the surface of the mud pit. These jets serve to keep the mud well dispersed and also aid in the release of gas which may have become entrained in the mud.

In developing equations suitable for mud-mixing calculations, the following symbols will be used:

W_0 = weight of the original mud, before a particular material is added, lb.
W_A = weight of the added material, lb.
W_F = weight of the final mud, after a particular material is added, lb.
V_0 = volume of the original mud, gal.
V_A = volume of the added material, gal.
V_F = volume of the final mud, gal.
D_0 = density of the original mud, lb./gal.
D_A = density of the added material, lb./gal.
D_F = density of the final mud, lb./gal.
w = pounds of material added per barrel of original mud.
$\tilde{w}$ = pounds of material added per barrel of final mud.

Since density is defined, in engineering terms, as weight divided by volume, the following relations are by definition:

$$D_0 = \frac{W_0}{V_0}$$

$$D_A = \frac{W_A}{V_A}$$

$$D_F = \frac{W_F}{V_F}$$

The following relations result from the definitions of w and $\tilde{w}$, since there are 42 U. S. gallons per barrel:

$$w = 42 \frac{W_A}{V_0}$$

$$\bar{w} = 42 \frac{W_A}{V_F}$$

Mud-weighting calculations are based on a material balance, or the Law of the Conservation of Mass, which states that the mass of the materials will remain unchanged. The final weight of the materials after mixing will be the same as the combined weights of the materials before they were mixed together. In terms of the symbols used, this is expressed as follows:

$$W_F = W_0 + W_A \tag{7-1}$$

All or most of the mud-weighting problems can be solved by substituting the known values into the material-balance equation and solving for the remaining unknown. However, some of the substitutions are rather difficult and the practice is not recommended. Rather, it is advantageous to develop a few special relations, based on the material balance, which can be used more conveniently.

In solving drilling mud-weighting calculations, it is common to use the following relation:

$$V_F = V_0 + V_A \tag{7-2}$$

This relation, which assumes that the volume of materials does not change upon mixing, is not strictly true. However, no appreciable error is involved in assuming that it is true in the water, oil, clay, and barite systems of drilling fluids. The common case in which a volume change of the mud causes concern is when the mud becomes gas cut, with small bubbles of gas entrained in the mud, thus decreasing the effective density of the mud and sometimes causing blowouts. The gas entrained in the mud usually escapes from the mud at the shale shaker, or its release is aided by jetting the mud pits or by passing the mud through a vacuum chamber. Accordingly, the volume relation given above will be used in developing further relations.

A series of special cases is considered in the following material, which illustrate the solutions of common mud-weighting problems.

1. Density resulting from adding solids to a mud.

This can be solved by the definition of density and by using appropriate substitutions.

DRILLING FLUIDS

$$D_F = \frac{W_F}{V_F}$$

$$D_F = \frac{W_0 + W_A}{V_0 + V_A}$$

$$D_F = \frac{V_0 D_0 + W_A}{V_0 + \dfrac{W_A}{D_A}}$$

$$D_F = \frac{D_0 + \dfrac{W_A}{V_0}}{1 + \dfrac{W_A}{V_0 D_A}} \tag{7-3}$$

Since $w = 42(W_A/V_0)$, Equation (7–3) may be expressed as

$$D_F = \frac{D_0 + \dfrac{w}{42}}{1 + \dfrac{w}{42 D_A}} \tag{7-4}$$

If the material to be added is barite, which has a specific gravity of about 4.3, or about 35 lb./gal., then $D_A = 35$ lb./gal., and Equation (7–4) reduces to

$$D_F = \frac{D_0 + \dfrac{w}{42}}{1 + 0.00068w} \tag{7-4a}$$

If the material to be added is a clay, the specific gravity will be taken as 2.5, so that $D_A = 20$ lb./gal.; then Equation (7–4) reduces to

$$D_F = \frac{1 + \dfrac{w}{42}}{1 + 0.0012w} \tag{7-4b}$$

Example: Ten tons of barite were added to 800 barrels of 9.2 ppg mud. What was the density after adding the barite? Also, calculate the increase in volume of the mud.

Solution:

$$w = \frac{(10 \text{ tons})(2000 \text{ lb./ton})}{800 \text{ bbl.}}$$

$$= 25 \text{ lb./bbl. of original mud}$$

Equation (7–4a) may be used directly.

$$D_F = \frac{D_0 + \dfrac{w}{42}}{1 + 0.00068w}$$

$$= \frac{9.2 \text{ ppg} + \dfrac{25 \text{ lb./bbl.}}{42}}{1 + (0.00068)(25 \text{ lb./bbl.})}$$

$$= 9.64 \text{ ppg}$$

The increase in volume equals the volume of the barite. This can be calculated from the known density of the barite.

$$\text{Volume of barite added} = \frac{(10 \text{ tons})(2{,}000 \text{ lb./ton})}{(35 \text{ lb./gal.})(42 \text{ gal./bbl.})}$$

$$= 13.6 \text{ bbl.}$$

Example: In mixing up a mud system, a basic bentonite-water suspension containing 6 per cent bentonite by weight was mixed before adding the barite weight material. What was the density of the basic bentonite-water suspension?

Solution: This problem is best solved by taking as a basis 100 pounds of total materials.

$$\text{Density} = \frac{\text{Total Weight}}{\text{Total Volume}}$$

$$= \frac{\text{Weight of Clay} + \text{Weight of Water}}{\text{Volume of Clay} + \text{Volume of Water}}$$

$$= \frac{\text{Weight of Clay} + \text{Weight of Water}}{\dfrac{\text{Weight of Clay}}{\text{Density of Clay}} + \dfrac{\text{Weight of Water}}{\text{Density of Water}}}$$

$$= \frac{6 \text{ lb. bentonite} + 94 \text{ lb. water}}{\dfrac{6 \text{ lb. bentonite}}{20 \text{ lb./gal.}} + \dfrac{94 \text{ lb. water}}{8.33 \text{ lb./gal.}}}$$

DRILLING FLUIDS

$$= \frac{100 \text{ lb.}}{0.30 \text{ gal.} + 11.28 \text{ gal.}}$$

$$= 8.64 \text{ ppg}$$

2. Density resulting from adding liquids to a mud.

The derivation of suitable equations for this operation, as well as for most mud-mixing relations, is best originated with the material-balance equation,

$$W_F = W_0 + W_A$$

In this case, no solids will be handled, and no weights, as such, will be required in the solution. All weight terms are therefore substituted out by the products of volume times density.

$$V_F D_F = V_0 D_0 + V_A D_A$$

The amount of liquid added to the mud may be expressed as a fraction of either the final volume or the original volume. If the original volume is eliminated as the difference of the final volume and the added volume, the derivation proceeds as follows:

$$V_F D_F = (V_F - V_A) D_0 + V_A D_A$$

$$D_F = \frac{V_F D_0 - V_A D_0 + V_A D_A}{V_F}$$

$$= D_0 - \frac{V_A}{V_F}(D_0 - D_A) \tag{7-5}$$

If it is desired to express the added volume in terms of the original volume, suitable relations can be developed from the material balance equation or by replacing the V_F term in Equation (7–5) by the sum of V_0 and V_A. However, Equation (7–5) can be used directly in most problems.

Example: Calculate the mud density resulting from adding 100 barrels of 42-degree API oil to 800 barrels of 11.3 ppg mud.

Solution:

The specific gravity of the oil may be found as follows:

$$\text{sp. gr.} = \frac{141.5}{°\text{API} + 131.5}$$

$$= \frac{141.5}{42 + 131.5}$$

$$= 0.816$$

Since water has a density of 8.33 ppg,

$$D_A = (0.816)(8.33 \text{ ppg})$$
$$= 6.80 \text{ ppg}$$
$$V_F = 800 \text{ bbl. original mud} + 100 \text{ bbl. added}$$
$$= 900 \text{ bbl.}$$

Equation (7–5) may now be used directly:

$$D_F = D_0 - \left(\frac{V_A}{V_F}\right)(D_0 - D_A)$$
$$= 11.3 \text{ ppg} - \left(\frac{100 \text{ bbl.}}{900 \text{ bbl.}}\right)(11.3 \text{ ppg} - 6.80 \text{ ppg})$$
$$= 10.8 \text{ ppg}$$

3. Solid material required per barrel of original mud.

The most common mud-weighting problem occurs when the density of the mud system must be increased to some predetermined value. The original volume of the mud is known approximately, as well as its density. The increase in the volume of the mud is often accepted as beneficial. The necessary relations are developed directly from the material balance.

$$W_F = W_0 + W_A$$
$$V_F D_F = V_0 D_0 + W_A$$
$$(V_0 + V_A)D_F = V_0 D_0 + W_A$$
$$\left(V_0 + \frac{W_A}{D_A}\right)D_F = V_0 D_0 + W_A$$
$$\frac{W_A}{V_0} = \frac{D_F - D_0}{1 - \dfrac{D_F}{D_A}}$$

and, since $w = 42(W_A/V_0)$

$$w = \frac{42(D_F - D_0)}{1 - \dfrac{D_F}{D_A}} \tag{7-6}$$

In the foregoing derivation, it may be noted that the weight term for the added material is retained, for solid mud materials are packaged by weight and will therefore be added according to weight.

DRILLING FLUIDS

Barite weight material and most clays are commonly shipped in 100-pound sacks. However, the original and final mud terms represent fluids, and they are replaced by volume-density products. The volume of the final mud represents an unknown quanity which is replaced by the sum of the volumes of the original and added materials. However, the volume term for the added material cannot be retained, since this material is already expressed as a weight term in the equation. Accordingly, the volume of the added material is expressed as a weight divided by the density. This completes all substitutions required in the derivation.

Example: The mud system at a drilling well contains 750 barrels of 10.4 ppg mud. How many sacks of barite will be required to increase the mud density to 12.4 ppg? If barite costs \$35.00 per ton at the well, and if the mud can be weighted during drilling operations so that no rig time costs will be involved, what will be the cost?

Solution: Equation (7–6) may be used directly:

$$w = \frac{42(D_F - D_0)}{1 - \dfrac{D_F}{D_A}}$$

$$= \frac{42(12.4 \text{ ppg} - 10.4 \text{ ppg})}{1 - \dfrac{12.4 \text{ ppg}}{35 \text{ ppg}}}$$

$= 131$ lb. of barite per bbl. of original mud

$$\text{Barite required} = \frac{(750 \text{ bbl.})(131 \text{ lb./bbl.})}{100 \text{ lb./sack}}$$

$= 983$ sacks

$$\text{Cost of barite} = \frac{(740 \text{ bbl.})(131 \text{ lb./bbl.})(\$35.00 \text{ per ton})}{2{,}000 \text{ lb./ton}}$$

$= \$1{,}720.$

Example: An emulsion mud is to be made by adding 100 barrels of 42-degree API oil to 800 barrels of 11.3 ppg mud. If the density of the mud is to be maintained at its original value, how much barite will also have to be added to the mud?

Solution: In a previous example it was determined that 900 barrels of 10.8 ppg mud would result from adding the oil to the mud. The amount of barite required to restore the density from 10.8 to 11.3 for the 900 barrels of mud may be calculated by Equation (7-6) as follows:

$$w = \frac{42(D_F - D_0)}{1 - \dfrac{D_F}{D_A}}$$

$$= \frac{42(11.3 \text{ ppg} - 10.8 \text{ ppg})}{1 - \dfrac{11.3 \text{ ppg}}{35 \text{ ppg}}}$$

$$= 31 \text{ lb./bbl.}$$

$$\text{Barite required} = \frac{(900 \text{ bbl.})(31 \text{ lb./bbl.})}{2,000 \text{ lb./ton}}$$

$$= 13.8 \text{ tons}$$

An alternate solution is available in this special case, in which the original density is to remain unchanged. It may be reasoned that the density of the mud will remain unchanged if, theoretically, sufficient barite is added to the oil to increase its density to that of the mud. In practice, of course, both must be added to the mud, for the barite would settle if added directly to the oil. It was previously determined that the density of the oil is 6.80 ppg. The alternate solution is as follows.

$$w = \frac{42(11.3 \text{ ppg} - 6.80 \text{ ppg})}{1 - \dfrac{11.3 \text{ ppg}}{35 \text{ ppg}}}$$

$$= 279 \text{ lb./bbl.}$$

$$\text{Barite required} = \frac{(100 \text{ bbl.})(279 \text{ lb./bbl.})}{2,000 \text{ lb./ton}}$$

$$= 13.8 \text{ tons}$$

4. Solid material required per barrel of final mud.

When mud is to be mixed to a predetermined final volume and density, it is convenient to use a relation expressing the pounds of solid material per barrel of final mud required to achieve a desired (known) final density. The density of the original mud must be known, but the original volume of the mud need not be known. A suitable relation may be developed from the material-balance equation.

DRILLING FLUIDS

$$W_F = W_0 + W_A$$

$$V_F D_F = V_0 D_0 + W_A$$

$$V_F D_F = (V_F - V_A) D_0 + W_A$$

$$V_F D_F = \left(V_F - \frac{W_A}{D_A}\right) D_0 + W_A$$

$$\frac{W_A}{V_F} = \frac{D_F - D_0}{1 - \dfrac{D_0}{D_A}}$$

and since

$$\bar{w} = 42 \frac{W_A}{V_F}$$

$$\bar{w} = \frac{42(D_F - D_0)}{1 - \dfrac{D_0}{D_A}} \tag{7-7}$$

In the preceding derivation, the weight term for the added material is retained, since the added material is a solid. The original and final mud terms represent fluids and are replaced by volume-density products. The volume of the original mud represents an unknown quantity and is replaced as the difference between the final volume and the added volume. The volume term for the added material cannot be retained, since this material is expressed in terms of its weight. Therefore, the volume term of the added material is replaced as a weight divided by the density. This completes all substitutions required in the derivation.

Example: A mud system consisting of 800 barrels of 12.5 ppg mud is to be mixed by adding barite to a 5 per cent bentonite-water suspension. How much barite will be required?

Solution: The density of the bentonite-water suspension must first be determined.

$$D = \frac{5 \text{ lb. bentonite} + 95 \text{ lb. water}}{\dfrac{5 \text{ lb. bentonite}}{20 \text{ lb./gal.}} + \dfrac{95 \text{ lb. water}}{8.33 \text{ lb./gal.}}}$$

$$= 8.59 \text{ ppg}$$

Equation (7-7) can now be used directly:

$$\bar{w} = \frac{42(D_F - D_0)}{1 - \dfrac{D_0}{D_A}}$$

$$= \frac{42(12.5 \text{ ppg} - 8.59 \text{ ppg})}{1 - \dfrac{8.59 \text{ ppg}}{35 \text{ ppg}}}$$

$$= 218 \text{ lb. per bbl. of final mud}$$

$$\text{Barite required} = \frac{(800 \text{ bbl.})(218 \text{ lb./bbl.})}{2,000 \text{ lb./ton}}$$

$$= 87.2 \text{ tons}$$

5. Relations between volumes and densities.

Several useful relations between volumes and densities can be obtained by replacing all weight terms in the material-balance equation with the equivalent volume-density products. If the volume of the added material is then eliminated as the difference between the final and original volumes, the relation between the original and final volumes, in terms of the densities involved, may be obtained as follows:

$$W_F = W_0 + W_A$$
$$V_F D_F = V_0 D_0 + V_A D_A$$
$$V_F D_F = V_0 D_0 + (V_F - V_0) D_A$$
$$V_0 D_A - V_0 D_0 = V_F D_A - V_F D_F$$

$$\frac{V_0}{V_F} = \frac{D_A - D_F}{D_A - D_0} \tag{7-8}$$

By similar algebraic operations, either the original or final volumes may be eliminated to give the two following relations:

$$\frac{V_A}{V_F} = \frac{D_0 - D_F}{D_0 - D_A} \tag{7-9}$$

$$\frac{V_A}{V_0} = \frac{D_0 - D_F}{D_F - D_A} \tag{7-10}$$

Example: Referring to the preceding example, how many barrels of water should be run into the mud pits before commencing to mix the mud? Also, calculate how much bentonite will be required for the mud.

Solution: With respect to adding barite weight material, the bentonite-water suspension is the original mud, with a density of 8.59 ppg. The density of the final mud is 12.5 ppg, and its volume is 800 barrels. The added material is barite with a density of 35 ppg. Equation (7–8) may be applied in this sense as follows:

$$\frac{V_0}{V_F} = \frac{D_A - D_F}{D_A - D_0}$$

$$= \frac{35 \text{ ppg} - 12.5 \text{ ppg}}{35 \text{ ppg} - 8.59 \text{ ppg}}$$

$$= 0.852$$

$$V_0 = 0.852 \, V_F$$

$$= (0.852)(800 \text{ bbl.})$$

$$= 682 \text{ bbl. of bentonite-water suspension}$$

With respect to adding the clay to the water, the water represents the original fluid with a density of 8.33 ppg. The clay is the added material with a density of 20 ppg. The suspension is the final material with a density of 8.59 ppg and a volume of 682 bbl. Equation (7–8) may again be used:

$$\frac{V_0}{V_F} = \frac{D_A - D_F}{D_A - D_0}$$

$$= \frac{20 \text{ ppg} - 8.59 \text{ ppg}}{20 \text{ ppg} - 8.33 \text{ ppg}}$$

$$= 0.978$$

$$V_0 = 0.978 V_F$$

$$= (0.978)(682 \text{ bbl.})$$

$$= 667 \text{ bbl.}$$

The amount of bentonite required may be calculated as 5 per cent of the total weight of the 682 barrels of bentonite-water suspension.

$$\text{Bentonite required} = \frac{(0.05)(682 \text{ bbl.})(42 \text{ gal./bbl.})(8.59 \text{ lb./gal.})}{2{,}000 \text{ lb./ton}}$$

$$= 6.15 \text{ tons}$$

Example: It is desired to decrease the density of a mud from 10.1 to 9.7 ppg by adding 40-degree API oil to the mud. There are 900 barrels of mud in the system. How much oil will be required? What will be the percentage of oil in the total mud?

Solution:

$$\text{Density of the oil} = \left(\frac{141.5}{131.5 + 40° \text{ API}}\right)(8.33 \text{ ppg})$$

$$= 6.87 \text{ ppg}$$

Equation (7–10) can be used for the solution:

$$\frac{V_A}{V_0} = \frac{D_0 - D_F}{D_F - D_A}$$

$$= \frac{10.1 \text{ ppg} - 9.7 \text{ ppg}}{9.7 \text{ ppg} - 6.87 \text{ ppg}}$$

$$= 0.142$$

$$V_A = 0.142 V_0$$

$$= (0.142)(900 \text{ bbl.})$$

$$= 128 \text{ bbl. of oil}$$

$$\text{per cent oil} = \left(\frac{128 \text{ bbl. oil}}{900 \text{ bbl.} + 128 \text{ bbl.}}\right)(100\%)$$

$$= 12.4 \text{ per cent}$$

The Yield of a Clay

The yield of a clay refers to the number of barrels of mud which may be obtained from the clay mixed with water. One of the prime functions of the clay used in drilling mud is to impart viscosity to the mud. Fundamentally, the viscosity produced by the clay is a measure of the colloidal quality of the clay. Accordingly, the viscosity produced by mixing a certain clay into water affords a convenient index of its quality. It is customary to use a viscosity of 15 centipoises (Stormer viscosity at 600 rpm) as the standard for comparison of different clays. Viscosity data may be obtained in the laboratory, so that the percentage of clay in the suspension required to produce a viscosity of 15 centipoises (or other standard used) may be ascertained. In order that uniform comparisons can be made, an exact procedure is given in *API RP 29*. After the percentage of clay in suspension required to produce the standard viscosity has been determined by laboratory experiment, the determination of the number of barrels of such mud which can be mixed per ton of clay becomes a mud-mixing calculation.

Since the yield of a clay is based upon a ton of 2,000 pounds of clay, and if x represents the percentage of clay in the total (final) suspension, it follows that

$$\frac{x}{100} W_F = 2{,}000 \text{ lb.}$$

$$\frac{x}{100} V_F D_F = 2{,}000 \text{ lb.}$$

On the basis of one pound of total suspension, where S represents the specific gravity of the clay, the density of the clay-water suspension is given by the following expression:

$$D_F = \frac{1}{\dfrac{x}{(100)(8.33)S} + \dfrac{100-x}{(100)(8.33)}}$$

$$= \frac{(100)(8.33)}{100 - x\left(1 - \dfrac{1}{S}\right)}$$

If the expression for D_F is substituted into the foregoing equation, the equation may be solved for V_F to yield

$$V_F = \frac{(2{,}000)(100)}{8.33x} - \frac{2{,}000\left(1 - \dfrac{1}{S}\right)}{8.33}$$

In the preceding equation V_F is expressed in gallons. If the yield, Y, is expressed in barrels per ton of clay, then $Y = V_F/42$, so that

$$Y = \frac{(2{,}000)(100)}{(42)(8.33)x} - \frac{2{,}000\left(1 - \dfrac{1}{S}\right)}{(42)(8.33)}$$

$$= \frac{572}{x} - 5.72\left(1 - \frac{1}{S}\right)$$

If a value of S equals 2.5 is assumed, then,

$$Y = \frac{572}{x} - 3.4$$

where

Y = yield, barrels of mud per ton of clay

x = percentage of clay, by weight, in the clay-water suspension.

Example: From laboratory data, it is found that 7 per cent clay is required in a clay-water suspension to produce a viscosity of 15 centipoises. What is the yield of the clay, in barrels of suspension per ton of clay?

Solution:

$$Y = \frac{572}{x} - 3.4$$

$$= \frac{572}{7} - 3.4$$

$$= 77.4 \text{ bbl. of suspension}$$

Much has been written on the subject of drilling fluids, and some good references for additional reading are listed below:

API RP 29, "Recommended Practice for Standard Field Procedure for Testing Drilling Fluids."

Principles of Drilling Mud Control, published by the University of Texas Extension Division.

W. F. Rogers, *Composition and Properties of Oil Well Drilling Fluids*, 2d ed. (New York, McGraw-Hill Book Company, Inc.).

B. Q. Green, *Training Course for Mud Engineers*, published by Magnet Cove Barium Corporation.

Drilling Mud Reference Manual, published by Baroid Division, National Lead Company.

Drilling Mud Data Book, published by Baroid Division, National Lead Company.

G. R. Gray, M. Neznako, and P. W. Gilkerson, "Some Factors Affecting the Solidification of Lime-Treated Muds at High Temperatures," *API Drilling and Production Practice* (1952), 73.

R. F. Burdyn and L. D. Wiener, "Calcium Surfactant Drilling Fluids," a paper presented at the 31st Annual Fall Meeting, Petroleum Branch, A.I.M.E., Los Angeles, October 14–17, 1956.

CHAPTER 8

Planning the Well

The function of an oil or gas well is to provide a conduit from the subsurface petroleum-bearing formations to the surface through which the petroleum can flow and be recovered at the surface. In order to provide this conduit, a hole must be bored or drilled to the petroleum-bearing formation. This hole is then cased with pipe, usually steel or steel alloy, and cemented, and production equipment is placed in the cased hole and at the surface to control and regulate the fluid-withdrawal rates. All of the foregoing must be accomplished with a minimum of capital outlay. The drilling system used to bore this hole will depend upon a number of factors: (1) cost per foot for the different drilling systems, (2) character of the formations to be penetrated, and (3) character of the producing formation. The drilling system will normally be either the cable-tool system or the rotary system. With a few exceptions, the cost per foot is less when the rotary system is used, regardless of the depth of the well. The time required to drill the hole to the desired depth is almost always less when the rotary system is employed. As a result, supervisory time is usually much less for a rotary-drilled well than for a cable-tool-drilled well. Therefore, unless other circumstances alter the situation, the rotary-drilled well is the more economical.

Something must be known of the character of the formations to be penetrated in reaching the producing horizon in order to select the proper drilling system. Where consideration of cost alone might indicate the use of a cable-tool drilling system, the potential presence of a formation containing fluids at abnormal pressures might dictate the use of a rotary-drilling system instead, so that the circulating fluid can be properly weighted to control the abnormal pressure formation. By the same token, where consideration of cost alone might indicate the use of the rotary-drilling system, the presence of a zone of lost circulation might alter costs so much that the cable-tool system would be preferred.

Another important consideration in the selection of the drilling system is the character of the producing formation. This has been found to be especially important in many shallow low-pressure formations. Many of these formations are water-sensitive; that is, when water is exposed to the formation, the water will be absorbed into the structure of the clay minerals, thus increasing the size of the clays present and drastically reducing the formation permeability. This phenomenon, coupled with an increased water saturation around the well bore and the low formation pressure, may seriously retard the movement of petroleum into the well bore. When the conventional rotary drilling system is used with a water-base drilling fluid, all of the factors requisite for the foregoing sequence of events are present. For these reasons, the cable-tool system of drilling is preferable in many shallow producing areas. The use of oil, or the so-called inverted oil-emulsion muds, where oil is the continuous phase, has eliminated the principal disadvantage of the rotary system for water-sensitive formations. However, other problems, such as interpretation of formation fluid content and increased cost, are introduced. In some cases the rotary system is used to drill the well to the top of the producing formation and the cable-tool system to drill into the pay zone. The use of air and natural gas for drilling these shallow water-sensitive formations also eliminates the basic disadvantage of the rotary system.

After the drilling system has been selected, the optimum size of drilling rig must be determined. Several factors enter into the selection of the rig, including casing requirements, casing-bit-size programs, nature of the formations penetrated, cost of the rig, and speed and safety of operation. It should be remembered that at all times the principal consideration is the end result, or the producing well. Drilling operations usually require only a few days or a few weeks, but once the well has been completed, it may produce for many years.

After the drilling system has been selected, much planning and coordination is still necessary in order that the drilling and completion program can progress efficiently. The details of the entire program must be planned before the rig is selected for actually drilling the well.

Planning usually begins at the bottom of the well, where the formation fluids will enter the bore hole. The size of the oil-string casing, which provides a conduit from the surface to the producing formation, is determined principally by subsurface considerations, such as (1) subsurface artificial lift equipment required; (2) multiple-zone completions requiring several different strings of tubing isolated from each other by packers; (3) type of completion method to be used, whether open hole, perforated casing, screened open hole, or screened perforated casing; and (4) prospects of deepening the well at a later date. The size of the oil string will usually be the minimum consistent with the demands of the factors just mentioned. Ample working room should be provided inside the oil-string casing for all types of production equipment which may be used, plus working space for fishing tools to recover any lost pieces of equipment. If the well is likely to be deepened at a later date, the same consideration must be given to the future deeper completion as to the present completion interval. In fact, where future deeper completions are considered likely, these deeper completions will probably be the principal controlling factors. It should be borne in mind that in order to deepen the well, the drilling bit must be run through the previous oil string; therefore, the bit size will be limited by the size of this casing, which will in turn limit the size of the casing and production equipment for the deeper horizon.

After the minimum acceptable size of oil-string casing has been selected, attention can be directed to mechanical requirements involved in boring the hole to the desired depth. However, there remains one other requirement which usually takes precedence over the actual drilling operations: the statutory regulations requiring the setting of surface casing to protect fresh-water sands from possible invasion by hydrocarbons or salt water from deeper horizons. These fresh-water formations occur near the surface, and only a few hundred feet of surface casing are usually needed. Knowing that the fresh-water sands must be protected with a string of casing and keeping in mind the requirements of the oil string, the planners of the well now must accomplish its drilling and completion with a minimum expenditure of captial. Something of the nature of the formations to be penetrated must be known in order to plan the drilling program intelligently. Important considerations are abnormal pressure formations, zones of lost circulation, heaving shales, dense zones where drilling is unusually slow, and high formation temperatures. From an analysis of all of these factors, a casing-bit-size program is formulated. This is probably the most important program in planning the actual drilling of the well. In determining the casing-bit-size program, the first consideration is the number of strings of casing which will be required. There will be a minimum of two strings, the surface casing and the oil-string casing, and possibly more. The nature of the formations may be such that one or more intermediate strings of casing may be necessary. These intermediate strings are not normally required for the proper functioning of the well during its producing life, but are strictly a part of the drilling phase. Intermediate strings of casing may be required where heaving shales are present in deep wells and it is more economical to isolate this troublesome condition by setting casing below the problem zone. Another condition which may dictate the use of an intermediate string of casing is the presence of a zone of lost circulation at a shallow depth and the presence of an abnormal pressure formation at a greater depth, requiring heavier than normal drilling mud. The use of intermediate strings of casing is usually an economic factor dictated by the actual drilling phase of the well. Whether or not the well owner plans to set a string of intermediate casing, the surface casing should be large enough to provide clearance for running a suitable intermediate string in case of an unforeseen drilling hazard, and the intermediate string must be large enough to accommodate the oil string. This is especially true in the case of wildcat wells.

In certain areas where the surface soil is so incompetent that, when drilling begins, caving around the sides of the hole becomes excessive, it is necessary to install another string of casing after drilling through the surface soil. This casing is called *conductor casing*, and is usually from ten to twenty feet in length.

Once the casing requirements have been determined, the size of each of the casing strings must be fixed. The weight of the casing determines its cost, and for a given strength of casing, as the diameter of the casing increases, the weight will increase correspondingly. Thus from a cost standpoint it is desirable to use the smallest diameter of casing possible. However, the minimum size of oil-string casing has already been determined and is usually the limiting factor in the design of the other casing requirements. Economic considerations are presently dictating the use of smaller casing in many oil wells. Savings usually result not only from the use of less steel but also from the fact that the smaller hole can usually be drilled at a lower cost with smaller equipment. One oil operator has already requested permission from a state regulatory body to set and cement 10,000 feet of $2\frac{7}{8}$-inch OD tubing as the oil-string casing in a specific well.

Another factor to be considered in planning the casing-bit-size program is clearance. Sufficient clearance should normally be allowed around the outside of the casing to provide a satisfactory thickness of cement to form a bond between the casing and the wall of the hole, and also to allow free passage of the casing into the hole.

After each string of casing has been run and cemented, in order for drilling to continue, the drill bit must pass through the cemented casing. Thus the bit, and consequently the next string of casing, must be smaller than the previous string. The maximum diameter of the bit should be less than the drift diameters of the casing through which it must pass. A typical casing-bit-size program for a 10,000-foot well is shown in Table 8–1.

TABLE 8–1
Typical Casing and Bit Programs

Casing String	Bit Size (inches)	Casing Size (inches [OD])
Conductor	24	20
Surface	$17\frac{1}{2}$	$13\frac{3}{8}$
Intermediate	$12\frac{1}{4}$	$9\frac{5}{8}$
Oil string	$8\frac{5}{8}$	$5\frac{1}{2}$

Figure 8–1 shows the relationship of the various strings of casing and the production equipment of a typical oil well.

After the well owner has completed his plans, it is essential that

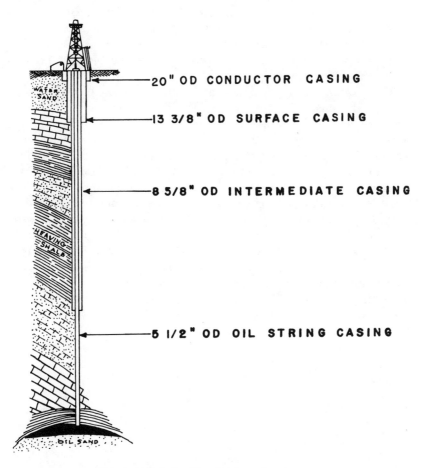

Fig. 8–1. Relationship of strings of casing.

the drilling contractor plan a drilling program. Considerations involved in developing the drilling program include:
1. The casing program and other specifications as provided by the owner.
2. Moving in and rigging up operations.
3. Bit sizes and types which will be required for the various formations.
4. Rotary speeds, bit weights, jet-bit hydraulics, pump speeds, mud-circulation rates, etc.
5. Drill-string equipment required; also any specialized equipment such as stabilizers and reamers.
6. Rig-maintenance program and equipment inspections.
7. Hole-deviation plans if crooked-hole conditions exist.
8. Suggested remedies for loss circulation and other hole problems which might be encountered.
9. Special equipment, such as instruments and blowout preventers which should be used.
10. Assignment of personnel, and familiarization of tool pusher and drillers with the terms of the drilling contract.

Proper advance planning of the drilling of the well will result in considerable savings of time and money. The drilling of a 10,000-foot well requires a large amount of equipment and manpower, and proper planning is requisite to insure that time will not be lost on account of the lack of necessary equipment or services. For example, when the total depth has been reached, an electric log is usually run, and in order to save valuable rig time, the logging equipment should be immediately available when it is needed. If the decision is made to run casing, the casing should be available on very short notice, if it is not already on location. Then, as soon as the casing has been run, cement and pumping equipment must be available to perform the cementing operation. The entire drilling operation is actually a series of co-ordinated activities, and proper advance planning will result in a smooth operation with a minimum number of delays.

CHAPTER 9

Power Plants

The power plant is the heart of the drilling rig. The power developed by the rig power plant is used principally for three operations: (1) rotating, (2) hoisting, and (3) drilling-fluid circulation. In addition to these major functions several auxiliary operations may also be powered by the rig power plant. Some of these auxiliary operations are (1) mud vibrating screen, (2) boiler-feed water pumps, (3) rig-lighting system, and (4) power for hydraulically operated blowout preventers. The power plant may be called upon to perform all of these jobs simultaneously, or it may be necessary to perform only one function or a combination of functions at any one time. In any event, a principal requirement of a power plant is flexibility. The power plant must be designed so that, whenever required, either of the last two principal recipients of power can receive essentially the entire power output of the plant.

Power for a drilling rig is normally furnished by steam, internal combustion engines, electricity, or any combination of these plants. The free-piston engine and the gas turbine may have some application in drilling-rig power plants, although they have not been used on rigs up to the present time.

To perform an efficient job, the drilling rig must be designed so that maximum efficient use is made of all its component parts. A rig which was designed to drill wells 20,000 feet deep could also drill a well 5,000 feet deep, but the cost of drilling the shallower well would probably be greater than if a smaller rig were used. The individual components of the rig must also be balanced so that each unit will perform its allotted task efficiently. For example, a rig would not be properly balanced if the power plant were large enough to drill to 20,000 feet and yet the mast or derrick could handle only 5,000 feet of drill pipe. All of the component parts of the drilling rig must have equivalent ratings and power in order for the rig to perform its function satisfactorily.

Another item of major importance in recent drilling-rig design is portability. The present trend is to assemble the rig in a number of unitized packages which will facilitate moving, rigging up, and rigging down. The number of these packages is kept to a minimum consistent with weight and space requirements. Consideration must be given to the fact that these unitized packages must, in most cases, be moved over public highways, where size and weight may be restricted.

In order to select properly the various parts of the rig, the designer must have a thorough knowledge of the power requirements for hoisting, rotating, and circulating. He must also determine the proper size of all the main and auxiliary equipment, such as derrick, mud pumps, rotary tables, transmission, crown blocks, traveling blocks, hooks, light plants, etc. In selecting the proper size of power plant to perform a specific type of drilling job, he must refer to certain power units. Therefore, a short discussion of the basic units used in describing power plants follows.

Basic Units of Power-Plant Design

When an object is displaced or moved a certain distance by the application of a force, then *work* is done on the object. For example, when drill pipe is removed from the hole being drilled, work is being done. Mathematically, work is equal to the product of force times distance, or

$$\text{Work} = \text{Force} \times \text{Distance}$$

or

$$W = F \times d \tag{9-1}$$

Example: Find the amount of work necessary to raise a weight of 100 pounds a distance of 7 feet.

In this case, the force is equal to the weight of 100 pounds; therefore,

$$\text{Work} = \text{force} \times \text{distance} = 100 \times 7 = 700 \text{ ft.-lb.}$$

The rate of doing work is defined as *power*. Power is the number of units of work performed in a unit time.

$$\text{Power} = \frac{\text{Work}}{\text{Time}}$$

Example: Find the power required to perform the work of the preceding example in five seconds.

$$\text{Power} = \frac{\text{work}}{\text{time}} = \frac{700 \text{ ft.-lb.}}{5 \text{ sec.}} = 140 \text{ ft.-lb. per sec.}$$

The most common unit of power is *horsepower*, which is an arbitrary rate of doing work. One horsepower is equivalent to 550 ft.-lb. per second, or 33,000 ft.-lb. per minute. The horsepower required to perform the work in the second example would be

$$\text{Horsepower} = \frac{140 \text{ ft.-lb. per sec.}}{550 \text{ ft.-lb. per sec./horsepower}} = 0.255 \text{ hp}$$

The power developed by an engine must be transmitted in order to be utilized effectively. In the transmission of the power originally developed by the engine, there will be some loss on account of friction developed in the moving parts of the engine. The work done on the piston of a reciprocating engine is called the *indicated horsepower*. The work delivered at the output end of the engine is called the *brake horsepower*. The indicated horsepower minus the brake horsepower is called the *friction horsepower*, or

$$\text{ihp} - \text{bhp} = \text{fhp}$$

Figure 9–1 shows the development of indicated horsepower and brake horsepower for a simple engine. To find the horsepower developed by

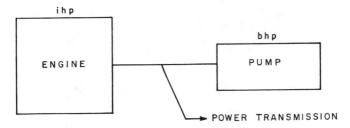

FIG. 9–1. Indicated and brake horsepower.

a reciprocating engine, it is necessary only to know the work developed in one power cycle of the engine, and then multiply this work by the number of power cycles developed in a unit time.

Expressed mathematically this is

$$\text{hp} = \frac{NW}{33{,}000} \tag{9-2}$$

where

hp = horsepower developed
N = number of power cycles per minute
W = work developed in ft.-lb. per cycle

The work developed in the power cyclinder, or piston, of an engine can be calculated if the pressure inside the cylinder is known, along with the dimensions of the power cylinder. If the mean effective pressure, Pm, is acting on a piston area, A, then the total force on the piston will be PmA. If the piston moves a distance, L, then the work developed will be

$$W = PmAL \qquad (9\text{--}3)$$

Combining Equations (9–2) and (9–3) yields

$$\text{hp} = \frac{PmLAN}{33{,}000} \qquad (9\text{--}4)$$

Equation (9–4) is a development of the indicated horsepower. In order to determine the brake horsepower of this engine, the friction horsepower would have to be subtracted from the indicated horse-

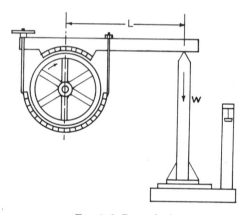

FIG. 9–2. Prony brake.

power. Since it is impractical to determine directly the friction horsepower developed by an engine, the brake horsepower, which is of prime importance to the user of the engine, is usually measured experimentally. The brake horsepower of an engine can be determined experimentally by measuring the power output of the engine. The word "brake" has come into common use to describe the output power of an engine because a brake is the usual method of measuring the horsepower developed by an engine. The *prony brake*, shown in Fig. 9–2 absorbs the power output from an engine through the friction of a number of wood blocks. At a given engine speed, the friction developed between the wheel and the wooden blocks is transmitted through a lever arm to a scale, where this friction is measured in pounds. The horsepower developed by the engine is equal to the net force observed on the scale multiplied by the distance through which the net force acts. On the prony brake, the net force is found by subtracting the weight of the lever arm when the engine is stopped, called the *tare*, from the weight on the scale when the engine is running. The distance through which the force acts in one revolution of the engine is equal to the circumference of a circle whose radius is the length of the lever arm. The total power developed in one minute would be

Power = Force on Scale × Distance Moved in One Revolution × rpm

Converting to horsepower,

$$\text{Brake horsepower} = \text{bhp} = \frac{F \times 2\pi R \times N}{33,000} \tag{9-5}$$

where

F = net weight on scales, lb.
R = length of lever arm, ft.
N = speed of engine, revolutions per minute.

The mechanical efficiency of an engine is

$$Em = \frac{\text{brake horsepower}}{\text{indicated horsepower}} \times 100 \tag{9-6}$$

where

Em = mechanical efficiency.

The electrical unit of power is the *watt*. One horsepower is equivalent to 746 watts. A more common expression of electrical power is the kilowatt, which is equal to 1,000 watts.

As shown by Equation (9-4), the horsepower developed by an engine is directly proportional to the number of power cycles developed per unit of time. Therefore, as the speed of the engine increases, the horsepower will increase.

The power rating of engines is by no means standard, although most industrial engines are rated at approximately 60 per cent of their maximum horsepower. Thus the manufacturer of an engine that will develop 1,000 horsepower at a maximum speed of 1,800 rpm will rate the engine at 600 horsepower. The manufacturer will normally guarantee that the engine can develop 600 horsepower for prolonged periods. It should be pointed out that all engines are not rated according to this 60 per cent factor. For instance, automobile engines are generally rated at their maximum power, although they are not designed to operate continuously at this maximum. Engines are usually described by the number of power cylinders and the size of these power cylinders. Thus a four-cylinder 10×10 engine would have four cylinders, and each cylinder would have a 10-inch-diameter piston with a travel, or stroke, of 10 inches.

An important principle in power-plant operation is *torque*. Torque is a measure of the ability of an engine to do work, and is distinguished from horsepower in that horsepower is a measure of the rate of doing work. Torque is important because it determines the ability of an engine to perform a specific job, while horsepower determines the rate at which the job can be performed. Torque is a product of the amount of force developed by an engine multiplied by the distance through which this force must operate to accomplish the desired work. The units of torque are therefore weight-length, the most common units being pound-feet. Thus, in removing drill pipe from the hole, the engine torque will determine whether the drill pipe can be removed, and the engine horsepower will determine the rate at which the drill string can be hoisted. Referring again to Equation (9–5), which developed the brake horsepower equation,

$$\text{bhp} = \frac{F 2\pi R N}{33,000}$$

it can be seen that the torque developed by an engine would be equal to

$$\text{Torque} = T = FR \tag{9-7}$$

where

$$T = \text{torque, lb.-ft.}$$

Therefore,

$$\text{bhp} = \frac{T N 2\pi}{33,000} \tag{9-8}$$

or

$$T = \frac{33,000 \text{ bhp}}{2\pi N} \tag{9-9}$$

Example: Calculate the torque developed by an engine which develops 1,200 bhp at 1,800 rpm.

$$\text{Torque} = T = \frac{33,000 \text{ bhp}}{2\pi N} = \frac{(33,000)(1,200)}{(2\pi)(1,800)}$$

$$= 3,501 \text{ lb.-ft.}$$

A high torque at low engine speeds indicates an ability to pick up large starting loads. Engines on a drilling rig may be required to perform one or all of the following jobs at the same time: (1) circulating, (2) hoisting, and (3) rotating. Engines can be built for high-starting torque or for low starting torque, depending upon what will be required of them in service. Normally, in order to increase the starting torque of an engine, the size of the engine must be in-

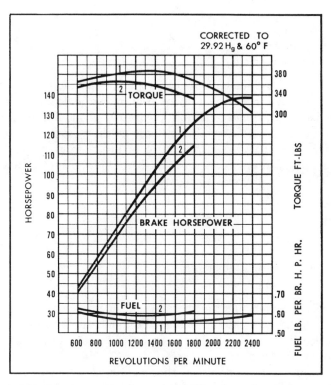

FIG. 9-3. Performance curves for typical drilling-rig engine. 1—bare engine; 2—complete power unit.

creased. Because of the nature of the loads involved, power plants suitable for drilling operations must normally have a high-starting torque. Performance curves for a typical drilling-rig engine are shown in Fig. 9-3.

In engineering terms, the efficiency of an operation is defined as the ratio of output to input. Of considerable importance to the operator of a drilling rig is the thermal efficiency of his power plant, for this term determines the amount of fuel required to obtain the desired work. Thermal efficiency is defined as the ratio of work output to heat input. As efficiency is usually expressed as a percentage, the above expression must be multiplied by 100. The general expression for thermal efficiency is

$$e_t = \frac{W}{Q} \times 100 \qquad (9\text{--}10)$$

where

W = work output
Q = heat input
e_t = thermal efficiency

As mentioned previously, the work output can be measured at several places. If the work used in the equation is the indicated work, then the efficiency as found by Equation (9–10) would be the indicated thermal efficiency. Of more importance to the power-plant owner is the *brake thermal efficiency*, the ratio of brake work to heat input, because he is primarily concerned with the power output of a power plant. The brake thermal efficiency is

$$e_b = \frac{W_b}{Q} \times 100 \qquad (9\text{--}11)$$

where

e_b = brake thermal efficiency
W_b = brake work

In order to determine the thermal efficiency of a power plant, the units of work and heat input must be identical. The units of heat and work will be reviewed briefly in order to establish a foundation for unit conversions.

The common unit of work is the *foot-pound*, with one foot-pound being the work, or energy, required to move a weight of one pound through a distance of one foot, where there is no friction involved.

The *British Thermal Unit* (Btu) is the energy required to raise the temperature of one pound of pure water by one degree Fahrenheit, when the original temperature and pressure of the water are 32 degrees Fahrenheit and 14.696 psia respectively. It has been found that one Btu is equal to 778 ft.-lb.

The *horsepower-hour* is the transfer of energy of one horsepower during a one-hour period. One horsepower has been defined previously as 33,000 ft.-lb./min. Therefore, the energy transferred by one horsepower in one hour is

POWER PLANTS

$$33{,}000 \frac{\text{ft.-lb.}}{\text{min.}} \times 60 \frac{\text{min.}}{\text{hr.}} = 1{,}980{,}000 \frac{\text{ft.-lb.}}{\text{hp-hr.}}$$

From the above definition of horsepower-hour, it can be seen that one horsepower-hour is equivalent to 1,980,000 ft.-lb. Converting the above energy from ft.-lb. to Btu's,

$$1{,}980{,}000 \text{ ft.-lb.} \div 778 \frac{\text{ft.-lb.}}{\text{Btu}} = 2{,}544 \frac{\text{Btu}}{\text{hp-hr.}}$$

Therefore, one hp-hr. is equal to 2,544 Btu.

The determination of thermal efficiency can best be determined by an example problem:

Example: Determine the brake thermal efficiency of an engine which uses 100.0 lb. of fuel in one hour while developing 500 brake horsepower. The heating value of the fuel is 20,000 Btu per lb.

From Equation (9–11),

$$e_b = \frac{W_b}{Q} \times 100 = \frac{500 \text{ hp} \times 1 \text{ hr.} \times 2{,}544 \frac{\text{Btu}}{\text{hp-hr.}}}{100 \text{ lb.} \times 20{,}000 \text{ Btu/lb.}} \times 100$$

$$= \frac{1{,}272{,}000 \text{ Btu}}{2{,}000{,}000 \text{ Btu}} \times 100$$

$$= 63.6 \text{ per cent}$$

DRILLING-RIG POWER PLANTS

Many types of power plants have been used or have been considered for use in oil well drilling operations. These include steam, internal combustion, electric, turbine, free piston, and combinations of some of these types. The operating principles and characteristics of each of these types of power plants will be studied briefly.

Steam Engine

The steam engine is one of the oldest types of power plant used in the drilling industry. The steam power plant was used almost exclusively in early-day drilling operations as it was the only type of plant available which had the power and flexibility to perform the required operations. Gradually, as larger, lower-cost internal combustion engines were developed, the steam power plants were replaced. The shortage of water and the high cost of fuel in many areas were largely responsible for this replacement; however, another factor of major importance was the lack of portability of the steam power plant. Although in 1958 there were still several large steam rigs in operation, these rigs were being gradually replaced with the more portable internal combustion or electric rigs. Not only is fuel cost important in the determination of the economics of rig change-over, but also the cost of rigging down, moving, and rigging up on a new location, which constitute a major consideration in over-all rig operating expenses.

The boiler is the heart of the steam power plant. The boiler generates steam from water, and the steam furnishes the power to run the mud pumps, rotate the drill pipe, and perform the required hoisting operations. When the steam reaches the prime mover and performs its work, the spent steam may either be released to the atmosphere or recovered for reuse in the system. If the spent steam is recovered, it must be condensed to water in order to be handled satisfactorily by the boiler feed pump.

Rating of Boilers

Boilers are often rated according to the number of pounds of steam delivered per hour, but this is not a satisfactory method since the energy contained in one pound of steam will depend upon the pressure and temperature of the steam. Another method formerly used in rating boilers was based on the heating surface. In early boiler design, it had been determined that 10 square feet of heating surface would produce one boiler horsepower. Therefore, a boiler with 800 square feet of heating surface would be rated at 80 boiler horsepower. Boiler efficiency has been increased to the point that many boilers will now develop one boiler horsepower for each two to four square feet of heating surface. Thus the amount of heating surface is an unsatisfactory method of rating boilers, as efficiency is not dependent on the heating surface alone, but also on the arrangement of the heating surface with respect to the boiler water. One boiler horsepower is defined as the heat required to evaporate 34.5 pounds of saturated water at 212 degrees Fahrenheit into steam at 212 degrees Fahrenheit in a period of one hour. The heat required to evaporate one pound of 212-degree Fahrenheit water to steam at 212 degrees Fahrenheit is 970.3 Btu. The heat required to evaporate 34.5 pounds of water is $34.5 \times 970.3 = 33,475.3$ Btu.

Therefore

$$\text{One boiler horsepower} = 33,475 \text{ Btu/hr.}$$

The total transferred heat of a boiler is

$$\text{Output of boiler} = W_s(h_s - h_{fw}) \text{ Btu/hr.}$$

where

W_s = weight of steam delivered per hour, lb.
h_s = enthalpy of steam leaving, Btu/lb.
h_{fw} = enthalpy of feed water, Btu/lb.

The boiler horsepower can now be calculated as follows:

$$\text{Boiler horsepower} = \text{bo. hp} = \frac{Ws(h_s - h_{fw})}{33{,}475.3} \quad (9\text{–}12)$$

The values h_s and h_{fw} can be obtained from published steam tables. *Enthalpy* may be defined as the total energy possessed by a system exclusive of kinetic and potential energy. It is a function of temperature and pressure. Any work performed by or on the system, or heat transferred to or from the system, results in a corresponding change in enthalpy. For this reason, enthalpy provides a "yardstick" to measure changes in energy.

A very important item in the rating of boilers is the over-all boiler efficiency. This is defined as a ratio of output heat to input heat. The over-all boiler efficiency can be calculated as follows:

$$e_b = \frac{\text{output heat}}{\text{input heat}} = \frac{W_s(h_s - h_{fw})}{W_f\, g_h} \times 100 \quad (9\text{–}13)$$

where

e_b = efficiency of the steam-generating unit
W_f = total weight of fuel fired, lb./hr
g_h = higher heating value of the fuel, Btu/lb

Fuel and Water Requirements

One boiler horsepower has been defined as 33,475.3 Btu/hr. In order to calculate the daily fuel requirements for a steam power plant, the following equation is used:

Daily fuel requirements

$$= \frac{33{,}475.3 \text{ Btu/hr.} \times 24 \text{ hr./day} \times \text{boiler hp}}{\text{heating value of fuel} \times E} \quad (9\text{–}14)$$

where

E = efficiency of the plant.

Example: Calculate the fuel requirements for three 150-horsepower boilers using natural gas with a heating value of 1,150 Btu/cu. ft. Assume 50 per cent efficiency.

$$\text{Fuel} = \frac{33{,}475.3 \times 24 \times 3 \times 150}{1{,}150 \times 0.50} = 627{,}000 \text{ SCF/day}$$

The above calculations probably represent minimum fuel requirements and would probably apply to optimum operating conditions. However, the fuel requirements during peak loads would be considerably greater because of reduced efficiency and also because of increased horsepower output for short periods. The rate of fuel consumption during short periods may be two or three times the rate of fuel consumption during normal operating periods.

Boiler Water Requirements

When the horsepower requirements of the rig are known, it is then possible to calculate the amount of boiler water which will be required. One boiler horsepower has previously been defined as 33,475.3 Btu/hr., which is the heat required to evaporate 34.5 pounds of saturated water at 212 degrees Fahrenheit into steam at 212 degrees Fahrenheit. Therefore, the barrels of water required for each boiler horsepower are

$$W = \frac{34.5 \text{ lb./hr./hp} \times 24 \text{ hr./day}}{8.33 \text{ lb./gal.} \times 42 \text{ gal./bbl.}}$$

$$= 2.33 \text{ bbl./day/boiler horsepower} \qquad (9\text{--}15)$$

The above equation will show how much boiler feed water will be required only if the spent steam is not recovered. If the spent steam is recovered for reuse, then, of course, the boiler feed water requirements will be materially reduced. When the boiler is charged initially, the only additional water required is makeup water for losses occurring in the circulating cycle.

The steam power plant is ideally suited to drilling-rig application for two basic reasons: (1) Maximum torque is developed at stall, or zero speed of the engine, and (2) there is extremely flexible power distribution. Development of maximum power at zero engine speed is ideal from a hoisting standpoint because, in lifting heavy loads off bottom, it is desirable to begin the lift slowly. Furthermore, in fishing operations, where attempts are being made to recover stuck equipment, the need is for maximum power at zero speeds. When drilling-fluid circulation is resumed after a shutdown period, high pump pressures are sometimes required to initiate, or "break," circulation. In this case, maximum power at zero speed is also desirable.

Power transmission, which is a major problem on the internal combustion engine-driven rig, is relatively simple on a steam rig.

All of the steam is generated at a central boiler-plant location, and the steam is piped to the various locations where it is required. The principal problems encountered in steam distribution are pressure and temperature losses, which can be reduced by large insulated steam lines.

Improving Steam-Power Efficiency

The demand for more horsepower has caused the following changes in steam power-plant design: (1) a more rugged construction which allows an increase in the working pressure of the boilers, (2) a change in boiler design of such a character that by relocation of heating surfaces more heating capacity is obtained, (3) the use of insulating material to coat the boilers and steam lines, and (4) the practice of preheating the boiler feed water.

Internal Combustion Engines

Although the steam engine provides excellent flexibility in operating characteristics, its lack of portability and high fuel cost where low-priced natural gas was unavailable led to the development of internal combustion engines for oil well drilling. The lack of portability was probably the major factor in the replacement of the steam rig. In deep drilling operations where five or more boilers are used, each weighing in excess of 30,000 pounds, the cost of moving from one location to another may be a major factor in the over-all operation. A majority of the wells are drilled by contractors; to be competitive, it is essential that their rig equipment be portable and that the cost and time involved in rig moves, for which they are responsible, be kept to a minimum. The development of compact internal combustion engines has increased tremendously the mobility of drilling rigs.

The internal combustion engine derives its power from the combustion of a fuel-air mixture, with the products of combustion directly providing the motive energy by performing work on movable pistons. The power output from the internal combustion engine is utilized through connecting links with the moving pistons.

There are two basic types of internal combustion engines, the *spark-ignition engine* and the *compression-ignition engine*. The spark-ignition engine develops power by igniting, with a closed *cylinder*, a combustible fuel-air mixture with a flame, or spark. One end of the cylinder is enclosed by a movable *piston*. As the fuel-air mixture is ignited, it expands, causing displacement of the piston. Connecting rods and cranks convert this linear motion of the piston to the desired rotary motion of the output shaft of the engine.

There are two major classifications of spark-ignition engines, the four-stroke cycle and the two-stroke cycle. The four-stroke-cycle engine was first built in 1876 by a German engineer named Otto. The basic principles of the Otto engine are still used in spark-ignition engines, and for this reason engines operating on the principle developed by Otto are still called Otto-cycle engines. The four-stroke Otto cycle consists of:

1. The *intake stroke*, which brings the fuel and air into the engine cylinder.
2. The *compression stroke*, which compresses the fuel-air mixture and increases the temperature of the mixture.
3. The *power stroke*, wherein the combustible mixture is ignited, forcing the piston toward the end of the cylinder, thereby developing the motive power of the engine.
4. The *exhaust stroke*, which clears the cylinder of the exhaust gases, preparing it for a repetition of the preceding series of events.

Figure 9–4 depicts the events occurring in the four-stroke Otto cycle. It should be noted that in the Otto cycle two complete revolutions are required to complete one power stroke from the cylinder.

In order to increase the power output of the four-stroke-cycle engine, an engine was developed which produced a power stroke from the piston with only two cycles of operation, thereby developing a power stroke for each revolution of the crankshaft. In the two-cycle engine, the air-fuel mixture is brought into the cylinder on the upstroke of the piston. The combustible mixture is compressed and ignited on the upward stroke, and power is produced on the downward stroke of the piston. Near the end of the down stroke, the combustion products are scavenged from the cylinder, this preparing the

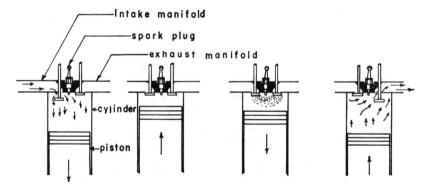

Fig. 9–4. Four-stroke Otto cycle.

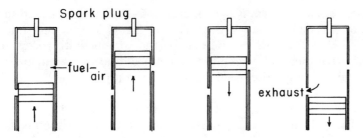

Fig. 9–5. Two-stroke Otto cycle.

cylinder for the next cycle. The series of events occurring in the two-stroke cycle are shown in Fig. 9–5.

In order to increase the power output of the engine, it is common practice to use several cylinders instead of only one. The use of several cylinders also has the desirable effect of providing smoother engine operation.

As fuel, spark-ignition engines commonly use gasoline, natural gas, liquefied petroleum gases such as butane or propane, or combinations of these.

The compression-ignition engine is based on an entirely different principle of operation from the spark-ignition engine. The compression-ignition engine operates on the principle of compressing air sufficiently that when fuel is injected into this compressed and hot air, spontaneous combustion will occur. This type of engine, which does not require a spark for ignition, was developed by a German engineer named Rudolf Diesel. Diesel engines may operate on either a two-stroke cycle or a four-stroke cycle. The four-stroke diesel cycle consists of:

1. The *intake stroke*, which brings air into the cylinder.
2. The *compression stroke*, which occurs as the piston moves to decrease the effective volume of the cylinder, thereby increasing the temperature and pressure.
3. The *power stroke*, which is obtained by injecting oil into the hot compressed air, causing spontaneous combustion of the mixture. This burning of the fuel-air mixture increases the pressure in the cylinder, thus forcing the piston to move.
4. The *exhaust stroke* which removes the spent gases from the cylinder.

The principal difference between the spark-ignition engine and the combustion-ignition engine is that the combustion-ignition engine does not utilize a spark in the combustion process. Instead it uses a *fuel injector*, which sprays the fuel into the cylinder in the form of a fine mist. The injection of the fuel into the cylinder in very fine particles reduces the time required for the combustion process to be completed, thereby increasing the power output of the engine.

Combustion-ignition engines are commonly referred to as *diesel engines*, in honor of their inventor.

As the diesel engine must compress the air-fuel mixture to the point where it will ignite spontaneously, it must be built to withstand higher pressures than a spark-ignition engine. As a result, the diesel engine is more expensive than a comparable spark-ignition (gasoline) engine. Although important, the factors of size and weight are not as critical in oil well drilling operations as they are in some other applications, such as in automobiles and airplanes, where weight and size are of major importance. Diesel engines are used throughout the oil industry because they are sturdy, have a small unit fuel consumption, and run economically at light loads.

Diesel engines are rated by the net brake horsepower they can deliver continuously when in good operating condition. The standard rating must be such that the engine will deliver an output 10 per cent in excess of the rating for two out of any twenty-four hours of operation with safe operating temperatures. Standard rating methods have been developed by the Diesel Engine Manufacturers Association and the API, and for complete information on rating tests, the DEMA tests should be examined. Other methods sometimes used in the power rating of diesel engines are (1) intermittent output, (2) rated output, and (3) continuous output. The *intermittent output* shows the brake horsepower which the engine will develop with a clear exhaust for a period of one hour or less. The *rated output* is the brake horsepower which the engine will develop for a period of twelve hours. The *continuous output* is the maximum recommended load in brake horsepower for continuous service of more than twenty-four hours.

The amount of fuel which is consumed in each power stroke of a diesel engine is governed by the fuel injection. In order to burn the fuel, there must be a supply of oxygen (air), and as the amount of air which can enter the cylinder is more or less fixed, it is entirely possible to inject an excessive amount of fel into the cylinder. When this occurs, the exhaust gases will be colored by the products of incomplete combustion. The color of the exhaust gases therefore provides a convenient and reliable index to the efficiency of combustion.

Conventional diesel engines in oil field service use a nonvolatile *middle distillate* fuel having an API gravity of approximately 37 degrees. This fuel is considerably less expensive than gasoline, the liquid fuel used in spark-ignition internal combustion engines.

Diesel engines can be adapted to use natural gas instead of diesel fuel, which results in considerable fuel savings in areas where natural gas is available. These engines are called *gas-diesel* or *dual-fuel* engines. The natural gas is injected into the cylinder filled with hot compressed air at top dead center. A small amount of oil, called *pilot oil*, injected at the same time, serves the function of igniting the hot compressed air, which in turn ignites the gas. In a different modification, the gas is brought into the cylinder on the suction stroke.

As mentioned previously, the weight of air which is drawn naturally into the cylinder is more or less limited, being governed by the prevailing temperature and pressure. In order to increase the amount of fuel consumed on the power stroke, which would increase the power output of the engine, it is necessary to increase the amount of air which enters the cylinder. If the air pressure outside the cylinder is increased, then a larger weight of air will enter the cylinder on each stroke. A *supercharger* can be used to accomplish this objective. A supercharger is an air pump which feeds the air to the inlet cylinder parts at elevated pressures. However, it should be noted that some power source must be used to run the supercharger; therefore, all of the additional power developed as a result of the combustion of more fuel and air is not usable as net increased power output.

Originally the internal combustion engine was not basically adaptable to oil well drilling operations, principally because of a lack of suitable power-transmission devices. However, the compactness of this type of power plant was extremely attractive. This fact, coupled with the scarcity of water and the high cost of boiler fuel in many areas, led to the development of power-transmission devices which have overcome the basic disadvantages of the internal combustion power plant. Prior to the development of fluid-type transmissions, when an internal combustion engine power plant was connected to a unit requiring power, such as a rotary table, the shock of mechanically connecting the rotating power source to a stationary body was quite great. With the application of the fluid-type transmissions to oil well drilling applications, the internal combustion engine lost most of its disadvantages as an oil well drilling power plant. As a result, at the present time a great majority of rig power plants are of the internal combustion type.

Electric Power

Electric power can be devised to provide the flexibility required in oil well drilling operations. To be adaptable to drilling operations, the electric power must provide (1) a dependable power source, (2) a wide range of speed-torque characteristics, and (3) a power source which is competitive, when all factors are considered, with other available sources.

Basically, electric power is developed by a generator driving electric motors, which in turn supply the power to run the mud pumps, draw works, rotary table, and auxiliary equipment. Some power must be available to drive the generator, and normally this is provided either by an internal combustion engine, or, more rarely, by purchased electric power. Thus, a diesel-electric rig is one powered by electric motors driven by a generator, which in turn is powered by a diesel engine. An all electric rig would be one in which the generator would be powered by an electric motor driven by purchased power. Obviously, the all electric rig can be used only in areas where reliable electric power is available.

The following report by Mr. Glenn Webb of the General Electric Company describes a simple control system for diesel-electric or gas-electric drilling-rig drives.

A Simple Control System for Diesel-Electric or Gas-Electric Drilling-Rig Drives

Basic Principles of DC Electric Machines

A DC motor is based on the principle that a force will be exerted on an electrical conductor carrying current if it is placed in a magnetic field. This force will be proportional to the amount of current flowing in the conductor and to the strength of the magnetic field. The stationary part of a DC motor consists of electromagnets to produce the magnetic field, and the rotating portion consists of copper wires or bars to carry the current. Thus the torque produced at the rotating shaft is proportional to the product of current and magnetic-field strength:

$$\text{Torque} \simeq \text{Current} \times \text{Field Strength} \qquad (9\text{--}16)$$

Stated another way, the current required by a motor is directly proportional to the torque demanded by its connected load and inversely proportional to its field strength.

$$\text{Current} \simeq \frac{\text{Torque}}{\text{Field Strength}} \qquad (9\text{--}17)$$

A generator is built the same way a motor is except that it performs the opposite function of converting mechanical to electric power. If a

POWER PLANTS

conductor is moved through a magnetic field, a voltage will be generated proportional to the speed with which the conductor is moved and to the strength of the magnetic field:

$$\text{Generator Voltage} \simeq \text{Speed} \times \text{Field Strength} \quad (9\text{--}18)$$

Voltage is electrical pressure which causes current to flow, just as hydraulic pressure causes a fluid to flow.

This relationship can be turned around and used to predict the speed of a motor:

$$\text{Motor Speed} \simeq \frac{\text{Voltage}}{\text{Field Strength}} \quad (9\text{--}19)$$

This shows that the speed of a motor is directly proportional to the applied voltage, and inversely proportional to the strength of its field. The generator output voltage is proportional to the speed at which it is driven and to the strength of its field. These relationships show the function of controls, because, to change the motor speed, the generator voltage must be changed, either by changing its speed or its field strength, or both.

Generator Selection

Obviously, if a generator and motor are selected which possess the characteristics required for drilling machinery, the requirements for controls can be much simplified. It is universally agreed that a drilling-rig drive should automatically slow down and pull harder as the load increases and speed up as the load decreases, in order to utilize engine power effectively. Since generator voltage governs motor speed and current supplies motor torque, it is desirable to have a generator which automatically trades volts for amperes. Starting with a simple generator, Fig. 9–6, whose shunt field is excited from a constant separate source, it is seen that load current has little effect on output voltage, which remains fairly constant as load current increases.

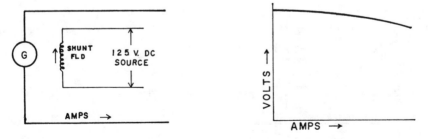

Fig. 9–6. Simple generator with volt-amp relationship.

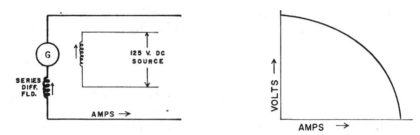

FIG. 9–7. Generator with second field winding in series—volt-amp relationship.

By adding a second field winding in series with the load current, however, the output voltage is reduced as load current increases, Fig. 9–7. This series field is arranged to oppose the constant shunt field, so that, since the series field carries more load current, the net effect of the two fields is reduced, until the series field finally completely counteracts the shunt field at a given value of load current. This means zero voltage (and speed) at the motor with maximum current (or torque), the stalled condition.

The addition of one more field to the generator provides two additional advantages, Fig. 9–8. The generator itself now furnishes most of its own excitation, reducing the capacity of the separate 125 V DC source, and it automatically becomes unable to overload or lug its driving engine. This engine-overload protection is explained as follows.

Starting at point *A* on the volt-amp curve of Fig. 9–8, the engine is idling at governed speed and the generator is producing maximum voltage, but there is no load on the connected motor, so that no current is required from the generator. Now, the effect of applying an increasing torque load to the motor will be examined. As it draws more current from the generator to develop more torque, the governor opens the engine throttle to maintain engine speed and produce the required power. At point *B* on the curve, full engine rating is reached

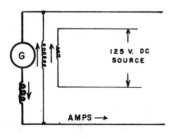 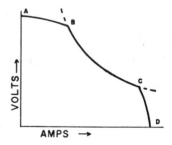

FIG. 9–8. Effect on volt-amp relationship of field windings.

and the throttle is full open. As the motor demands more torque or current, the engine can only slow down. But a slight decrease in engine and generator speed reduces the generator voltage because

$$\text{Generator Voltage} \simeq Speed \times \text{Field Strength}$$

When generator voltage reduces, the voltage applied to its self-excited field is reduced, and consequently its net field strength is reduced. Thus, as engine speed reduces, both generator speed and field strength are reduced, with the result that its voltage (and the motor speed) reduce as the square of the reduction in engine speed. This means that the electric transmission would drop about 10 per cent of its load if engine speed and load-carrying capacity dropped 5 per cent. The result is that a slight decrease in engine speed unloads the generator faster, and a balance condition is quickly reached where the generator adjusts its load to just match engine capability with no tendency to overload the engine. This condition exists as current increases to point C, where voltage and motor speed are low enough no longer to require full engine power even though torque is still increasing.

From point C to point D the governor closes the throttle, and at point D the engine is again at governed idle speed. The motor is developing maximum torque in the stalled condition, but the series field has completely counteracted the shunt fields so that only sufficient voltage is produced to circulate stall current. The process occurs in reverse if the motor load is decreased and then allowed to accelerate to maximum no-load speed corresponding to the voltage at point A.

Motor Selection

A look at the motor equation,

$$\text{Motor Speed} = \frac{\text{Voltage}}{\text{Field Strength}}$$

shows that a motor whose field strength is held constant will always run at a speed dictated by its applied voltage, regardless of its load, and that speed can be limited by limiting voltage.

The other equation

$$\text{Motor Torque} \simeq \text{Current} \times \text{Field Strength}$$

shows that such a motor will demand current directly in proportion to its torque load and that torque can be limited by limiting current.

The only motor whose field can be held constant is a separately excited shunt motor. It permits maximum simplification of controls because no extra devices are needed to obtain stable speeds at light loads or to prevent overspeeding at no load. This is an important safety consideration where chain drives are used. Because the motor always operates with full field, it also has inherently superior deceleration characteristics without additional control devices.

Controls

Now, in Fig. 9–9, the separately excited shunt motor is connected to the generator through a switch or contactor. Contactors are also provided to disconnect the fields, and a rheostat is used to control the generator field strength and output voltage. By turning this

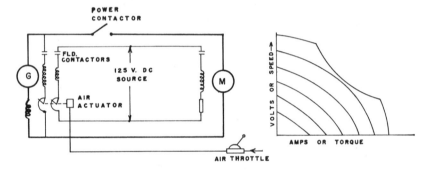

FIG. 9–9. Separately excited shunt motor connected to generator.

rheostat, an infinite number of operating curves can be obtained, as shown. For a given setting of the rheostat, the motor operates up and down the selected speed-torque curve, depending on the load imposed on it.

With no load, the motor would run at a stable speed at the left end of one of the curves, depending on the setting of the rheostat. This speed could be varied at will from about 10 per cent to 100 per cent of rated speed.

With an excessive load, the motor would stall at a point at the lower end of one of the curves, developing a torque in proportion to the rheostat setting. This stalled torque can also be varied from about 10 per cent to 100 per cent of maximum generator current by the operator.

Thus the operator has full control of torque and speed at all times. He can apply torque gradually until the load starts to move, then accelerate as slowly or as rapidly as desired. Furthermore, a standard ammeter will provide a continuous indication of torque, line pull, or pump pressure, since amps are directly proportional to torque in a shunt motor. And a standard voltmeter will indicate rotary rpm, line speed, or pump strokes per minute, since voltage is directly proportional to the speed of a shunt motor. The product of volts and amps is watts, which may be converted to horsepower by dividing by 746.

Figure 9–9 illustrates the basic control scheme where the generator is driven at a constant speed by an engine. Voltage control is obtained by varying generator field strength. But from the basic equation for generator voltage,

$$\text{Generator Voltage} \simeq \text{Speed} \times \text{Field Strength}$$

it is seen that output voltage can also be controlled by changing the generator (or engine) speed. The common method of doing this is with an air-actuated engine governor. If generator speed is con-

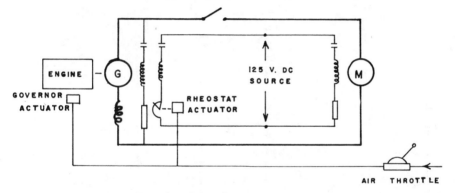

Fig. 9–10. Controlling output voltage.

trolled, there is no need for as much field control to obtain the same family of output curves; therefore, the rheostat plate in the self-excited field is eliminated, and in its place an air actuator is put on the other plate. This permits the control of both engine speed and rheostat setting simultaneously from a common throttle valve at the driller's position. See Fig. 9–10.

Actually, remote air operation is a convenient and simple way to control the rheostats in either system. Many draw works already have provisions for such throttles for hand or foot operation or both, and drillers are familiar with them. They can be combined with air-clutch operators or with each other in various ways.

The basic controls have been shown for one generator and one motor. The same system is used with any required number of machines, provisions being made to switch generators from one motor to another, and to put two or more generators on one motor for maximum power. Figure 9–11 illustrates a switching arrangement for four 600 hp generators and five motors each rated 625 hp continuously for pumping, and capable of delivering 1,000 hp intermittently for hoisting. Two generators driven by different engines are available to each motor to provide standby power for any function. All four generators could be used to develop 2,000 hp at the draw works, two generators in parallel on each motor.

An assignment switch for each generator is provided on the opera-

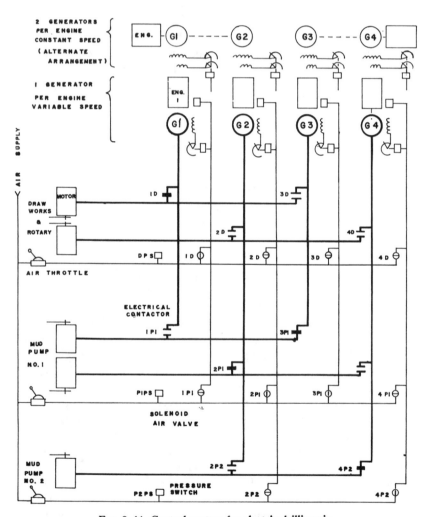

Fig. 9–11. Control system for electric drilling rig.

tor's console. This sets up both the speed control and power circuits and puts full field on all motors to which power is assigned. Thus, if power assignments were made as shown in Fig. 9–11, Generator 1 would be assigned to drive the rotary, Generators 2 and 3 to Mud Pump 1, and Generator 4 to Mud Pump 2. If the throttle of Mud Pump 1 were advanced, the power contactors $2P1$ and $3P1$ would close, field would be applied to Generators 2 and 3, and the throttle would control the output of these two generators and, consequently, the pump speed.

The throttle system for this arrangement is also shown in Fig. 9–11. When Generators 2 and 3 assignment switches are set on Pump 1 ($P1$) position, the electrically operated air valves ($2P1$) and ($3P1$) are opened, connecting throttle ($P1$) to control Generators 2 and 3. When this throttle is advanced, the initial air pressure in the throttle line causes pressure switch ($P1PS$) to close its contacts, which in turn closes power contactors ($2P1$) and ($3P1$), starting the pump. Similarly, Generator 1 would be controlled by the draw works throttle, and Generator 4 by Pump 2 throttle.

The power contactors are locomotive type, pneumatically operated and rated to carry 1,000 amps continuously. They close on initial throttle movement, and are opened by a spring when the throttle is turned off. But they include blowout coils and arc chutes, and are capable of repeated openings while carrying maximum stalled current. They have been proved on severe locomotive duty, and offer long, maintenance-free service on drilling duty.

Reversing of the draw works or rotary-drive motors is accomplished by turning a switch on the operator's panel which merely reverses the shunt fields of these motors. This can be done only with these throttles in the *OFF* position.

This describes all the basic controls required for this system. It can be adapted to any combination of motors and generators with any desired degree of power-assignment flexibility to suit mobile platforms, platform-tenders, and small or large land rigs. When two generators are driven in tandem by one large engine, as shown by the alternate arrangement at the top of Fig. 9–11, the generator fields only are controlled by the throttle, since a change in engine speed would change the output of both generators, which may be connected to different loads. But separate controls can be provided to preselect definite engine speeds, such as idle, half, three-quarters and full, depending on power demand at the time. With one generator per engine or one generator driven by two engines, either system may be used, depending on whether it is desired to reduce engine speed automatically at reduced throttle settings or to provide separate engine-speed controls.

A separate 125-volt DC source is required for power to generator and motor fields and control devices. This can be a single small auxiliary generator of 25 to 50 kw capacity depending on the size of the rig. It can be driven by one of the main engines, by a light-plant engine, or by an AC motor if sufficient AC power is available. It can be a constant or variable speed machine. Normally a second such auxiliary generator is provided as a standby.

Some protective features are included, as follows:

Emergency stop button. This is located on the operator's panel to shut down all power in emergency.

Electrical fault protection. One side of the power circuit is grounded through a sensitive ground relay. When an electrical fault occurs, this relay operates to remove all power.

Cutout switches. Switches are provided at each pump, draw works, etc., which can be opened and padlocked if desired to prevent the application of power to machinery during maintenance or inspection.

Warning devices. Lights are provided on the operator's panel to signal loss of ventilation or ground relay operation. An alarm horn also sounds during an abnormal condition, if any throttle is opened.

Engine shutdown. If an engine shuts down for any reason, its associated generators are automatically isolated from the system to prevent running the engine by "motoring" action of its generator.

The operator's control panel includes the following devices: an assignment switch for each generator, an ammeter for each motor, a voltmeter for each motor where a speed indication is desired, a reversing switch for the draw works and rotary if separately driven, an emergency stop button, and indicating lights as required.

The speed-control air throttles may be mounted in the electrical operator's panel or separately on the draw works console.

The control system is purposely as simple as can be devised for controlling DC machines. It contains no complicated circuits or devices, but only contactors, relays, resistors, and rheostats, plus the standard air devices which are familiar to the industry. All the devices have been proven by years of locomotive service, and the system has operated in drilling service with outstanding success.

The preceding report shows how maximum torque is obtained at zero speed, which is a desirable element in a drilling-rig power plant. For example, in removing drill pipe from the hole, or in working with stuck drill pipe, it is highly desirable to develop maximum torque (or power) at a zero speed in order to reduce stresses on the equipment. Also, when attempting to rotate the drill pipe, it is desirable to develop maximum torque at very low rotational speeds for the same above-mentioned reasons.

Development of maximum torque at zero speed results in a smooth pickup of hoisting loads.

Electric power plants have the same flexibility of steam, yet with much greater compactness and portability. The principal drawback of the electric power plant is its high initial cost.

Turbines

The gas turbine as a prime mover is not a new development. In fact, it is one of the oldest forms of combustion engines. Shaft work from a turbine is developed by using moving gases (or steam) to propel a row of blades or "buckets" which are rigidly connected to a shaft.

A centrifugal compressor furnishes air to a combustion chamber, into which fuel is injected. The combustion gases are then directed to a turbine where the hot gases impinge upon the blades of the turbine, causing them to rotate the shaft, producing rotational, or mechanical, energy. The combustion of the fuel and air results in a large increase in gas volume and pressure. As the hot gases enter the turbine, the pressure is reduced and the velocity of the gas is thereby greatly increased.

In practice the gas turbine is used to drive the air compressor; therefore, it is necessary to have some auxiliary source of starting power to initiate the cycle. This is normally provided by a small electric motor or internal combustion engine.

The gas turbine provides a continuous flow of power to the output shaft, and there are no reciprocating elements, as there are in piston-type engines. On account of this absence of reciprocating motion, the gas turbine can be operated at much higher speeds than conventional piston-type engines. Consequently, the unit weight per brake horsepower is quite small.

Although the cycle of operation of the turbine eliminates many complicated parts of the combustion engine, this same principle has retarded its practical use as a prime mover. Because of the turbine's continuous operation, the operating temperature of the combustion chamber and turbine are much higher than the operating temperature of combustion engines, where combustion occurs in cycles. Until metals were developed which could withstand these high operating temperatures, the gas-turbine engine was not an important prime mover.

Within recent years, materials capable of withstanding these continuous high operating temperatures have been developed, and this type of engine may have some application in the drilling industry. A very promising area lies in the combination free-piston–gas-turbine plant, which will be discussed later.

Free-Piston Engines

The merits of a free-piston engine have long been recognized. It consists essentially of two pistons opposing each other. The power cycle begins with injection of fuel and air into the combustion chamber; after combustion occurs and pressure increases within the combustion chamber, the opposing pistons are forced out of the combustion chamber. However, in doing this, the bounce pistons, located on the opposite ends of the power pistons, compress air in the bounce chamber, and at the same time ports are opened in the combustion chamber permitting the combustion products to escape and the pressure to be relieved. The high pressure in the bounce chambers then forces the power pistons back into the combustion chamber, increasing the pressure until ignition occurs and the cycle is repeated. Thus the free piston operates on the diesel principle, in which ignition occurs by compression.

Ideally the free-piston engine has many advantages on account of its simplicity. The stroke and compression ratio of the free-piston engine are not fixed, but are determined by the amount of fuel and air injected, the mass of the moving parts, and the friction developed. As the mass of the moving parts is fixed and little control can be exercised on the friction developed, control of the engine is obtained by regulating the amount of fuel injected into the compression chamber. When more fuel is injected, a longer stroke is realized, which results in greater compression in the bounce chamber. This greater compression in the bounce chamber produces more thrust on the return stroke, and greater compression on the fuel-air mixture results.

Theoretically, the operation of the two pistons should be synchronous; however, because of difference in friction, they will not be completely synchronous. To insure proper operation, the two pistons must be synchronized, and for this reason a linkage is usually placed between the two pistons to assure synchronization.

The principal application of the free-piston engine has been as a compressor used to drive a gas-turbine. The combustion gases developed in combustion chambers of the free-piston gas generator are used to drive a turbine. The thermal efficiency of the combustion cycle and the flexibility of the turbine drive combine to provide a promising power-plant combination. The crankshaft, connecting rods, and bearings of the conventional internal combustion engine are eliminated.

Auxiliary Equipment

In addition to the three principal operations in drilling—circulating, rotating, and hoisting—several miscellaneous functions are performed which require power from some source. Careful consideration must be given to the placing of this auxiliary equipment in the power arrangement or else adequate power may not be delivered to one of the principal operations when it is needed. Some of these miscellaneous functions which require power are (1) light plant, (2) sand reel (for swabbing, etc.), (3) fans or blowers, (4) shale shakers, (5) mud centrifuges, (6) transfer pumps, and (7) air compressors.

In auxiliary equipment where steam power can be efficiently utilized, the power steam can be piped to the equipment without materially reducing the over-all power output, as the steam boilers provide a simple power source. Electric power or internal combustion power does not enjoy the complete flexibility of steam because more than one power unit is usually employed, and the linking together of the individual power units into a single component is not as simple as when steam power is used.

Cable-Tool Power Plants

Power requirements in cable-tool drilling are considerably less than that required for rotary drilling. In cable-tool operations, the fact that there is no fluid circulation system eliminates a major power requirement. The weight of cable tools in the hole is materially less than the weight of rotary tools. This is due to the use of a wire line to place the bit on the bottom of the hole instead of drill pipe as required in rotary operations. The horsepower requirements in cable-tool drilling will be more fully discussed in Chapter 11, because most of the power requirements for cable-tool drilling are for hoisting. The churn drilling action of the cable tools is accomplished by hoisting the drill stem a few feet and then dropping the tools on bottom.

Basically a cable-tool rig and a rotary rig use the same type of power plant, the principal difference being size. A cable-tool rig will require only a fraction of the horsepower required by the same size rotary rig because no circulating fluid and no drill pipe are used in cable-tool operations. Thus the need for circulating horsepower is completely eliminated, and the need for hoisting horsepower is greatly reduced. Principal uses of horsepower on a cable-tool rig are for the churning action of the drilling tools and for the removal of the drilling tools from the hole. These operations are similar in that they both involve the lifting of the drill string off bottom.

Two important considerations in the horsepower requirements of a cable-tool rig are the number of churning strokes per minute desired and the rate of removal of the drill stem from the hole. Sample calculations follow which illustrate the method of determining horsepower requirements.

Example: Determine the drilling horsepower required for a cable-tool operation where the maximum weight of drilling tools and lines will be 4,000 pounds. The length of stroke of the walking beam will be 4 feet and the beam will be operated at 30 strokes per minute. Mechanical efficiency is 90 per cent.

Solution: If the beam is operating at 30 strokes per minute, then the drilling tools will have to be lifted and dropped once each two seconds, or the tools will have to be lifted in one second if the assumption is made that the tools will fall as rapidly as they are raised. Therefore, the velocity of travel of the tools on the upstroke will be 4 feet per second.

The horsepower required is

$$\text{Theoretical horsepower} = \frac{4{,}000 \text{ lb.} \times 4 \frac{\text{ft.}}{\text{sec.}}}{550 \text{ ft.-lb./sec./hp}} = 29.1 \text{ hp}$$

$$\text{Actual horsepower} = 29.1 \div 0.80 \text{ mechanical efficiency}$$

$$= 36.4 \text{ hp required}$$

It should be noted that additional horsepower will actually be required because of the sticking action of the tools on bottom.

Example: Calculate the horsepower required to remove the drill stem from the hole at the rate of 500 feet per minute, if the maximum weight of the drill stem and line is 4,000 pounds. Assume a mechanical efficiency of 80 per cent.

$$\text{Actual horsepower} = \frac{4{,}000 \text{ lb.} \times 500 \frac{\text{ft.}}{\text{min.}}}{33{,}000 \frac{\text{ft.-lb.}}{\text{min.}} \times 0.80 \text{ efficiency}}$$

$$= 75.75 \text{ hp required}$$

Drilling horsepower requirements and hoisting horsepower requirements are not additive since these operations are never conducted simultaneously; however, certain auxiliary operations, such as the light plant and water pumps will require some horsepower. In determining the total horsepower requirements, therefore, it is necessary to add the auxiliary horsepower requirements to either the drilling or the hoisting horsepower requirements, depending on which is greater, and then apply a suitable reserve horsepower to allow for stuck tools or fishing operations which may require additional horsepower.

Steam, electricity, and internal combustion power have been used to power cable-tool rigs. Steam was the first type of power used and was readily adaptable to cable-drilling operations. Electricity was first used as early as 1909.

Power-Transmission Mechanisms

The power output of most power plants is developed in the form of a rotating shaft. In order to convert this rotating energy into useful work, power-transmission devices are required. These power-transmission devices can be subdivided into (1) gear devices for changing speeds of rotation and power output, (2) the actual transfer of the developed power from one point to another, (3) the utilization of this power after it has been transferred to the desired place, and (4) clutches which permit these speed and power changes.

Gears

The principal function of gears is speed manipulation and increasing output torque. A change in rotational speed of shafts connected to the gears is obtained by meshing two gears with different diameters. The change of rotational speed will be directly proportional to the circumferences of the respective gears. This is illustrated where two shafts are connected with gears. The gear on input shaft A has a circumference of C, while the gear on output shaft B has a circumference of $2C$, or twice the circumference of gear A. For each complete revolution of shaft A, shaft B will turn one-half revolution; therefore, the speed of rotation is cut in half. As there is also a direct relationship between circumference and diameter, i.e., $C = \pi d$, the change in rotational speed is directly proportional to the diameter of the gears. Gears are used on internal combustion engines to enable them to handle larger loads at a slower pace, which also changes the output torque. For example, in removing heavy loads of drill pipe from the hole, if the rate of removal of the drill pipe is reduced, the horsepower requirements are reduced. Or stated in another way, with a given horsepower output from an engine, by operating the engine at its maximum power output and then reducing the output work by means of gears, the rate of doing work, and thereby the horsepower requirement, is reduced, but the torque is increased.

Power Transmission

The principal users of power on a drilling rig are (1) circulating equipment, (2) draw works, (3) rotary table, and (4) auxiliary equipment. Therefore, power from the power plant must be available for all of these services. In a steam power plant, transmission of power is accomplished by the steam lines, making this type of power transmission the simplest of any type yet developed. In the internal combustion engine power plant—or power rig, as it is generally called—the transmission of power becomes more difficult. On small power rigs one engine may develop sufficient power for all of the four different power users, in which case suitable power-transmission devices would have to be developed to satisfy this need. The fact that several of the power users will have to be served simultaneously also complicates the transmission equipment. For example, circulation, rotation, and certain auxiliary items of equipment must normally have power simultaneously. In the larger power rigs, where two or more engines may comprise the power plant, the problem still is not simple, because in the interest of flexibility and maximum usability of developed horsepower, it may be desirable to change one or more of the engines to different power uses as the demand for power changes. During drilling operations, the major power requirement is for circulating, and a large percentage of the total power developed may be directed to the circulating equipment. However, during round-trip operations, the major power requirement is for the draw works. Thus, in the interest of flexibility, the power transmissions should be arranged to direct the developed power to any needed place. Transmission of power on electric-powered rigs is normally relatively simple. The generators can be located in a central area with the motors placed conveniently near the driven equipment.

Drilling-mud circulating pumps are usually driven by flexible belts or by chain drives. On the other hand, the draw works and rotary table are usually chain driven.

Power Utilization

After the power has been transmitted to the point of utilization; i.e., either the mud pumps, draw works, or rotary table, it must be effectively utilized. The transmitted power from the power plant is applied to the various users of power through mechanical couplings, fluid couplings, electrical couplings, and torque converters.

The mechanical couplings are usually the gear type. Fluid couplings have no direct mechanical connection between the power source and the driven equipment. Since this fluid coupling will not transmit shock loads from one piece of equipment to another, it helps increase the life of the equipment. The electric coupling is similar in operation to the hydraulic coupling, except that electricity rather than a liquid is utilized. The electric coupling utilizes two concentric rotors, one rotor being attached to the driving shaft and the other to the driven shaft. The coupling is designed so that current flowing through the driving rotor will set up a magnetic flux in magnetic circuits formed by the two rotors. Then when the driving rotor is set in motion, an electro-motive force is set up which causes rotation of the driven rotor.

The torque converter is an outgrowth of the fluid coupling. However, the torque converter is designed not only to absorb shocks from the power plant or the driven equipment, but also to multiply the input torque as required. When the driven equipment begins to slow down on account of an increase in loading, fluid movement within the torque converter is such that the torque on the input shaft is increased. Torque can be increased as much as five-fold on some torque-converter designs. Thus the torque converter serves the same purpose as changing to a lower gear ratio in a mechanical transmission. As this torque conversion is performed automatically, it is ideally suited to the hoisting operations on a drilling rig.

During the torque conversion process, the output horsepower is not changed. It is impossible to increase output horsepower by torque conversion.

Clutches

Some means must be provided to disengage the driven equipment from the power plant. The clutch is the equipment used to perform this function. Two basic types of clutches are in use, the mechanical clutch and the air clutch. Mechanical clutches are either jaw or gear-type clutches, and can be used satisfactorily on steam or electric rigs, because power can be applied in small increments. However, on internal combustion engine–powered rigs the mechanical clutch imposes shock loads on equipment, as there is a minimum speed of rotation below which the internal combustion engine cannot be run. The utilization of the air clutch has been one of the major factors responsible for the wide application of the internal combustion engine as a power plant for drilling rigs. The air clutch is actually a friction-type clutch; with the air causing the gripping surface to close around the power input shaft. A typical air clutch is shown in Fig. 11–25 in Chapter 11. The air clutch is used extensively in the drilling industry at the present time.

Typical arrangements of power plants, power transmissions, and power uses for a power rig are shown in Fig. 9–12.

TABLE 9–1

Typical Actual Fuel and Horsepower Requirements for Power Rigs

Depth (feet)	Required Horsepower	Fuel Consumption		
		Diesel (gal./hr.)	Gas (SCF/hr.)	Butane (gal./hr.)
1,000	175	11.7	1,750	15.15
2,000	250	15.5	2,500	21.61
3,000	275	17.2	2,750	23.80
4,000	300	18.8	3,000	26.55
5,000	350	23.4	3,500	30.25
6,000	375	24.3	3,750	32.40
7,000	500	32.4	5,000	43.25
8,000	600	36.8	6,000	51.80
9,000	750	46.0	7,500	64.90
10,000	850	61.6	8,500	73.50
11,000	950	68.8	9,500	82.20
12,000	1,000	72.5	10,000	86.50
15,000	1,250	92.3	12,250	108.00
20,000	2,000	145.0	20,000	173.00

TABLE 9–2

Water Requirements for Power Rigs*

Area	Water Requirements (bbl./day)
Oklahoma (Elk City Field)	350– 400
West Texas (Block 31 Field)	330– 380
Gulf Coast (Land Rig)	550– 650
Gulf Coast (Barge)	600– 800
Rocky Mountains	300– 400
Offshore (Tender)	650–1,000+
Slim Hole operations	80– 200

* Average water requirements for areas indicated.

POWER PLANTS

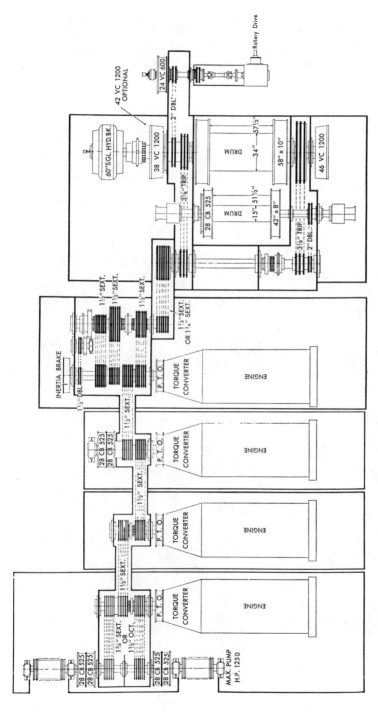

FIG. 9–12. Typical arrangements of power plants, power transmission, and power uses for a power rig.

Fig. 9–13. Deep-drilling rig in operation.

Fig. 9–14. Drilling-rig power plant.

POWER PLANTS

TABLE 9-3

CABLE-TOOL HORSEPOWER*

Depth (feet)	Horsepower
1,500	55
2,000	90
3,000	150
4,000	158
5,000	158
6,000	180
7,000	192
8,000	210

* Typical rigs operating in the United States.

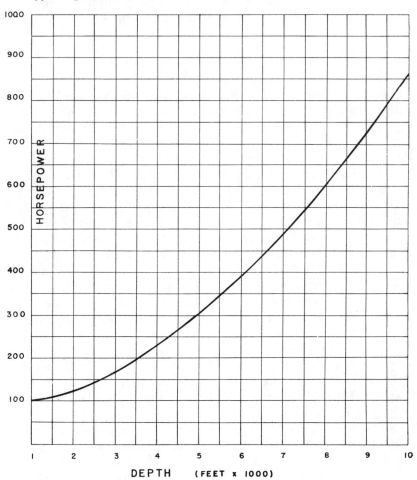

FIG. 9-15. Average horsepower required *vs.* depth.

CHAPTER 10

Rotary Operations, Drill Strings, and Bits

In conventional rotary drilling, a new hole is drilled by rotating the drill string and bit, and this is generally referred to as *rotary operations*. Other common rotary operations include *milling*, in which pipe is cut by rotating a cutting tool on the bottom of a string of pipe, and *reaming*, in which the hole is enlarged by rotating suitable tools in the section of the hole which is to be enlarged. The equipment concerned in rotary operations includes the swivel, kelly, rotary table, drill pipe, drill collars, and the bit.

SWIVELS

The swivel performs three functions: (1) it suspends the kelly and drill string; (2) it permits free rotation of the kelly and drill string; and (3) it provides a connection for the rotary hose and a passageway for the flow of drilling fluid into the top of the kelly and drill string. Accordingly, the chief operating parts of the swivel are a high-capacity-thrust bearing, which is often of tapered roller bearing design, and a rotating fluid seal consisting of rubber or fiber and metal rings which form a seal against the rotating member inside of the housing. The fluid seal is arranged so that the abrasive and corrosive drilling mud will not come in contact with the bearings.

The swivel is suspended by its bail from the hook of the traveling block. The fluid entrance at the top of the swivel is a gently curving tube, which is referred to as the *gooseneck*, which provides a downward-pointing connection for the rotary hose. In this manner the rotary hose is suspended between the upper nonrotating housing of the swivel and the standpipe, which extends part way up the derrick and conveys mud from the mud pump. The fluid passageway inside of the swivel is commonly about three or four inches in diameter so that there is no restriction to mud flow. The lower end of the rotating member of the swivel is furnished with left-hand threads of API tool-joint design.

Swivels are furnished by the several manufacturers in various load capacities. Since the load-capacity requirement is determined largely by the weight of the drill string, rigs capable of deep drilling operations require swivels of greater capacity than rigs used in shallower drilling.

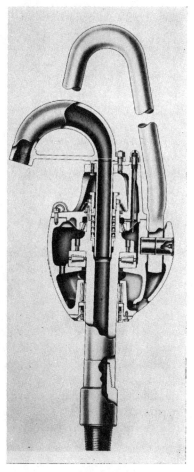

FIG. 10–1. Swivel (cutaway view).

KELLYS

The kelly, or *grief stem*, is a hollow section whose upper end is suspended from the swivel. The flow of drilling fluid passes downward through the kelly and into the top of the drill pipe. Internal diameters of about three inches are common. The outside cross-section of the kelly is usually square, although hexagonal kellys are sometimes used. The purpose of the square or hexagonal outside cross-section is to enable torque to be transmitted from the rotary table to the drill string. The tool-joint thread on the lower end of the kelly is right-handed (similar to the threads on the drill pipe) and the thread on the top is left-handed, so that normal right-hand rotation will tend to keep all joints made up tight. During drilling operations the kelly bushing remains on the kelly. The outside of the kelly bushing is commonly square, about sixteen inches on each side, so that when it is removed from the rotary table, the largest bits used in the well will pass through the rotary table. Torque is applied from the rotary table, through the kelly bushing, and thence to the kelly stem itself. The kelly is free to slide through the kelly bushing (or rotary bushing) so that the drill string can be rotated and simultaneously lowered or raised during drilling operations. The better bushings are provided with roller bearings for the purpose of facilitating the sliding of the kelly through the bushing. For use with Range 2 drill pipe, which makes up in lengths of about 31 feet per joint, kellys are made with a length of square section (length in the clear) of 38 feet; and the overall length, including the round end sections in which the tool joints are cut, is about 41 feet. Of the 38 feet "in the clear," about 2 feet are taken up by the kelly, or rotary, bushing. The original lengths of the round end sections on the kelly permit rethreading when the original tool-joint threads become worn.

A *kelly saver sub* is used between the kelly and the upper joint of the drill-pipe string. The use of this short section eliminates the necessity of unscrewing the lower end of the kelly itself during drilling operations and thereby prevents thread wear on the kelly joint. The tool joint on the kelly saver sub should be weaker than the tool joint on the kelly. This protective arrangement will cause failures to occur in the sub rather than in the kelly. The kelly saver sub also provides a space for mounting a rubber protector which will prevent the kelly from whipping against the inside of the casing, so that wear on both the casing and the kelly is prevented.

Kelly cocks are short sections containing a valve which may be manually closed. They are placed between the kelly and the swivel. The use of a kelly cock permits closing the top of the drill string so that flow cannot take place through the inside of the drill string. The flows of concern are usually those which might be associated with drill-stem testing or other operations wherein the pressures of the subterranean formations might otherwise be applied against the rotary hose.

Rotary Tables

The primary function of the rotary table is to transmit torque and impart a rotary motion to the kelly and drill string. The top of the housing of the rotary table customarily forms a portion of the derrick floor and is provided with a nonskid tread. The exposed top of the rotating table, as well as the exposed upper part of the rotary or kelly bushing, should be free of projections which might be hazardous to personnel. The rotating table is usually cast of alloy steel and fitted underneath with a ring gear which is shrunk onto the table proper. The table is supported either by ball or tapered roller bearings capable of supporting the dead load of the string of drill pipe or casing which may be run into the well. Provisions in the form of suitable bearings must also be made for holding the table in place against any possible tendency to upward movement imparted during rig operations. Suitable guards are arranged so that mud or water cannot get

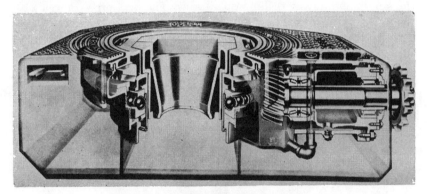

Fig. 10–2. Rotary table (cutaway view).

into the oil bath provided for the gears and bearings. The ring gear and its driving pinion gear are customarily of spiral bevel construction, which provides for smoother operation than straight bevel gears. The speed reduction from the pinion shaft to the table is of the order of 3 or 4 to 1. The pinion shaft is customarily furnished with antifriction bearings, often of tapered roller design. It is also equipped with oil seals and guards to prevent entry of mud or water into the interior of the equipment.

Power for driving the rotary table is most often taken from the draw works and transmitted to the rotary table by a chain and sprocket drive. In this arrangement the excess power available, as well as the momentum of the heavy moving parts, presents a hazard to twisting off the drill pipe if the bit should become stuck. The hazard is reduced where a single internal combustion engine can be used for driving the rotary. Independent drives are also used for the rotary tables; and these employ a steam engine, an internal combustion engine, or an electric motor, according to the type of power used in the drilling rig. Also, a hydraulically driven table has been developed. In

166 OIL WELL DRILLING TECHNOLOGY

common, all such drives rotate the horizontal pinion shaft, the spiral bevel pinion gear rotates the ring gear, which in turn rotates the vertical drill string.

Rotary tables are rated according to the size of the opening through the table, such as twenty inches, and the dead-load capacity of the table, such as 700,000 pounds

Drill Pipe and Tool Joints

The major portion of the drill string, or drill column, is ordinarily composed of drill pipe. The upper end of the drill pipe is supported by the kelly stem during drilling operations. The drill pipe rotates with the kelly, and the drilling fluid is simultaneously conducted down through the inside of the drill pipe and subsequently returned to the surface in the annulus outside of the drill pipe. In a deep well, the top

FIG. 10-3. Tool joint mounted on drill pipe. Note weld which increases strength and hard facing on box which reduces wear.

Fig. 10-4. Tool joint mounted on internal-external upset drill pipe.

portion of the drill pipe is under considerable tension while drilling, since most of the weight of the drill string is supported from the derrick.

The drill pipe in common use is hot-rolled, pierced seamless tubing. API Grade D drill pipe has a minimum yield strength of 55,000 psi, and Grade E drill pipe has a minimum yield strength of 75,000 psi. Drill pipe made of stronger steel is also available.

The drill pipe most commonly used is Range 2 pipe, which has an average length of 30 feet per individual length (joint) of pipe. The addition of tool joints produces an average made-up length of about 31 feet per individual length of pipe.

The individual lengths of pipe are fastened together by means of tool joints, which means that there are complete tool joints spaced at 31-foot intervals throughout the length of the drill pipe. The male half of a tool joint is fastened to one end of an individual piece of pipe and the female half is fastened to the other end. Ordinary pipe proved to be unsuitable for drill pipe, and pipe made of higher grades of steel failed in the threaded connections. Upsetting the ends of the pipe to compensate for metal removed in cutting threads proved to be a partial answer. In internal upset pipe, the pipe wall near the end of the length is made thicker by decreasing the internal diameter. In external upset pipe, the pipe wall near the end is made thicker by increasing the external diameter. In internal-external upset pipe, greater thickness at the ends of the pipe is achieved by both decreasing the ID and increasing the OD. Threads of ordinary dimensions will not stand up under the repeated uncoupling and making-up which is required every time a round trip is made with the drill pipe for such purposes as replacing a worn bit. Therefore, tool-joint threads, which are large, tapered threads, were cut into short lengths of alloy steel with sufficient OD, six inches or more, to accommodate such threads. These short pieces of alloy steel were then threaded on their other end so that they in turn might be screwed on the upset ends of the drill pipe. Satisfactory performance has been achieved by shrink fitting the tool joint to the pipe and welding a bead around the end of the tool joint so that it is firmly fastened on the pipe. Before this practice became widespread, failures were common at the last exposed thread of the pipe where it fastened into the tool joint. In addition to threaded connections, tool joints are butt-welded to drill pipe and some pipe is sufficiently upset that an integral tool joint is cut on the pipe.

Tool joints are subject to abrasion on their exterior surface since they invariably rub against the rock wall of the hole. Welding beads of hard surfacing material around the outside of the tool joints often doubles their useful life. The threads and shoulders of the tool joints represent surfaces which slide over each other with extreme pressures as the joints are made up. If foreign particles are present or too much friction develops, parts of the metal will tear away and roll up very much like a snowball, which results in "galling" of the thread or shoulder surface and eventual failure of the joint. It is therefore very important that the threads and shoulders be maintained clean and covered with a suitable thread lubricant.

The drill string in a deep well has the relative dimensions of a length of thread; the long, slender shape gives the drill string certain inherent weaknesses. A new piece of drill pipe could fail from the application of excessive tension or torque, although such failure is rare. The yield strength of the various types and sizes of drill pipe is listed in tables giving the properties of tubular goods, and drill pipe will be permanently stretched and thereby weakened if the yield strength is exceeded in a direct pull on the pipe. Most drill-pipe failures are caused through the process of corrosion-fatigue. Such failures usually start on the inside of the pipe. Minute cracks in the surface open and close as the pipe works in rotation and tension or compression, and corrosive fluid is pumped in and out by such working so that the crack

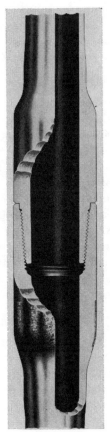

FIG. 10-5. Integral tool joint with hard-faced box.

is enlarged by both mechanical failure and corrosion until eventually the piece of pipe is discarded or fails. The inside surface of some drill pipe is plastic coated to retard such action. In some instances, sodium dichromate has been added to the drilling mud to inhibit corrosion, particularly where the mud is salty. Where a small hole has extended through the pipe during drilling operations, the hole has been rapidly enlarged by mud erosion to the extent that the pipe twisted off. An alert driller will be able to detect a washout almost immediately by a change in pump pressure, thereby eliminating a fishing job or damage to casing. Where such holes have occurred inside of the cased portion of the well hole, the casing has been damaged by erosion of the mud. Any wobbling action in the rotating drill string is conducive to fatigue failures. Such wobble may be caused at the top of the string by a bent kelly. Formerly, wobble was common in the bottom portion of the drill pipe when part, or all, of the weight on the bit was supplied by the drill pipe. Shock such as may occur from carrying too much weight on a drag type bit is also conducive to pipe fatigue and failures.

Drill Collars

The lower section of the rotary drill string is composed of drill collars. The name derives originally from the short sub which was used to adapt the threaded joint of the bit to the drill string. However, modern drill collars are each about 30 feet in length, and the total length of the string of drill collars may be from about 100 to 700 or more feet. With the exception of the Gulf Coast area and the shallow drilling in the Mid-Continent area, most drilling utilizes twenty to thirty drill collars, or a length of 600 to 900 feet. The purpose of the drill collars is to furnish weight and stiffness in the bottom portion of the drill string. During drilling, all of the drill pipe should be in tension, since drill pipe is essentially a tube of medium wall thickness and has but little resistance to bending by column action. This means that the total weight of the string of drill collars should be determined by the weight carried on the drill bit. During the drilling of any interval of rock, the weight on the bit varies instantaneously with the action of the bit, so that the effective weight of the drill collars immersed in mud should exceed the average weight on the bit as shown by the weight indicator. Practice varies; however, one of the earliest recommendations was that the effective weight of the drill collars should be such that 70 per cent of the drill collars would be in compression and furnishing the weight on the bit.

Drill collars are usually made with an essentially uniform OD and ID. A tool-joint pin is cut on the lower end and a tool-joint box is cut into the upper end; however, when the individual lengths are coupled together, they present a smooth exterior. The internal bore of the drill collars is usually $2\frac{1}{4}$ inches or $2\frac{7}{8}$ inches, but the smaller bore tends to give excessive fluid-flow pressure losses, and very little difference in weight results from using the larger bore. The OD of the drill collars is limited by the size of the hole being drilled. Most of the development of the use of long strings of drill collars took place in the Permian Basin of West Texas and New Mexico, which is hard rock country requiring high weights on the bit for satisfactory penetration rates. The practice there was originally to use drill collars of 6- to $6\frac{1}{4}$-inch diameter, which was about the same OD as the tool joints, for drilling $7\frac{7}{8}$- and $8\frac{3}{4}$-inch diameter holes. If such drill collars, or the bit, became stuck, they could be "washed over," since sufficient annular space remained between the drill collars and the hole that a tube through which mud could be circulated could be lowered down over the outside of the drill collars to aid in their recovery. It is, of course, a decided advantage, in case of trouble, to have nothing in the hole which cannot be washed over, the bit excepted. However, the requirements of weight and rigidity have prompted operators to assume the calculated risk of using drill collars of large OD. In the period from 1953 to 1955, $6\frac{1}{4}$-inch drill collars were largely replaced by $6\frac{3}{4}$-inch and 7-inch OD drill collars for drilling $7\frac{7}{8}$- and $8\frac{3}{4}$-inch holes.[1] Wash pipe was available which could be used with such drill collars. In addition, $7\frac{7}{8}$-inch holes were being drilled in some instances with 7-inch OD drill collars, and some $8\frac{3}{4}$-inch holes were being drilled with 8-inch OD drill collars, with both of the latter termed *oversize collars* because of the fact that washing over was originally impossible, with fishing operations restricted to the use of tapered taps or similar devices which could take their hold on the inside of the lost, or disengaged, drill collars. However, wash pipe for the 7- and 8-inch drill collars has subsequently been made available, although overshots are not available which can operate in the small annular clearance.

Better bit performance has been reported with the use of oversize drill collars, and it was suggested by Bromell that the larger collars have less buckling, or rather buckle to a less degree, because of their closer fit in the drilled hole, and consequently hold the bit more squarely on bottom, thereby reducing any walking or alternate loading of one bit cone at a time.[1] One purpose of using such larger OD drill collars is to produce a straighter hole. There is some evidence that the use of larger drill collars tends to reduce the sticking of drill collars in key seats.[2] Key seats are usually formed at a shoulder where a hole which had deviated some from the vertical straightens downward, and a groove which tends to assume the size of the OD of the tool joints on the drill pipe wears into the area on the side of the hole which must support part of the weight of the drill string as drilling

[1] R. J. Bromell, "Drilling Practices in the Permian Basin," *The Petroleum Engineer*, Vol. XXVII, No. 4 (April, 1955), p. B55.
[2] Personal communication from G. C. McDonald, Gulf Oil Corporation, Oklahoma City.

proceeds deeper. When pulling the pipe in such key seats, or grooves, there is presumably less tendency to stick a drill collar whose diameter appreciably exceeds the tool-joint OD than there is of sticking a drill collar just slightly larger than the tool joints, which continuously wear through the key seat during actual drilling.

Rotary Drill Bits

The bit which does the actual drilling is attached to the lower end of the drill collar. In the rotary system of drilling, hole is made by lowering the string of drill pipe and drill collars until the bit touches or approaches the bottom of the hole. Circulation of drilling fluid is established down through the drill pipe, and the fluid is discharged through ports or watercourses in the bit so that the bit and bottom of the hole will be kept clean. Rotation of the drill string is established by means of the rotary table. The top of the drill string is then gently lowered by means of the draw works or hoist until suitable weight for drilling is applied to the bit.

The type of bit which should be used in any particular instance is governed primarily by the characteristics of the rock to be drilled and the conditions under which the rock must be drilled. The softer shales of the younger sedimentary rocks are drilled very effectively with drag-type bits, particularly in the Gulf Coast area. Toothed wheel-type bits are generally used for drilling the harder shales, sands, and limestones. Rock strata which lie horizontally are more easily drilled than inclined beds, where it is often necessary to use less weight on the bit in order to maintain a vertical hole. Rocks of uniform character present a single problem, while interbedded soft and hard layers require a suitable compromise in order that the over-all interval can be effectively penetrated. Drilling conditions include the drill string, particularly the drill collar size and weight, the weight which may be carried on the bit while drilling, the straightness and vertical character of the hole, the nature of the drilling fluid, rate of circulation, jetting action employed, the rotary speed, and possible safety precautions demanded by the drilling equipment, caving formations, or the financial investment represented by the depth of the hole already drilled. Different types and variations of bits are therefore in common use in drilling wells.

Economic considerations involved generally require that the hole should be drilled at the lowest possible cost per foot. The total cost per foot depends upon, among other considerations, the average rate of drilling and the total feet drilled per bit. The total feet drilled per bit becomes increasingly important at greater depths where more rig time is required for a round trip to replace the worn-out bit.

Fig. 10-6. Relation of fluid nozzle discharge to chip generated by drag-type bit modified for high nozzle fluid velocity drilling.

Drag bits. Drag bits have no independently moving parts, and the term is applied specifically to bits of the blade type. The simplest of the drag bits are the fishtail bits with two blades spaced 180 degrees from each other. Bits having three or four blades are also used, but such bits are usually fingered so that the total length of the cutting edges does not exceed the length of the hole diameter by more than about 20 per cent. The blades on modern bits are furnished with tungsten carbide inserts or are otherwise hard surfaced to reduce wear. Short blades are preferred, for their use permits the mud-discharge nozzles to be positioned a short distance above the bottom of the hole so that maximum jet energy can be utilized in the drilling. The mud streams flowing out of the discharge nozzles are directed to the bottom a short distance ahead of each cutting edge. Drag bits are used in drilling soft formations, and under ideal conditions the drilling action probably resembles the turning of earth by a plow. The mud streams directed to the bottom of the hole break up material loosened by the bit and carry it upward to the surface. In many soft formations, hole can be made by the jetting action of the drilling fluid. In most cases, however, the chief function of the mud appears to be removing cuttings and keeping the bit and bottom of the hole clean. In a later section of this book the operating variables are discussed, and it is shown that the rate of drilling which can be maintained is related to the hydraulic horsepower of the mud streams which are jetted out of the bit housing and directed toward the bottom of the hole.

FIG. 10-7. Three-cone bit for softer formations with conventional mud-discharge ports.

Disk bits. These bits are interesting from a historical standpoint. They are a form of drag bit in which the cutting edges are mounted on disks. The disks are mounted off-center with respect to the axis of the drill string, so that as the drill string is rotated, the scraping action on the bottom causes the disks to rotate slowly. In this manner the total cutting edges available for drilling are increased by comparison with the stationary blades of the drag bit. Two or four disks are mounted in the bits. The bottom of the drilled hole is rounded, a form adapted to the flushing of cuttings by the mud stream. The disk bit inherently does not have the weight-bearing capacity of the drag bit and does not provide as much clearance at the bottom of the hole for removal of cuttings. It is related to the drag bit somewhat in the same manner that the Zublin differential bit is related to ordinary rolling cutter bits in that both have cutting elements which rotate and present successive surfaces against the bottom of the hole.

Rolling cutter bits. The most widely used bits are rolling cutter bits. Structurally the bits are classified as cone-type bits and as cross-roller bits. Rows of teeth are cut into the rolling members, so that these bits

FIG. 10-8. Jet-type cone bit illustrating bearing construction.

are also referred to as toothed-wheel bits. The teeth are hard surfaced with such material as tungsten carbide in order to obtain longer bit life. The toothed wheels rotate independently; as the drill string is rotated, the rolling cutters turn by virtue of their contact with the bottom of the hole. A recent adaptation for the drilling of very hard rock has been the substitution of rounded tungsten carbide inserts in place of the teeth ordinarily cut on the rollers, and these latter bits are referred to as button bits. With respect to the size of the mud fluid-discharge ports and their arrangement, the bits are generally classified as conventional or as jet bits.

FIG. 10–9. Cross-roller bit.

FIG. 10–10. Bottom view of cross-roller bit.

In the design of the toothed-wheel rolling cutter bits, the size of the rotating members is limited by the geometry of the bit and the diameter of the hole. A further design compromise must be reached between the size of the bearing in the center of the rotating member, the thickness of the hub of the member, and the depth of the teeth cut

FIG. 10–11. Hard-formation three-cone bit showing bottom-hole drilling pattern.

into the member. The strength of each of these parts, including also the stationary axle fastened to the bit housing, depends largely upon their size. The amount of metal which can be retained in each of these parts, and therefore the strength of the bit, tends to vary as the square of the linear dimension—the diameter of the drilled hole. Conversely, the contact between the rotating members and the bottom of the hole

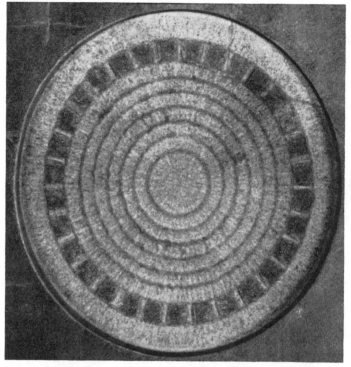

Fig. 10-12. Hard-formation bit and drilling pattern in hard rock. Webs between gauge (outside) teeth.

tends to be linear and varies directly as the diameter of the drilled hole. The larger bits are therefore capable of operating under greater loads per inch of hole diameter than are the smaller bits and give better rates of penetration, at least up to sizes of $7\frac{7}{8}$-inch hole diameter. The teeth on the larger bits can be made longer, and it therefore appears consistent with present knowledge that the larger bits are inherently capable of drilling more footage per bit.[3] Such comparisons

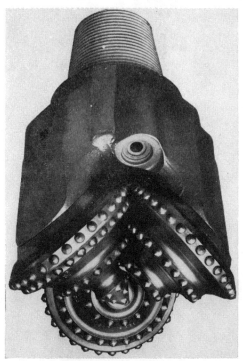

FIG. 10–13. Jet-type carbide insert or "button" bit.

are believed to be particularly valid between the ranges of about $4\frac{3}{4}$-inch and about $7\frac{7}{8}$-inch hole diameter.

Rolling cutter bits are designed for soft, medium, and hard formations. For drilling in softer formations such as the younger shales, it is desirable to have widely spaced, long, slim teeth so that any tendency for cuttings to pack between the teeth will be minimized. In some instances the teeth on one cutter wheel sweep through the space between the rows of teeth on adjacent cutter wheels. This permits the use of longer bit teeth in these areas and also provides for some clean-

[3] Personal communication from H. B. Wood, Hughes Tool Company, Houston.

ing action and removal of lodged cuttings. The long, slim teeth tend to remain relatively sharp. Although the teeth are widely spaced, the number must be sufficient to insure rotation. The different rotating elements, such as the cones, have slightly different total numbers of teeth on the different cones, or rollers. This provision prevents the

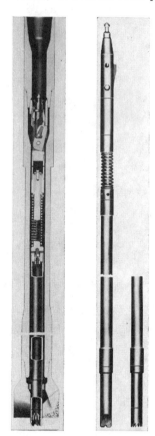

Fig. 10–14. Soft-formation coring assembly. Core barrel and bit are retrievable by wire line and can be replaced with the small drilling bit.

teeth on successive rollers from touching the same place in the bottom of the hole, and the size of the chips is less than the spacing between the teeth. A twisting action on bottom is attained by offsetting the axes of the cones forward of the center of the hole, so that a true rolling action is impossible. Thus the teeth on soft formation bits tend to tear and scrape at the bottom of the hole, and this scraping aids in the drilling action.

Rolling cutter bits for use in drilling hard formations have closely spaced and shorter teeth. This feature provides for the maximum strength and tooth surface that can be utilized in drilling the formation and therefore increases the footage drilled per bit. The rolling elements are designed so that they have a true rolling action on the bottom of the hole, since sliding on bottom tends to wear the teeth of the bit. The drilling action of such bits possibly consists of the re-

FIG. 10–15. Rolling cutter core bit.

moval of small fragments or grains of rock by crushing action with each tooth contact and the occasional removal of larger fragments such as might be formed by the ridges which develop in the drilling pattern. For example, where a bit is drilling at the rate of fifteen minutes per foot and the rotary table is turning at 60 rpm, the rate of penetration per revolution of the bit is

$$\text{Penetration per revolution} = \frac{12 \text{ in./ft.}}{(15 \text{ min./ft.})(60 \text{ rev./min.})}$$
$$= 0.013 \text{ in. per revolution}$$

Very little is actually known about the precise drilling mechanisms and manner of rock failure under the conditions of deep drilling. On the other hand, as noted under the discussion of the drill collar, there is some evidence that the action of the bit is not independent of the characteristics of the drill string—particularly the drill collars. The efficient drilling of hard formations requires suitable drill collars. Furthermore, the evidence must be considered conclusive that the drilling action is strongly influenced by the nature of the drilling fluid and the pressure relations in the bottom of the hole.

For drilling formations which are intermediate between the very soft and the very hard, the several bit manufacturers have a sufficient choice of bits of intermediate characteristics to drill economically the formations encountered. Occasionally bits have been made with one cone whose characteristics were different from those of the other two, and such mixing of cutting elements has occasionally given good results. Occasionally, more on an experimental basis, one cone has been cut off of a three-cone bit before running in the hole. Such bits are known as pegleg bits. Occasionally such bits have been tried in crooked holes. Often they have drilled faster than they would have with three cones present, but as the total footage drilled per bit is less, their use is questionable from economic and safety considerations.

The mud-discharge ports in the conventional type of bit are often referred to as *watercourses*. The theory of positioning the conventional watercourses so that the mud fluid is directed down onto the top of the cutting elements is obviously to maintain the cutters in a clean condition so that they will be free to drill into the bottom of the hole. Visual tests made by placing a bit inside a large lucite tube have shown that there is considerable fluid turbulence on the bottom of the hole even with this arrangement and that large particles simulating large drill cuttings are readily washed away. This finding is confirmed by the continued successful use of such bits. However, in some areas and in some formations, the use of jet bits has given greatly increased rates of penetration. In such bits the mud-discharge nozzles, or jets, are so arranged that the mud fluid is discharged directly on the bottom of the hole. The removal of cuttings from the bottom of the hole is thus insured. The fluid velocities required for effective jet drilling seem to indicate, however, that more than mere removal of the drill cuttings may be involved. Other factors may possibly be that the jetting velocity prevents incipient mud filter cake formation on the bottom of the hole or that the liquid fluid phase of the mud is forced by jet pressure into the formation so that shales are weakened or the pressure gradient on the face of the bottom of the hole is reduced. As implied, the matter must be considered speculative at the present time regarding the mechanisms whereby jet bits achieve advantages in drilling certain formations.

Zublin bit. This bit, named for its designer, is the only bit manufactured in recent years in which the cutting element has an articulating motion against the bottom of the hole. However, early patents were issued for articulating types of bits. As the drill string is rotated, the cutting element revolves at a slower rate while the teeth slide up

FIG. 10-16. Zublin bit.

and down along the rounded bottom of the hole with each revolution of the drill string.

Zublin differential bit. This bit has been used for drilling shales where the weight carried on the bit has been limited. It has sometimes been used for straightening a well and maintaining a vertical hole. It has also been used experimentally with the mud turbine. The cutting elements are small and mounted in the rim of a larger wheel. Rotation of the drill string causes the cutting elements to rotate on the bottom of the hole, which is rounded. Since the large wheel is mounted off-center, it also rotates slowly so that different cutting elements are brought to bear progressively against the bottom of the hole. The mud stream is divided so that one portion washes the cutting elements while they are in their upper position inside of the housing of the bit while the main portion of the mud streams sweeps the rounded bottom of the hole free of any drill cuttings.

Diamond bits. Diamond bits, similar to drag bits, have no independently moving parts. They drill by direct abrasion or scraping against the bottom of the hole. Diamonds are much harder than the

Fig. 10–17. Zublin differential bit.

common rock minerals and are the only material which has been used economically for such severe service. When diamonds are mined, the stones of gem quality are separated from the commercial diamonds, which are colored or imperfect in crystalline form. The better grades of these stones are used in cutting tools for lathes and milling machines.

In manufacturing a diamond bit, a steel bit blank is machined with threads suitable to attaching it to a core barrel or the drill collars.

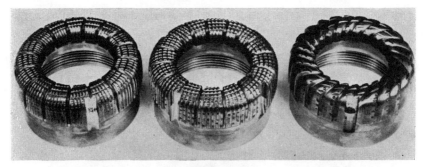

Fig. 10–18. Hard-, medium-, and soft-formation diamond core bits.

Diamonds are placed in a bit mold and covered with powdered metal. With the bit blank held in position, the assembly is heated until the powdered metal fuses, and then it is slowly cooled. In this manner, the diamonds are held in a matrix which is attached to the steel bit blank. The matrix is composed partly of hard particles such as tung-

Fig. 10–19. Diamond bit for medium-hard formation. Plug-type bit drills out the entire hole.

sten carbide, together with a softer alloy which acts as the binder material. The binder material must melt at a sufficiently low temperature that the diamonds will not be injured in the process. The hard particles in the matrix are necessary in order that the matrix will not erode or wear away, thus causing loss of the diamonds and failure of the bit. In the early 1940's, several attempts were made to develop solid diamond bits which would drill out the entire hole without re-

Fig. 10-20. Hard-formation plug-type diamond bit. Special type of bit for going through undersized sections of the hole.

quiring a periodic round trip to empty the core barrel. These bits failed in the center, for in that area the diamonds have little or no peripheral speed and consequently tend to drill more slowly and take more drilling weight than the diamonds at the outer edges of the bit. About 1950, diamond bit manufacturers began using a much harder matrix material, and subsequently solid diamond bits have been developed which give satisfactory service.

Diamond bits are much less sensitive to the nature of the drilling fluid than are the rolling cutter bits. It is generally believed that thinner drilling fluids are advantageous in diamond bit performance and economy; however, they will continue to drill with thick muds. They differ in another important respect, for small-diameter diamond bits perform as well and better than larger-diameter bits and are, of course, much less expensive. A disadvantage of the diamond bit is its fragility. If drill cuttings or broken chips from rolling cutter bit teeth

Fig. 10–21. Reamer.

remain on the bottom of the hole and roll under the bit as it is rotated, they may rapidly cause bit failure. Thus it is necessary to wash the bottom of the hole thoroughly with mud circulation each and every time the diamond bit is set back on bottom. In this operation care must be taken that the throbbing action of the mud pump and consequent longitudinal vibration of the drill string does not cause the bit to pound on bottom. While slightly higher rotary speeds might often be advantageous for faster drilling, the bits are customarily rotated in the same range of speed as rolling cutter bits at about 40 to 100 rpm, in order to avoid any irregular action of the drill string which would damage the bit.

Reamers. The primary use of a reamer is to enlarge the hole already drilled by the bit. Such enlargement is generally restricted to the practice of running a reamer above the bit while the hole is being drilled. In this connection, the reamer acts to stabilize the bit. In addition to a reamer directly above the bit, one or more reamers have been positioned in the string of drill collars at points of suspected buckling. The purpose of using reamers in this manner is to permit the use of greater weights on the bit and still maintain a straight hole.

CHAPTER 11

Hoisting Operations

The function of the hoisting equipment is to get necessary equipment in and out of the hole as rapidly as is economically possible. The principal items of equipment that will be used in the hole are drill string, casing, and miscellaneous well-surveying instruments such as logging and hole-deviation instruments. The drilling string is probably the most important consideration in the design of hoisting equipment, although casing loads may be the largest loads imposed on the derrick. In the conventional drilling operation, the drilling bit becomes worn at frequent intervals, necessitating its removal and replacement, which requires removal of the entire drilling string. The number of times the drilling bit will have to be replaced will depend on the depth of the well and the character of the formations encountered. In shallow wells, where bit wear is not excessive, only a few bits will be required, while in deep wells, where bits may become worn after having drilled only a few feet, several hundred bits may be required. Removal of the drill pipe for the purpose of replacing a worn bit is commonly referred to as "making a trip." A *round trip* therefore refers to both removal and insertion of the drill string. Drilling-rig design is influenced to a large extent by the economics of removing the drill string from the hole.

In order to decrease the amount of time required to make these round trips, individual joints of drill pipe may not be unscrewed when the bit is changed. Also, the drill pipe is stacked vertically in one corner of the derrick, materially reducing the time required to make the round trip. The drill string may be unscrewed at every other joint, thus reducing the work of connecting and disconnecting by half. Each length of pipe which is stacked in the derrick is called a *stand* of drill pipe. When each joint of pipe is unscrewed and stacked, this length of pipe is called a *single stand*. If the pipe is disconnected at every other joint, the resulting length of pipe is called a *double stand*. The drill pipe also may be disconnected at every third joint, which is called a *thribble stand*, or at every fourth joint, which is called a *fourble stand*. Obviously the height of the derrick will determine the length of each stand. As the height of the derrick increases, the permissible length of each stand increases correspondingly.

The principal components of the hoisting system are the (1) block-and-tackle system, (2) derrick, (3) draw works, and (4) miscellaneous hoisting equipment such as hooks, elevators, and weight indicators.

The block-and-tackle system is comprised of the (1) crown block, (2) traveling block, and (3) drilling line. The principal function of this block-and-tackle system is to provide the means for removing equipment from, or lowering equipment into, the hole. The block-and-tackle system develops a mechanical advantage, to be described later in more detail, which permits easier handling of large loads. The positioning of the drilling equipment in the hole is also a function of the block-and-tackle system, as well as providing a means of gradually

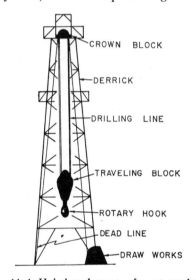

FIG. 11–1. Hoisting elements of a rotary rig.

lowering the drilling string as the hole is deepened by the drill bit. A schematic diagram of the hoisting elements of a rotary rig are shown in Fig. 11–1. The function of the derrick is to provide a structure for the removal and insertion of down-the-hole working equipment and also for properly positioning the drilling tools over the hole. The draw works is the power control center of the rig. The principal component of the draw works is a hoisting drum on which the drilling line is spooled. The rig power plant supplies motive power to the hoisting drum permitting the reeling or unreeling of drilling line from the hoisting drum. There are other auxiliary components of the draw works, such as small revolving drums, or *catheads*, which can be used for many purposes, including minor hoisting and moving operations, and *jerk chains* for the final tightening or initial unscrewing of joints of drill pipe as they are connected or unscrewed.

Hoists and Their Operation

The starting point in the design of hoisting equipment must be in the block-and-tackle system used in raising and lowering drill pipe, casing, and other equipment. As mentioned previously, very large weights must be handled by the hoisting system. Casing ordinarily imposes the largest load on the hoisting equipment; for example, 12,000 feet of $9\frac{5}{8}$-inch OD 47 #/ft. casing weighs 564,000 pounds, disregarding the effects of buoyancy, while 20,000 feet of $6\frac{5}{8}$-inch OD, 25.20 lb./ft. drill pipe weighs only 504,000 pounds, disregarding the effects of buoyancy. It is possible to reduce these large casing loads by placing a plug in the bottom of the casing to prevent entry of the drilling fluid inside the casing, thus increasing the effects of buoyancy. However, this technique presents other problems and will be discussed more fully in Chapter 19.

The horsepower required to lift these big loads may be quite large. For example, if it were desired to lift a drill string weighing 500,000 pounds at the rate of 100 feet per minute, the theoretical horsepower requirements would be

$$\text{hp} = \frac{500{,}000 \text{ lb.} \times 100 \text{ ft./min.}}{33{,}000 \text{ ft.-lb./min./hp}} = 1{,}515 \text{ hp}$$

This calculated requirement of 1,515 hp does not include any friction losses, and when the friction losses are included, it is obvious that the horsepower requirements would be quite large.

In order to reduce the force, or torque, required to remove the drill string from the hole, advantage has been taken of a device for mechanically reducing the horsepower requirements. It should be emphasized that the energy expended or the work performed does not change, for this, of course, is not possible. This mechanical device which is used to advantage is the block-and-tackle system. Figure 11–2 shows schematically two derricks with different block-and-tackle systems in place. Assume that the weight, W, attached to the lower end of the traveling block is the drill string. In view A of Fig. 11–2, the work done in moving this weight of 300,000 pounds, from point 1 to point 2, a distance of one foot is

$$\text{Work} = \text{Force} \times \text{Distance}$$
$$= 300{,}000 \text{ lb.} \times 1.0 \text{ ft.}$$
$$= 300{,}000 \text{ ft.-lb.} \qquad (11\text{–}1)$$

Horsepower is defined as the rate of doing work. Therefore, if the load in this problem is raised the distance of one foot in one second, the horsepower requirements for a direct lift would be:

$$\text{hp} = \frac{300{,}000 \text{ ft.-lb./sec.}}{550 \text{ ft.-lb./sec./hp}} = 545.4 \text{ hp}$$

In view *B* of Fig. 11–2, the hoisting system is different, and it can be seen that this is not a direct lift, but the hoisting line is wound around three sheaves in the crown block and two sheaves in the traveling block. Therefore, in order to raise the load one foot, four

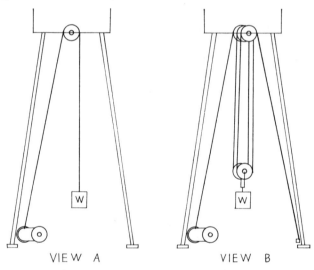

FIG. 11–2. Block-and-tackle systems.

feet of line must be wrapped around the hoisting drum. Since the work done must be the same, then the weight of the effective load lifted can be computed as follows:

$$\text{Work} = \text{Force} \times \text{Distance}$$

or

$$\text{Force} = \frac{\text{Work}}{\text{Distance}}$$

$$= \frac{300{,}000 \text{ ft.-lb.}}{4 \text{ ft.}} = 75{,}000 \text{ lb.}$$

Another way to analyze the forces present with a block-and-tackle system is to utilize the diagram shown in Fig. 11–3. If a weight of 300,000 pounds is attached to the traveling block, then each of the lines around the traveling block will support one-fourth of the weight. These forces would be present in the same proportion at the crown block. Therefore, from analyses of this figure, it is obvious that the pull on the line to the hoisting drum would also be one-fourth of the total weight, W, on the traveling block.

The horsepower required to lift the weight, W, at a rate of one foot per second, using the block-and-tackle system, can also be deter-

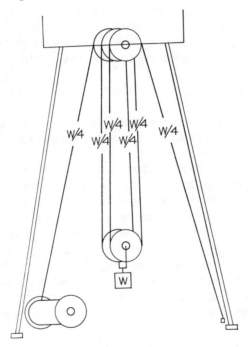

FIG. 11–3. Distribution of load on derrick lines.

mined. However, in order to achieve a net lift of one foot, the line to the drum must be moved a distance of four feet. The horsepower required is

$$\text{hp} = \frac{\text{Work}}{\text{Time}} = \frac{\text{Force} \times \text{Distance}}{\text{Time}}$$

$$= \frac{75{,}000 \text{ lb.} \times 4 \text{ ft.}}{1 \text{ sec.} \times 550 \text{ ft.-lb./sec./hp}} = 545.4 \text{ hp} \quad (11\text{–}2)$$

Thus the horsepower requirements are the same in both cases. The use of the block-and-tackle system has not reduced either the work or the horsepower required to perform the job.

An important characteristic of any power plant is torque. The torque of a power plant is an indication of the power which the plant is capable of developing. As discussed in Chapter 9, the units of torque are foot-pounds. Referring to Figs. 11–2 and 11–3, the line pull on the drum is 300,000 pounds in the case of the single-sheave system and only 75,000 pounds in the case of the double block-and-tackle system. Therefore, the power plant torque required to initiate movement in the hoisting drum when the double block-and-tackle system is employed would be only one-fourth of the torque required to initiate movement in the hoisting drum when only a single sheave was used. The block-and-tackle system has two distinct advantages in hoisting operations:

1. Actual horsepower requirements can be less because the rate of doing work can be reduced.
2. Engine torque requirements will be considerably less, depending upon the number and arrangements of lines in the block-and-tackle system.

The number of sheaves and the arrangement of the drilling line through these sheaves is important when derrick loads are considered. One phenomenon of the block-and-tackle system is that the actual load on the supporting structure, in this case the derrick, may be considerably larger than the weight of load actually lifted. Figure 11–4 shows three possible block-and-tackle combinations. View A shows a single sheave at the top of the derrick. View B shows three sheaves in the crown block and two sheaves in the traveling block, with the end of the line (called the *dead line*) attached to the derrick floor. View C shows the same sheave arrangement except that the dead line is now

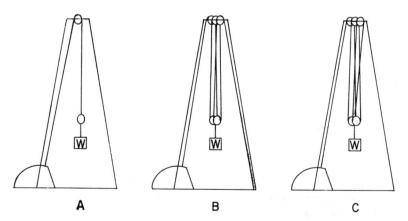

Fig. 11–4. Block-and-tackle combinations.

attached to the traveling block. Summation of the total pull on the derrick for the three arrangements is shown in Table 11–1.

TABLE 11–1

Loads on Derrick

View A	$2.0W$
View B	$1.5W$
View C	$1.2W$

Important conclusions developed from an analysis of the various possible loading combinations are:
1. Actual loads imposed on the derrick may be much larger than the weight to be lifted.
2. As the number of sheaves in the block-and-tackle system is increased, the actual load on the derrick is decreased.
3. Attaching the dead line to the traveling block rather than to the derrick floor reduces the actual load on the derrick.

From the preceding discussion it is obvious that when the block-and-tackle system is used, the force required to move a given weight is reduced. This ratio of the weight lifted to the amount of force required to lift the weight is called the *mechanical advantage*. The amount of mechanical advantage which will be obtained in any case will depend upon the number of lines supporting the load. In Fig. 11–4, the mechanical advantage for each of the three different line-and-sheave arrangements can be calculated:

View A:

$$\text{Mechanical advantage} = \frac{\text{weight lifted}}{\text{force exerted}}$$

$$= \frac{W}{W} = 1$$

View B:

$$\text{Mechanical advantage} = \frac{W}{W/4} = 4$$

View C:

$$\text{Mechanical advantage} = \frac{W}{W/5} = 5$$

From the foregoing calculations, it is evident that the mechanical advantage is exactly equal to the number of lines which are actually helping lift the load.

196 OIL WELL DRILLING TECHNOLOGY

The load on the drilling line decreases as the number of sheaves in the crown block and traveling block increases, as is evident from an examination of Fig. 11–3. This is of considerable importance in wire-rope design and requirements. However, since the wire-line load decreases as the number of sheaves is increased, the line must travel a correspondingly longer distance in order to raise the load on the traveling blocks a given distance.

Hoisting Horsepower

For a given hoisting speed the actual horsepower required to remove the drill string from the hole will be a maximum at the total depth of the well, and a minimum at the surface when drilling has just begun. The horsepower required in removing the drill string from the hole is often referred to as *hook horsepower*, as the horsepower calculation is made by determining the weight on the hook multiplied by the lift velocity of the hook. Figure 11–5 is a curve showing hook horsepower requirements versus number of stands in hole for a constant hoisting speed and the same-size drill pipe. Where electricity or torque converters are used, an essentially constant hook horsepower is available, and where mechanical drives are employed, for any given gear ratio, the hook horsepower is constant at maximum speed. Therefore, in hoisting applications, the graph shown in Fig. 11–5 is actually of little practical importance, once its significance has been grasped. As a constant horsepower source is available, this means that as the total load is reduced coming out of the hole, the speed at which the drill pipe is removed from the hole can be increased. For example, if 1,000 hook horsepower is available and the total weight of the drill stem is 200,000 pounds, then the time required to lift one stand of drill pipe 90 feet in length from the hole is

$$hp = \frac{\text{Force} \times \text{Distance}}{\text{Time}}$$

or

$$\text{Time} = \frac{\text{force} \times \text{distance}}{hp} = \frac{200{,}000 \text{ lb.} \times 90 \text{ ft.}}{1{,}000 \text{ hp} \times 550 \frac{\text{ft.-lb.}}{\text{sec./hp.}}}$$

$$= 32.7 \text{ seconds}$$

When half of the drill-stem load has been removed from the hole, the time required to remove one stand from the hole would be

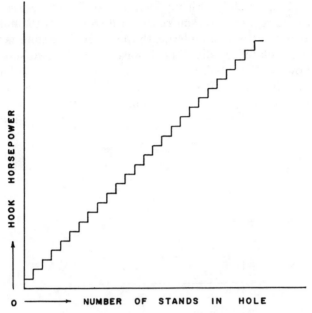

Fig. 11-5. Hook horsepower vs. depth.

$$\text{Time} = \frac{\text{force} \times \text{distance}}{\text{hp}}$$

$$= \frac{100{,}000 \text{ lb.} \times 90 \text{ ft.}}{1{,}000 \text{ hp.} \times 550 \frac{\text{ft.-lb.}}{\text{sec.}}} = 16.4 \text{ seconds}$$

However, for any rig there is always some limiting hoisting velocity above which it is unsafe to work. Therefore, regardless of the horsepower available, this maximum safe hoisting speed should not be exceeded. This relationship is shown graphically in Fig. 11-6. Using this information, it is possible to calculate the relative economics of two rigs having different horsepower, where the number of round trips at the various depths can be reasonably assumed and the costs of the rigs are known. It is assumed, of course, that all other factors, such as crew efficiency and condition of the rig equipment, are identical for both rigs. An efficient crew can in many cases more than offset a rather large difference in horsepower. However, if all other factors are equal, the time saved on each round trip can be calculated, and after this time saving has been converted to dollars, the more economical rig can be selected.

In Fig. 11–6, which is basically identical to Fig. 11–5, it is seen that the total time required to remove the drill stem from the hole at any depth is equal to the area underneath the curve. The time can be conveniently calculated by dividing this area into the two sub-areas, A and B, shown in Fig. 11–6. Both of these areas can be calculated quite rapidly by this method. It must be realized that the time required for each round trip made during the course of drilling the well will be different because of the difference in the length of drill pipe in the hole.

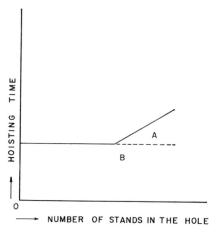

Fig. 11–6. Hoisting time per stand *vs.* number of stands in the hole for constant horsepower.

Component Parts of the Hoisting System

Derricks

The conventional derrick is a four-sided truncated pyramid ordinarily constructed of structural steel, although tubular steel is used infrequently for certain parts of the derrick. Derricks may be either portable or nonportable, the portable derrick being commonly referred to as a *mast*. The nonportable derrick is usually erected by bolting the structural members together. After the well has been drilled, the derrick can be disassembled, by unbolting, and erected again at the next location. As erection of the derrick may require several days, it is the practice of some companies to use two nonportable derricks with each rig. After a well has been completed, no attempt is made to dissemble derrick No. 1 immediately, because derrick No. 2 has already been erected at the next location. Considerable rig time can be saved by making use of the two derricks, because no time is

Fig. 11-7. A drilling mast.

lost by having to wait for the rig-building crew to disassemble the derrick, move it to the new location, and assemble it again. After the rig is drilling on well No. 2 and when the next location is known, then derrick No. 1 can be disassembled and erected at the next location. In flat terrain the derrick is sometimes skidded to the next location without dismantling

The principal considerations involved in the design of a derrick are these:

1. The derrick must be designed to carry safely all loads which will ever be used in wells over which it is placed. This is the collapse resistance caused by vertical loading, or the dead-load capacity

of the derrick. The largest dead load which will be imposed on a derrick will normally be the heaviest string of casing run in a well. However, this heaviest string of casing will not be the greatest strain placed on the derrick. The maximum vertical load which will ever be imposed on the derrick will probably be the result of pulling on equipment, such as drill pipe or casing, that has become stuck in the hole. The designer must consider that, sometime during the useful life of the derrick, severe vertical strain will be placed on it because equipment has become stuck in the hole. Several methods can be employed to provide for these maximum strains. One is to allow a considerable additional load above the maximum casing load. Another is to design the derrick to withstand loads which exceed the capacity of the wire line which will be used on the rig.

2. The derrick must also be designed to withstand the maximum wind loads to which it will be subjected. Not only must it be designed to withstand wind forces that will act on two sides at the same time (the outer surface of one side of the derrick, and the inner surface of the opposite side), but cognizance must also be taken of the fact that the drill pipe may be out of the hole and stacked in the derrick during periods of high winds. The horizontal force of the wind acting on the derrick and drill pipe is counteracted by the pyramidal design of the derrick, by bolting the derrick legs to their foundations, and by the use of from one to three guy wires on each leg of the derrick. These guy wires are attached to *dead men* located some distance from the derrick. A *dead man* is made from a short length of large pipe, a concrete block, or a short section of timber, which is buried in the ground to provide an anchor for the guy wire. Guy wires are small-diameter wire lines, usually less than one-half inch in diameter.

A typical derrick is shown in Fig. 9–13. The component parts of the derrick are gin pole, crow's nest, water table, derrick man's working platform, legs, girts, braces, and ladder.

The gin pole, located at the extreme top of the derrick, is used principally for hoisting the crown block into place. The crow's nest provides a safe working surface around the crown block. The water table is the opening in the top of the derrick into which the crown block fits. The derrick man's working platform is the working area from which the upper end of the drill pipe is handled as the drill pipe is removed from, or inserted into, the hole. The derrick legs are the principal structural members of the derrick. Each leg, of which there are four, is a continuous member extending from the base of the der-

HOISTING OPERATIONS

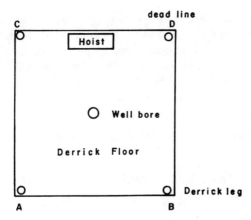

FIG. 11-8. Plan view of a derrick floor.

rick to the water table. The girts are the horizontal structural members connecting and supporting the four derrick legs. The braces are the structural members used to strengthen the derrick by proper bracing between girts. The derrick ladder is used to provide a convenient access to the upper parts of the derrick, principally the working platform for the derrick man and the crown block.

In view B of Fig. 11-4 the vertical loading of a derrick can be determined. The pull on each line is shown to be $W/4$, where W is equal to the weight to be lifted. The pull on each line will be called T for further development of the problem. In other words, for the specific example of view B, Fig. 11-4, $T = W/4$. The load on the derrick caused by the hanging weight, W, will be the sum of the downward line pulls. As there are six lines pulling downward, the total vertical load is $6T$. Figure 11-8 is a plan view of a derrick floor, showing the hole, the hoist, the derrick legs, and the position of the dead line. The loads on the individual derrick legs are shown in Table 11-2.

TABLE 11-2

DERRICK LOADS

	Total Load	Load on Individual Derrick Legs			
		A	B	C	D
Centered Load	$4T$	$1.0T$	$1.0T$	$1.0T$	$1.0T$
Hoist-line Load	T			$0.5T$	$0.5T$
Dead-line Load	T				$1.0T$
Total Load		$1.0T$	$1.0T$	$1.5T$	$2.5T$

Table 11-2 assumes that there are two sheaves in the traveling block. The maximum load on any leg would be $2.5T$ in this particular instance, and this maximum would occur only if the dead line were attached to one of the legs which also supported the hoist-line load. In practice, this is not a recommended procedure, as the maximum total load on any derrick leg could be reduced to $2.0T$ by attaching the dead line to the leg A or leg B. However, using the case illustrated in Fig. 11–8, the maximum equivalent load on the derrick would be the maximum leg load multiplied by four, as shown below

$$\text{Maximum equivalent load} = 4 \times 2.5T = 10T$$

However,

$$\text{Actual load on derrick} = 6 \times T = 6T$$

where

$6 = $ number of lines pulling down

The ratio of the equivalent load to the actual load is called the *derrick efficiency factor*. This is evident because of the various derrick-leg loading combinations and also because of the fact that the derrick leg is the principal load-supporting structure, and if one derrick leg fails, obviously the entire derrick will fail.

$$\text{Derrick efficiency factor} = \frac{\text{actual load}}{\text{equivalent load}} \times 100$$

$$= \frac{6T}{10T} \times 100 = 60\% \qquad (11\text{–}3)$$

If the dead-line load is shifted to leg A or leg B, then the maximum load on the individual derrick leg becomes $2.0T$. The equivalent maximum load will now be: $4 \times 2T = 8T$. The derrick efficiency factor will be altered from the previous calculation:

$$\text{D.E.F.} = \frac{\text{actual load}}{\text{equivalent load}} \times 100 = \frac{6T}{8T} \times 100 = 75\%$$

Since the derrick as a whole is no stronger than its weakest member, if one of the derrick legs is overloaded, the derrick may fail even though the total load is well below the calculated failure load with an evenly distributed load. Therefore, the derrick load should not exceed the safe equivalent load. The derrick efficiency factor, then, is a measure of the use made of the derrick strength, and proper placing of the dead line is a major factor in the most efficient utilization of the derrick design capacity. The required derrick capacity, based on vertical loading only, can be computed as follows:

Required Derrick Capacity

$$= \frac{\text{Hanging Load} \times \text{No. Lines Down}}{\text{D.E.F.} \times \text{No. Lines Up}} + \text{Wt. of Crown Block} \quad \textbf{(11-4)}$$

An example problem will illustrate the practical utilization of the equivalent load theory.

Example: What percentage of rated derrick capacity may be utilized when six lines to the traveling block and eight lines on the crown block are used, with the dead line being attached to a derrick leg opposite the hoisting drum (equivalent to attaching to leg A or leg B in Fig. 11-8).

Solution:

Tension in each line $= \dfrac{W}{6} = T$

Total load on derrick $= 8T$
  (disregarding weight of crown load)
Centered load $= 6T$

Portion of centered load absorbed by each derrick leg $= \dfrac{6T}{4} = 1.5T$

Hoist-line load on leg C or leg D $= 0.5T$
Total load on leg C or leg D $= 2.0T$
Dead-line load on leg A $= T$
Total load on leg A $=$ centered load $+$ dead-line load

$$= 1.5T + 1.0T = 2.5T$$

Total derrick load if all legs were carrying a load equivalent to leg A

$$= 4 \times 2.5T = 10T$$

Derrick efficiency factor $= \dfrac{\text{actual load} \times 100}{\text{equivalent load}}$

$$= \dfrac{8T}{10T} \times 100 = 80\%$$

Therefore, 80 per cent of the rated derrick capacity can be used.

The weights of the crown block, derrick, and derrick accessories were not considered in the previous development since they would have been an additive term in both the numerator and denominator of the derrick efficiency factor, and would not materially affect the result, especially where the load, T, is large compared to these weights.

The effective load-carrying capacity of a derrick can be increased by reinforcing the derrick legs. This reinforcement is usually in the form of tubular steel columns placed adjacent to the derrick legs.

The API has developed specifications for derricks. The size number of the derrick refers to the number of *panels* or *bays* between the uppermost and lowest girts. Each panel has a height of seven feet. The first, or lowest, girt is approximately ten feet above the derrick floor for all derricks except some very tall ones, in which the lowest girt is located approximately fourteen feet above the derrick floor. In developing standard specifications for derricks, the API has defined the most important components of the derrick. Derrick height is measured along the derrick leg from the top of the derrick floor to the bottom of the water table beams. The *base square* is the distance between derrick legs and is measured at the top of the derrick floor and inside the derrick legs a distance of approximately one-fourth the width of the leg angle. The water table opening is the open distance between flanges of the water table beams.

The first girt extends around three sides of most derricks. On the fourth side greater clearance must be provided in order to allow the handling of long joints of drill pipe, casing, and other pieces of equipment. The larger opening on the fourth side is called a V-window, because of its similarity to an inverted V.

The API safe-load capacity of a derrick is determined by computing the strength of the derrick leg at its weakest point and multiplying this result by four, since there are four legs. The API rating does not consider either the weight of the derrick itself or the wind or other loads which may be imposed on the derrick. This is an important fact to understand, because in the API specifications (Standard 4A) there are requirements to withstand certain wind velocities. These wind forces are used in the design of the braces and girts in the derrick.

Masts

The design and development of oil well derricks has kept pace with other fast-developing segments of the oil industry. The development of the portable derrick, or *mast*, as it is usually called, has resulted in a material reduction in drilling costs, primarily in the amount of time consumed in rigging up and tearing down.

The development of the portable well-service unit has paralleled the development of the portable derrick, for one of the principal requirements of a portable well-service unit is a mast or derrick of some kind. Before the advent of the portable well-service unit, it was necessary to leave the derrick in place over the well for use in future well-service operations. The portable well-service unit eliminated the need for a permanent derrick and thus materially reduced over-all oil well costs.

A mast, shown in Fig. 11–9, is defined as a structure which can be moved as a unit, without dismantling. The masts used in oil well drilling operations may vary from a simple pole structure for drilling shallow wells to a four-leg structure closely resembling a standard derrick. Masts have been developed with heights as great as 146 feet and load capacities exceeding 1,000,000 pounds. These masts can be used safely to depths approaching 15,000 feet in many areas. The masts require a simple, reliable method of raising and lowering, which is generally accomplished by using the drilling line in conjunction with an A-frame of some type.

One of the problems associated with the development of the portable derrick is providing sufficient working room on the derrick floor while limiting the derrick base to dimensions that will permit its being transported across public highways. Most designs which are available telescope or unbolt to provide the short lengths required for highway travel.

Two basic types of masts have evolved as a result of the increased demand for portable derricks: (1) the free-standing mast and (2) the guyed mast. The *free-standing mast*, as the term indicates, uses no guy wires for support. If properly designed, it will withstand loads as great as any guyed mast available.

Construction of Derricks and Masts

In the early days of the oil industry, oil derricks were constructed of wood, but the advent of structural steel has eliminated wooden derricks. The few remaining are those which were left standing over old wells and are still in a safe enough condition to use for well-servicing work.

The material most commonly used for derricks today is structural steel of one form or another. Angle, channel, and I-beam in various sizes are consistently used in derrick designs. Tubular steel is commonly used to reinforce derrick legs. Single-pole masts are generally made of tubular steel, many of them being fabricated of oil well casing or line pipe. As the problem of weight is paramount in portable mast design, some masts have been fabricated of structural aluminum, which materially reduces the weight.

Derrick Substructures

In drilling operations, a working space must be provided below the derrick floor. The actual space required will depend upon the type of rig being used and the formation pressures that will be encountered

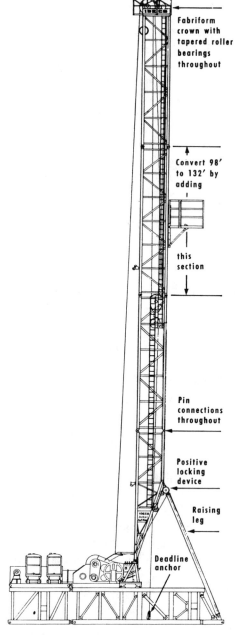

Fig. 11–9. A mast.

in the drilling of the well. In order to control excessive formation pressures and prevent the wells from getting out of control, blowout preventers must be attached to the innermost string of casing in order that they can be closed around the drill pipe, or, if the drill pipe is out of the hole, can be used as a valve to close the casing completely. Some blowout-preventer designs will close the opening at the top of the casing regardless of whether the drill pipe is in or out of the hole. The closing element is comprised of a hydraulically expanded rubber sleeve designed to withstand any formation pressures which might be encountered. The rubber element in this type of blowout preventer is sufficiently flexible to permit the rotation or even the complete withdrawal of the drill pipe while maintaining a positive seal. Other types of blowout preventers are fitted with *rams*, which are made of steel faced with hard rubber. The closing elements in this type of blowout preventer are not flexible and therefore will close around only one geometrical configuration. If these blowout preventers are equipped to close around the drill pipe and it becomes necessary to close the blowout preventers while the drill pipe is out of the hole, then at least one joint of drill pipe must be placed in the hole before an effective seal can be maintained. The rams, or closing elements, of these blowout preventers can be changed to fit the size of drill pipe being used, or they can be fitted with *blank rams*, which give complete closure with the drill pipe removed from the hole. In many areas, at least two blowout preventers are used in series, and in this case several feet of space beneath the derrick floor are required. To provide ample work room, the derrick floor is elevated above the ground level by placing it on a substructure. The substructure is normally fabricated from structural steel, and the loads it must bear are greater than those borne by the derrick, since the substructure must support not only the derrick with its load, but other loads, such as the rotary table and draw works, as well.

The derrick and substructure must be placed on foundations that will safely support all the loads to which the derrick and substructure may be subjected. Concrete provides an excellent foundation, but it is expensive, and, further, in areas where foundations must be removed after the drilling has been completed, removal of concrete becomes a problem. In efforts to eliminate concrete foundations with their objectionable features, extensive use has been made of the bearing capacity of the soil to support the required loads. Footings for each derrick leg must, of course, be larger when concrete foundations are not used. Spreading the load of each derrick leg over a relatively large area can be accomplished by placing each corner of the substructure on a base which provides the proper bearing area. These bases may be made of structural steel members laid on a wooden timber matting. The steel bases may be welded and the timber mats bolted to reduce the set-up and tear-down time when the rig is moved. To determine the size of each mat required, it is necessary to know the load-bearing capacity of the soil and the maximum derrick loads anticipated.

Example: Calculate the size of the corner mats required for a sandy loam soil condition and a maximum derrick load of 500,000 pounds. Use a safety factor of 3.

$$\text{Load on each leg} = \frac{\text{derrick load}}{\text{number of supports}} \times \text{safety factor}$$

$$= \frac{500,000}{4} \times 3$$

$$= 375,000 \text{ lb.} \tag{11-5}$$

Load-bearing capacity of soil = 10,000 lb./ft.^{2}

$$\text{Mat size} = \frac{\text{load}}{\text{load-bearing capacity of soil}} = \frac{375,000}{10,000} = 37.5 \text{ ft.}^2$$

Miscellaneous Hoisting Equipment

In addition to the traveling block, other items of equipment are necessary in the proper handling of the drill string and other equipment which may be used in the hole. Four other important elements are hooks, links, swivel, and elevators. The *hook* provides a connection between the traveling block and the swivel. The *swivel* is a device that permits the drill pipe to rotate without causing rotation of the traveling block and lines. *Links* and *elevators* are used in removing or inserting drill pipe and casing in the hole. The conventional hook may be replaced by a *rotary connector*.

Weight Indicators

A *weight indicator* is a device for measuring the weight of equipment hanging from the traveling block. This device, as originally designed, was intended principally to show the load on the derrick at all times, thus preventing derrick overload. The instrument has gained wide use, for, in addition to showing visually, at all times, the load on the derrick structure, it helps to determine other important information, such as weight on the bit during drilling and friction loads when removing drill pipe from the hole. A very important use of weight indicators occurs when maximum pulls are being applied to recover stuck equipment during fishing operations. In fact, in any fishing operation to recover stuck drill pipe, casing, or tools, the weight indicator is indispensable.

Three basic types of weight indicators have been developed and used to some extent: (1) an instrument attached to the traveling

Fig. 11–10. Swivel.

block or hook which measures the load directly; (2) an instrument placed on the dead line which causes a bend in the line and, by measuring the tendency of the line to straighten when placed under a load, converts this measurement into an approximate load on the line itself; and (3) an instrument composed of two elements, a pressure transformer and an indicating gauge, connected by a hydraulic hose. The pressure transformer may be one of several different designs, and it may be attached to either the dead-line anchor or the dead line itself. In either case the indicating gauge is actuated by the forces on the pressure transformer. This latter type of instrument is generally accepted as the most reliable and rugged weight indicator available at

Fig. 11-11. Traveling block.

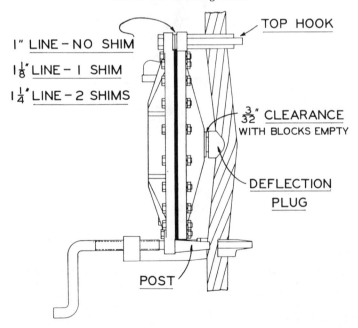

Fig. 11-12. Weight-indicator pressure transformer, located on dead line.

the present time, principally because of certain limitations in the other two types.

A weight indicator attached to the traveling block has the obvious disadvantage of not always being in full view of the driller, although it does measure the hook load directly. This disadvantage alone seriously limits the usefulness of the instrument. The second type of in-

Fig. 11-13. Pressure transformer for weight indicator attached to dead-line anchor.

strument discussed has two serious limitations: (1) since it must be attached to the dead line, it is in many cases too far away from the driller for accurate and easy reading, and (2) the dead line is often subjected to severe "whipping," which is detrimental to the continued accuracy of any instrument attached to it. Figure 11-12 shows a typical pressure transformer of the type attached to the dead line, with the indicating gauge placed in a position which is convenient for the driller to observe at all times. Figure 11-13 shows the type of pressure transformer which is attached to the dead-line anchor. The latter type of pressure transformer is normally preferred for continuous drilling operations.

A weight indicator indicating gauge is shown in Fig. 11–14. More than one dial may be read simultaneously on the indicating gauge.

FIG. 11–14. Weight, rotary-speed, and drill-string torque-indicating gauges.

Information which may be obtained includes (1) derrick load, (2) weight on bit, and (3) vernier readings for amplifying slight weight changes. The net weight on the bit is obtained by setting the zero mark of this scale opposite the pointer just before the bit reaches bottom. The bit-weight dial readings 'ncrease in a direction opposite to the hook-load scale. When the bit reaches bottom and the indicator pointer begins to reflect this reduced hook load, the weight on the bit can be read directly from the appropriate dial.

An important consideration in reading and interpreting weight-indicator readings is that the reading of the weight indicator is a direct function of the number of lines strung in the blocks. For a specific hook load, as the number of lines strung in the blocks is increased, the effective load on the pressure element of the weight indicator is decreased. Depending on the design of the pressure transformer, the size of the line may also be a factor in actual deflection of the weight-indicator pointer. Therefore, in order for the weight indicator to indicate the actual weight on the derrick, the indicating mechanism must be calibrated for the size of the line and the number of lines strung in the block. As these factors are subject to change from time to time, most weight indicators are furnished with several different dial faces, each calibrated for a specific line size and a specified number of lines strung in the blocks. Extreme caution should be exercised to insure that the proper dial face is used.

Buoyancy

Archimedes' Principle states that a body submerged either wholly or partially in a liquid is buoyed up by a force equal to the weight of the liquid displaced. This is an important principle in many phases of drilling operations, and for this reason the basic principle will be discussed briefly. A rectangular block of solid material immersed in a liquid will be used as an example. In Fig. 11–15, it is seen that the cube is immersed a distance, L_1, beneath the surface of the liquid. The horizontal forces on all sides of the block are balanced; however, the vertical forces are not balanced. The total force acting on the top of the block is equal to the weight of the liquid acting on the cube. If the weight per unit volume of the liquid is ρ_L, then the total weight of the liquid above the block is $\rho_L L_1 A$. Where A = cross-sectional area of cube. The force acting on the lower face of the cube will be $\rho_L L_2 A$. The net upward force is equal to

$$\text{Net force up} = \rho_L L_2 A - \rho_L L_1 A = \rho_L h A \qquad (11\text{--}6)$$

Referring again to Fig. 11–15, it is seen that the volume of the block is equal to hA; therefore, the buoyant force of the liquid is exactly equal to the volume of the liquid displaced.

The buoyant force of the drilling mud will reduce the actual weight of anything suspended in it. This may be a very important factor when heavy strings of casing are being run in a well. The magnitude of this reduction in actual weight can be examined by referring to

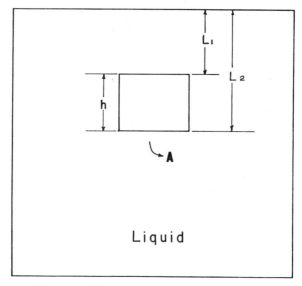

Fig. 11-15. Illustration of Archimedes' Principle.

Fig. 11-16, which shows a one-foot section of casing immersed in drilling mud. According to Archimedes' Principle the actual weight of the casing will be reduced by the weight of the liquid which the casing displaces. The weight of one foot of this casing is

W_{ca} = volume of one foot of the casing × density of casing steel

$$= (\pi r_1^2 1 - \pi r_2^2 1)\rho_s \qquad (11\text{-}7)$$

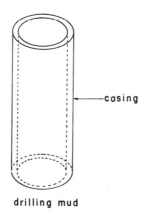

Fig. 11-16. Illustration of buoyancy.

where

W_{ca} = weight of casing in air, lb./ft.
r_1 = inside radius of casing, ft.
r_2 = outside radius of casing, ft.
ρ_s = density of casing steel, lb./ft.³

The weight of an equivalent volume of drilling mud is

$$W_{dm} = (\pi r_1^2 1 - \pi r_2^2 1)\rho_m \qquad (11\text{--}8)$$

where

W_{dm} = weight of drilling mud, lb./ft.
ρ_m = density of drilling mud, lb./ft.³

The effective weight of the casing when immersed in the drilling mud is

$$W_{cm} = W_{ca} - W_{dm} \qquad (11\text{--}9)$$

where

W_{cm} = effective weight of casing in drilling mud, lb./ft.

Substituting Equations (11–7) and (11–8) in Equation (11–9) yields

$$W_{cm} = (\pi r_1^2 1 - \pi r_2^2 1)\rho_s - (\pi r_1^2 1 - \pi r_2^2 1)\rho_m \qquad (11\text{--}10)$$

Multiplying and dividing the right-hand side of Equation (11–10) by W_{ca} yields

$$\begin{aligned} W_{cm} &= W_{ca} \frac{(\pi r_1^2 1 - \pi r_2^2 1)\rho_s - (\pi r_1^2 1 - \pi r_2^2 1)\rho_m}{(\pi r_1^2 1 - \pi r_2^2 1)\rho_s} \\ &= W_{ca} 1 - \frac{(\pi r_1^2 1 - \pi r_2^2 1)\rho_m}{(\pi r_1^2 1 - \pi r_2^2 1)\rho_s} \\ &= W_{ca}\left(1 - \frac{\rho_m}{\rho_s}\right) \end{aligned} \qquad (11\text{--}11)$$

As the density of steel is more or less uniform, Equation (11–11) can be further simplified:

$$\begin{aligned} W_{cm} &= W_{ca}\left(1 - \frac{7.48B}{489}\right) \\ &= W_{ca}(1 - 0.015B) \end{aligned} \qquad (11\text{--}12)$$

where

$7.48 = $ gal./ft.^{3}
$B = $ drilling mud weight, lb./gal.
$489 = $ average density of steel, lb./ft.^{3}

Equation (11–12) can be used to determine the effective weight of casing or drill pipe when suspended in drilling fluid.

Wire-Line Usage

The wire line used for drilling purposes is fabricated by assembling individual small-diameter wires into *strands*. Several strands are then wrapped around a *core* for the completed wire line. The *lay* of a wire

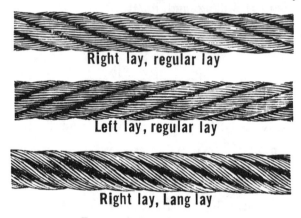

Fig. 11–17. Lay of wire lines.

line describes the direction in which the wires and strands are wrapped with respect to each other. Figure 11–17 shows the three standard construction arrangements. View A of Fig. 11–17 shows "right lay, regular lay" wire rope. *Right lay* mean that the strands are wrapped in a right-hand direction when looking away from one end of the line. *Regular lay* describes the individual wire arrangements of a single strand, and means that the wires in each strand are formed into that strand by turning in the reverse direction to that in which the strands were laid to form the line. In the case of right lay, regular lay, the individual wires of each strand are turned in a left-hand direction. View B of Fig. 11–17 shows "left lay, regular lay" wire line. The *left lay* designation means that the strands have been wound in a left-hand direction. *Regular lay* indicates that the wires forming each strand were wound in the opposite direction to the winding of the strands. Thus, in this instance, the term *regular lay* means that the wires were wound in a right-hand direction. The term *Lang lay* means that the strands and the wires in each strand are laid in the same direction. View C of Fig. 11–17 shows a "right lay, Lang lay" wire line. In Lang lay wire lines the strands and the individual wires in each strand are wound in the same direction.

The *core* of the drilling line, around which the strands are wrapped, can be either fiber or wire rope. The fiber used in fiber-core wire ropes manufactured under API specifications must be hard-twisted, best-quality Manila or sisal, or their equivalent. Jute cores cannot be used in API line. The fiber is thoroughly impregnated with a lubricating compound to increase the useful life of the rope. The principal advantage of the fiber-core wire line lies in its greater flexibility. It can be wound over smaller drums and sheaves without damage than a similar wire-rope-core line. The principal advantage of the wire-rope-core wire line is its greater strength. It is considerably more rigid than the fiber-core wire rope, and therefore its use is normally restricted to deep-drilling operations where equipment is large and small-diameter sheaves and drums are not used.

Wire rope is described by (1) lay of the strands, (2) lay of the individual wires forming the strand, (3) core material, (4) number of strands used in the line, and (5) number of individual wires used in each strand. Thus, a right-lay, regular-lay, 6×19 wire line with fiber core will adequately describe a wire line. The right-lay, regular lay terms describe the lay of both strands and individual wires in each strand; the number 6 designates the number of strands used in fabricating the line, the number 19 refers to the number of individual wires used to form a strand, and the core is shown to be fiber.

The individual wires used to form a strand may have different diameters in some wire-rope designs. In order to increase the strength and useful life of the line, some wire-rope designs use a smaller filler wire in those areas of a strand which would normally be void of wire. This filler wire, if used, will be designated in the wire-line description, usually by the term *filler*, or *seale*. Figure 11–18 shows several different wire-rope designs. From a study of this figure, which also gives a description of each of the different wire ropes, the meanings of the various terms in wire-rope description will become apparent.

The diameter of a wire rope is the diameter of the smallest circle which can be circumscribed about the line. The correct and incorrect methods of measuring wire-line diameter are shown in Fig. 11–19.

The drilling line used in rotary operations is always above ground; it is never lowered into the hole. Its principal function is to lower, raise, or otherwise manipulate the drill pipe, casing, or other equip-

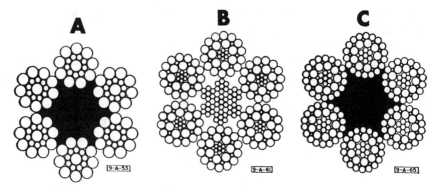

Fig. 11-18. Wire-rope designs. A. 6×19 seale with fiber core. B. 6×35 filler wire with independent wire-rope core. C. 6×43 filler wire with fiber core.

ment used in drilling and completing the well. A relatively short length of line is required, and the length does not depend on the depth of the well, but on the height of the derrick and the number of lines strung in the crown and traveling blocks. Although the length of drilling line is relatively short, the stresses imposed on it may be great especially when it is considered that the drilling line must pass around sheaves of relatively small diameter, thus placing severe bending as well as tension stresses in the line. Failure of the drilling line can cause expensive fishing jobs, side-tracking operations, or even abandonment of the hole; therefore, it is essential to maintain the line in satisfactory condition as long as it is being used as a drilling line. The drilling line cannot, however, be replaced until its condition is such that it is no longer considered safe, because it is an expensive maintenance item in over-all drilling-rig costs.

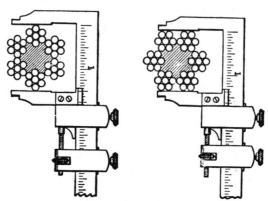

Fig. 11-19. Measuring wire-line diameter. *Left*, correct way to measure the diameter of wire rope. *Right*, incorrect way to measure the diameter of wire rope.

As the drilling line cannot be used to destruction, some method of determining the time for replacement is necessary. Of course, visual inspection of the line at all times is required in order to locate immediately any points of weakness. Probably the best method yet devised for evaluating the service life of drilling line is the *ton-mile usage* of the line. A *ton-mile* refers to the lifting of a weight of one ton (2,000 pounds) through a distance of one mile (5,280 feet). Although it is realized that the number of ton-miles of work a line has performed is not an absolute indication of its useful life, it is nevertheless a usable method for determining the approximate life of the line. Unusually severe stresses or conditions accelerating wear on the line will cause variations in the useful life of a drilling line. For this reason, continuous visual inspection is a necessity.

The principal work of the drilling line is confined to three distinct operations: (1) drilling, or hole-making, (2) round trips to replace worn bits, (3) running casing, and (4) miscellaneous operations such as coring and fishing.

In order to determine the number of ton-miles of work the line has performed in the course of normal drilling operations, it is necessary to evaluate the cycle of operations involved in drilling the length of the kelly. Beginning with the kelly down, the cycle of operations are as follows:

1. Pick up entire drill string the length of the kelly, plus a few additional feet.
2. Break out kelly and lower into rat hole the length of kelly. The weight involved will be the weight of kelly plus the weight of block, hook, elevators, etc.
3. Raise traveling block the length of one joint of drill pipe in order to pick up the joint of drill pipe. The weight in this case will be the weight of a joint of drill pipe plus the weight of the blocks.
4. Lower the new joint of drill pipe into the hole. This involves the weight of the entire drill string and blocks.
5. Raise the kelly out of the rat hole and attach to drill pipe. The weight lifted will be only that of the kelly and blocks.
6. Drill ahead the length of the kelly. The weight on the drilling line is equal to the weight of the entire drill string and blocks, minus the weight carried on the bit.
7. Pick up entire drill string the length of the kelly.
8. Ream ahead the length of the kelly. The weight on the drilling rig is equal to the weight of the entire drill string plus the weight of the blocks.

Analysis of the cycle of operations shows that for any one hole steps 1 and 6 will be equal to one round trip of the drill pipe, steps 7 and 8 will also be equal to one round trip, and steps 4 and 5 will be equal to one-half round trip. Steps 2 and 3 involve only small weights and are insignificant when compared to the other steps in the cycle, especially in deep wells. Therefore, the work performed by the drilling line in drilling the entire hole is actually equal to approximately two and one-half round trips to bottom. It should be pointed out that the work involved in steps 2 and 3 could be a substantial percentage of the total wire-line work when the wells being drilled are shallow. In order to take this work into consideration and at the same time retain a simple method of calculating wire-line work, the *API Recommended Practice 9B* recommends that the wire-line work done in drilling be assumed to be three times the work performed in a round trip at total depth, which, in effect, assumes that the wire-line work done in steps 2 and 3 is equal to one-half round trip with the drill string. Expressed in equation form, the total wire-line work done by drilling a well is approximately

$$T_D = 3T_{TD} \tag{11-13}$$

where

T_D = ton-miles drilling, for entire well
T_{TD} = round-trip ton-miles at total depth

The wire-line work done, by drilling, between round trips is found by the following equation:

$$T_d = 3(T_2 - T_1) \tag{11-14}$$

where

T_d = ton-miles drilling
T_1 = ton-miles for round trip at depth 1
T_2 = ton-miles for round trip at depth 2

A summation of the ton-miles drilling between round trips, as determined by Equation (11-14) will equal the ton-miles drilling for the entire well. This is evident from analysis of the following equation:

$$\begin{aligned}T_D &= 3(T_n - T_{n-1}) \\ &= 3T_1 + 3(T_2 - T_1) + 3(T_3 - T_2) + 3(T_4 - T_3) \\ &\quad + \cdots 3(T_{TD} - T_{TD-1}) \end{aligned} \tag{11-15}$$

where

T_1 = ton-miles during first round trip in the course of drilling the well
T_{TD} = ton-miles work done in round trip at total depth

In Equation (11–15), all of the terms will be canceled from the right-hand side of the equation except the term $3T_{TD}$.

In "hard rock country," it may be unnecessary to ream the hole after having drilled the length of the kelly. In this case, steps 7 and 8, in the cycle of operations previously shown, will be eliminated. Thus, the wire-line work equivalent to one complete round trip will be eliminated, and the wire line work done in drilling will be

$$T_D = 2T_{TD} \qquad (11\text{–}16)$$

and

$$T_d = 2(T_2 - T_1) \qquad (11\text{–}17)$$

A large proportion of the work done by the drilling line is in the round-trip operations. The cycle of operations performed during a round trip with the drill pipe is briefly described below:

1. The drill string is picked up the length of the kelly. This involves the weight of the entire drill string, including the kelly, and the traveling block assembly.
2. The kelly is removed and lowered into the rat hole. The weight involved in this operation is only the weight of the kelly and block assembly.
3. The drill string is raised the length of one stand, the stand is disconnected and stacked in the derrick. The entire weight of the drill string and block assembly is lifted by the drilling line.
4. The block assembly is lowered to the derrick floor, where it is attached to the drill string in preparation for another lift. The weight handled by the drilling line is the weight of the block assembly only.
5. Steps 3 and 4 are repeated until the entire drill string has been removed from the hole.

In returning the drill string to the hole, the previously outlined procedure is repeated in reverse. However, the weights handled and the distances traveled will be identical.

An analysis of steps 1 through 5 shows that the traveling block–elevator assembly moves a distance approximately equal to twice the length of a stand in removing the drill pipe from the hole, and also a distance of approximately twice the length of a stand in running the drill pipe back into the hole. Therefore, for the round trip, the wire-line work done in moving the block assembly will be equal to the weight of the block assembly multiplied by four times the depth of the hole, when the drill pipe is assumed to extend to the bottom of the hole. The ton-miles wire-line work done in raising and lowering the traveling block–elevator assembly in one complete round-trip operation is

$$T_b = \frac{4DM}{5{,}280 \text{ ft./miles} \times 2{,}000 \text{ lb./ton}} \qquad (11\text{–}18)$$

where

T_b = ton-miles work in moving block assembly
D = depth of hole, ft.
M = total weight of traveling block–elevator assembly, lb.

In removing the drill pipe from the hole, two assumptions are made in order to develop a simple formula for computing the wire-line work. These assumptions are:
1. The kelly is assumed to have weight equivalent to one stand of drill pipe.
2. The drill pipe is assumed to extend to the bottom of the hole. A correction factor is then included which takes into account the additional weight of the drill collars.

The wire-line work done in removing the drill string from the hole is equal to the sum of the work required to remove each stand of drill pipe from the hole. Mathematically this can be expressed as follows:

$$T_t = \frac{L_s n W_s + L_s(n-1)W_s + L_s(n-2)W_s + \cdots L_s[n-(n-1)]W_s}{5{,}280 \times 2{,}000} \qquad (11\text{–}19)$$

where

T_t = ton-miles work in removing drill pipe from hole
L_s = length of one stand of drill pipe, ft.
n = total number of stands of drill pipe in hole.
W_s = weight of one stand of drill pipe, lb.

Equation (11–19) can also be expressed as

$$T_t = \frac{L_s \times n \times W_{avg}}{5{,}280 \times 2{,}000} \qquad (11\text{–}20)$$

where

W_{avg} = average weight lifted, lb.

However, $L \times n = D$, and a critical examination of Equation (11–19) shows that the average weight lifted is equal to one-half the sum of the weight of the initial load plus the weight of the final load. The weight of the initial load is the effective weight of the entire drill stem, corrected for bouyancy; and the weight of the final load is the effec-

HOISTING OPERATIONS

tive weight of one stand of drill pipe, also corrected for bouyancy. Therefore the average weight lifted is

$$W_{avg} = 1/2(W_s + nW_s) \tag{11-21}$$

For convenience, drill pipe lengths are usually expressed in pounds per foot of length. Using the drill pipe weight on a pounds per foot basis, Equation 11-21 will now be

$$W_{avg} = 1/2(L_sW_m + L_snW_m)$$

where

W_m = effective weight per ft. of drill pipe, corrected for buoyancy of drilling mud, lb.

Combining Equations (11-20) and (11-21) yields

$$T_t = 1/2(L_sW_m + L_snW_m)D \tag{11-22}$$

Assuming that the friction losses going in the hole are equal to those coming out of the hole, the wire-line work performed in going back into the hole can also be found by Equation (11-22). The total wire-line work performed in manipulating the drill pipe in one round-trip operation will then be equal to Equation (11-22) multiplied by two, or

$$T_p = \frac{2 \times 1/2(L_sW_m + L_snW_m)D}{5{,}280 \times 2{,}000}$$

$$= \frac{(L_sW_m + L_snW_m)D}{5{,}280 \times 2{,}000}$$

$$= \frac{DW_m(L_s + D)}{5{,}280 \times 2{,}000} \tag{11-23}$$

where

T_p = ton-miles of wire-line work done in handling drill pipe during one complete round trip.

As has been said, for convenience the drill pipe is assumed to extend to the bottom of the hole. As the unit weight of the drill collars and bit are greater than the unit weight of the drill pipe, the additional wire-line work done in removing and lowering this extra weight is equal to the additional weight multiplied by the distance moved. The total additional wire-line work done in a round-trip operation is

$$T_e = \frac{2 \times C \times D}{5{,}280 \times 2{,}000} \tag{11-24}$$

where

T_e = ton-miles work done in moving excess weight of drill collar–bit assembly
C = excess weight of drill collar–bit assembly over drill pipe, lb.
D = depth of hole, ft.

The total wire-line work done in making a round trip is equal to the algebraic sum of the work done in moving (1) traveling block assembly, (2) drill pipe, and (3) extra weight of drill collar assembly. In many cases a simple equation combining the above operations can be developed. If the cycle of operations is such that Equations (11–18), 11–23, and 11–24 properly define the wire-line work for the over-all round-trip operation, then these equations can be combined as follows:

$$T_r = T_b + T_p + T_e$$

$$= \frac{4DM}{5{,}280 \times 2{,}000} + \frac{DW_m(L_s + D)}{5{,}280 \times 2{,}000} + \frac{2CD}{5{,}280 \times 2{,}000}$$

$$= \frac{4DM + DW_m(L_s + D) + 2CD}{5{,}280 \times 2{,}000}$$

$$= \frac{DW_m(L_s + D)}{10{,}560{,}000} + \frac{4D(M + 0.5C)}{10{,}560{,}000}$$

$$= \frac{D(L_s + D)W_m}{10{,}560{,}000} + \frac{D(M + 0.5C)}{2{,}640{,}000} \qquad (11\text{–}25)$$

A slightly different approach can be used to calculate wire-line work, which, although somewhat simpler in the method of solution, results in an equation identical to Equation (11–25). In Fig. 11–20, it can be seen that, in order to remove the drill pipe from the hole, the center of gravity of the drill pipe must be moved a distance, h, which is equal to one-half the depth of the well (assuming that the drill pipe extends to the bottom of the hole), plus one-half the length of one stand of drill pipe. The ton-miles of wire-line work required to remove the drill pipe are

$$T_p = \frac{1/2D(W_m \times D) + 1/2L_s(W_m \times D)}{5{,}280 \times 2{,}000}$$

$$= \frac{1/2W_m(L_s + D)D}{5{,}280 \times 2{,}000} \qquad (11\text{–}26)$$

The center of gravity of the additional weight of the drill collar assembly must be raised a distance, d, equal to the depth of the well

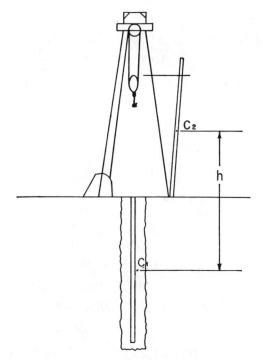

Fig. 11-20. Calculating wire-line work by determining the change in center of gravity of the pipe.

minus one-half the length of the drill collar assembly, plus one-half the length of a stand of drill pipe. Expressed as an equation this is

$$T_e = \frac{(D - 1/2a)(C) + 1/2L_s C}{5{,}280 \times 2{,}000} = \frac{C(D - 1/2a + 1/2L_s)}{5{,}280 \times 2{,}000} \quad (11\text{-}27)$$

where

a = length of drill collar assembly, ft.

In bringing each stand of pipe out of the hole, the traveling block assembly moves a distance equal to twice the length of a stand of drill pipe. Therefore, in the process of removing the entire drill string from the hole, the traveling block assembly will move a distance equal to twice the depth of the hole, or

$$T_b = 2D \times M \quad (11\text{-}28)$$

The total wire-line work done in a round-trip operation is equal to twice the work done in removing the entire drilling string from the hole

$$T_r = 2(T_p + T_e + T_b)$$
$$= 2\left[\frac{1/2 W_m(L_s + D)D + C(D - 1/2a + 1/2L_s) + 2DM}{5{,}280 \times 2{,}000}\right] \quad (11\text{--}29)$$

If the length of the drill collar assembly, a, and the length of a stand of drill pipe, L_s, are approximately equal, or if the algebraic sum of the two quantities is small when compared to the total depth of the hole, D, then Equation (11–27) becomes

$$T_e = \frac{CD}{5{,}280 \times 2{,}000}$$

and Equation 11–29 becomes

$$T_r = 2\,\frac{1/2 W_m(L_s + D)D + CD + 2DM}{5{,}280 \times 2{,}000}$$
$$= \frac{W_m(L_s + D)D + 4D(M + 1/2C)}{5{,}280 \times 2{,}000}$$
$$= \frac{D(L_s + D)W_m}{10{,}560{,}000} + \frac{D(M + 1/2C)}{2{,}640{,}000} \quad (11\text{--}30)$$

Equations (11–30) and (11–25) are identical. However, Equation (11–30) involves an assumption which is probably never exactly correct, and for this reason Equation (11–29) is normally preferred.

Running Casing

The cycle of operations involved in running casing is identical to the operations performed in running the drill string back into the hole. Therefore, the ton-miles of work done in running casing is equal to one-half the wire-line work done during a round-trip operation. As there is no excess weight on account of the drill collars, Equation (11–30) can be modified as follows

$$T_s = \tfrac{1}{2}\,\frac{W_{cm}(L_{sc} + D)D + 4DM}{5\,280 \times 2.000} \quad (11\text{--}31)$$

where

T_s = ton-miles wire-line work done in setting casing
W_{cm} = effective weight per foot of casing in mud, lb.
L_{cs} = length of joint of casing, ft.

Equation (11–31) is correct when only one weight of casing is run. When combination strings of casing are used—i.e., casing with the same OD, but with varying unit weights—this equation becomes somewhat tedious to handle and the meaning of the individual terms becomes less clear. However, when it is analyzed from the standpoint of the distance the center of gravity of the casing must be moved, the solution of the problem becomes relatively simple, because the center of gravity of each section of the same casing weight is easily determined. When three different weights of casing are run in a well, the total wireline work done in running the casing is the sum of the work required to run each segment:

$$T_s = T_A + T_B + T_C \qquad (11\text{-}32)$$

where

T_A = ton-miles wire-line work done in setting section A
T_B = ton-miles wire-line work done in setting section B
T_C = ton-miles wire-line work done in setting section C

The wire-line work can be subdivided into the work done in moving the center of gravity of the various sections of casing plus the work done by the traveling block assembly. During the casing setting operation, the traveling block assembly will move a distance equal to approximately twice the total depth of the well. Then the total wire line work will be

$$T_s = (1/2L_A + 1/2L_{cs})W_{cm}L_A + (L_A + 1/2L_B + 1/2L_{cs})W_{cm}L_B$$
$$+ (L_A + L_B + 1/2L_C + 1/2L_{cs})W_{cm}L_B + 2DM \qquad (11\text{-}33)$$

where

L_A = length of section A, ft.
L_B = length of section B, ft.
L_C = length of section C, ft.

Coring Operations

The wire-line work performed in coring operations is practically identical to the wire-line work done in drilling; and as in the case of drilling, coring wire-line work can very conveniently be expressed in terms of round-trip wire-line work. A typical cycle of coring operations is as follows:
1. With the core barrel on bottom, core ahead the length of the kelly;
2. pick up length of kelly;
3. lower kelly in rat hole;
4. pick up joint of drill pipe and add to drill string;
5. lower this joint into the hole;

6. pick up kelly;
7. repeat steps 1 through 6 until the desired length of core has been obtained, or until the total length of the core barrel has been drilled.

Analysis of the cycle of coring operation shows that steps 1 and 2 are equivalent to one round trip, step 5 is equivalent to one-half round trip, and steps 3, 4, and 6 are approximately equal to one-half round trip. Therefore, the total wire-line work would be

$$T_c = 2T_r \qquad (11\text{-}34)$$

Since few if any wells are cored from surface to total depth, the work done in coring a particular interval may be determined by

$$T_c = 2(T_{r1} - T_{r2}) \qquad (11\text{-}35)$$

where

T_{r1} = ton-miles wire-line work for one round trip at depth 1, where coring was started

T_{r2} = ton-miles wire-line work for one round trip at depth 2, where coring ended.

Evaluation of Wire-Line Service

Determining the useful life of a wire line may be very difficult because no way has yet been devised to determine the actual magnitude of the stresses and strains imposed on the line. As has been said, up to the present time the best method developed of evaluating wire-line usage is ton-miles of service. The sum of the ton-miles of work performed by drilling, coring, round trips, and running casing yields the total wire-line work. The principal objective in keeping a record of the total ton-miles of wire-line work is to permit maximum utilization of the wire line. Certain parts of the line are subjected to considerably more wear than others. Probably the point of most severe wear on a drilling line is the point where the dead line passes over the first crown block sheave. This sheave does not turn, and therefore all the stresses applied to the line during all of the operations are acting on this one point. In order to increase the effective life of the wire line and reduce over-all wire-line costs, drilling lines are usually purchased several hundred feet longer than the actual length required. This excess length is then reeled on the drum a few feet at a time, so that the points of maximum wear will vary. The optimum time to change the point of severest wear is difficult, if not impossible, to determine. The API has developed several methods for determining when the line

HOISTING OPERATIONS

should be moved, and these techniques are fully outlined in *API Recommended Practice 9B*.

CALCULATING LENGTH OF LINE ON A DRUM

Quite often it may be necessary to calculate the line-carrying capacity of a drum, or the capacity of a specific layer of line around the drum. The capacity of the first layer around the drum will be equal to the length of line used in one wrap around the drum, multiplied by the number of times the line can be wrapped around the drum. The actual diameter around which the first layer is wrapped is equal to the diameter of the drum plus the diameter of the line. The capacity, in feet, of the first layer on the drum will be

$$L_1 = \frac{\pi}{12}(D + d) \times \frac{W}{d} \qquad (11\text{-}36)$$

where

L_1 = wire line capacity of first layer, ft.
D = drum diameter, in.
d = line diameter, in.
W = drum width, in.

From Equation (11-36), the term $\pi(D+d)$ is the actual circumference around which the first layer is wrapped, and the term W/d is the number of wraps which can be placed on one layer of the drum. The capacity of the second layer is

$$L_2 = \frac{\pi}{12}(D + 3d) \times \frac{W}{d} \qquad (11\text{-}37)$$

Likewise, the capacity of the third layer is

$$L_3 = \frac{\pi}{12}(D + 5d) \times \frac{W}{d} \qquad (11\text{-}38)$$

The capacity of the nth layer

$$L_n = \frac{\pi}{12}[D + (2n - 1)d] \times \frac{W}{d} \qquad (11\text{-}39)$$

where

n = total number of layers on the drum.

The total capacity of the drum is equal to the sum of the capacities of each layer:

$$L_T = L_1 + L_2 + L_3 + \cdots L_n \qquad (11\text{--}40)$$

Combining Equations (11–36) through (11–40) results in

$$L_T = \frac{\pi}{12}(D+d)\frac{W}{d} + \frac{\pi}{12}(D+3d)\frac{W}{d} + \frac{\pi}{12}(D+5d)\frac{W}{d}$$

$$+ \cdots \frac{\pi}{d}\left(D + (2n-1)d\frac{W}{d}\right)$$

$$= \frac{\pi}{12}\frac{W}{d}[D + d + D + 3d + D + 5d + \cdots D + (2n-1)d]$$

$$= \frac{\pi}{12}\frac{W}{d}[nD + (\Sigma n's)(d)]$$

However, $\sum n's = n^2$

$$L_T = \frac{\pi}{12}\frac{W}{d}[nD + n^2 d] \qquad (11\text{--}41)$$

also

$$n = \frac{h}{d} \qquad (11\text{--}42)$$

where

$h =$ flange depth, in.

Combining Equations (11–41) and (11–42) yields

$$L_T = \frac{\pi}{12}\frac{W}{d}\left[\frac{h}{d}D + \frac{h^2}{d}\right]$$

$$= \frac{\pi}{12}\frac{Wh}{d^2}(D+h) \qquad (11\text{--}43)$$

Draw Works

The draw works has often been referred to as the power-control center of the drilling rig, because on it are located most of the controls required to run the rig. The principal parts of the draw works are (1) hoisting drum, (2) catheads, (3) brakes, and (4) clutches. The hoisting drum is probably the most important single item of equipment on the draw works because it is through this drum that power is transmitted to remove the drill string, and on this drum the drilling line is wound and unwound as equipment is manipulated in the hole.

Power for the drum comes from the power plant through suitable transmission devices, either mechanical, hydraulic, or electric. These transmission devices have been discussed at length in Chapter 9. The proper design of a hoisting drum requires careful balancing of several objectives, some of which are in conflict with each other. From the standpoint of supplying the motive power for hoisting, the ideal drum would have a diameter as small as possible and a width as great as possible. This is true because the input power required to rotate the drum will be a function not only of the line pull but also of the diameter of the drum. This becomes more obvious if specific cases are illustrated. In Fig. 11–21, the input power required to overcome the mo-

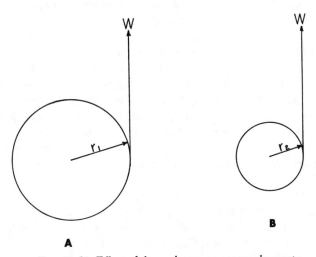

FIG. 11-21. Effect of drum size on power requirements.

ment of inertia of the drum in view A will be r_1W, while the input power required to overcome the moment of inertia of the drum in view B is r_2W. Thus it is seen that for a given line pull, W, the input power required to cause rotation of the drum against the line pull will be directly proportional to the radius of the drum. As the drum radius increases, the power requirements will increase. And for a given drum radius it is desirable to have the largest drum length possible in order to limit the number of wraps on the drum to the minimum, for as additional wraps are placed on the drum, the effective diameter of the drum is decreased.

Solely from the standpoint of the drilling line, the drum should be as large in diameter as possible and also as wide as possible. The bending stresses in the drilling line are decreased as the diameter of the drum around which they are wrapped is increased. Also line crossover reduces the effective life of the drilling line.

There are maximum safe speeds at which the line can be wrapped around a drum, which vary directly with the drum diameter, large drum diameters permitting greater line speeds.

Another major factor to be considered in drum design is the over-all size of the draw works. Strictly from a size standpoint, the drum should be as small as practical. In the consideration of economy, the drum should be large enough to permit fast line speeds, yet small enough to present no insurmountable moving problems.

In the final analysis, in drum design all of the aforementioned factors must be considered, with the end result a compromise compatible with all the various considerations.

Catheads are small rotating drums located on both sides of the draw works. They are used as a power source for many routine operations around the rig, such as screwing and unscrewing ("making up" and "breaking out") drill pipe and casing, and pulling single joints of drill pipe and casing from the pipe rack up to the rig floor. Most of the earlier "catheading" was done manually; and was accomplished by wrapping a Manila line several turns around the cathead and applying the necessary power by the friction hold of the Manila rope on the rotating cathead. This operation is extremely hazardous, because once a rope becomes fouled on the cathead, with a resulting loss of the slipping action of the line, the rotating cathead then immediately becomes a powerful force, rapidly winding the rope on the drum. Special catheads have been developed which eliminate much of the manual catheading originally done. Special break-out and make-up catheads now automatically perform the make-up and break-out operations without requiring the services of a roughneck on the cathead drum. The location of the catheads with respect to the other parts of the draw works is shown in Fig. 11-22.

The brakes are important units of the draw-works assembly, as they are called upon to stop the movement of large weights being lowered into the hole. When a round trip is being made, the brakes are in almost continuous use. Therefore, the brake must have a relatively long life, and be so designed that the heat generated by braking can be dissipated rapidly. The principal brake on a hoisting drum is usually a friction-type mechanical brake. A diagram of this brake is shown in Fig. 11-23. It is essentially comprised of a flexible band (made of asbestos or other heat- and wear-resistant material) wrapped around a drum. One end of this flexible band is permanently anchored, while the other end is movable and attached to a lever by

FIG. 11-22. Drilling-rig draw works.

means of which the band can be either loosened or tightened against the rotating drum. The brake lever is designed to provide a mechanical advantage in its action. To minimize the heat generated by continual use of the friction brake as equipment is being lowered into the hole, auxiliary brakes have been developed which perform a large part of the "slowing" action required to prevent excessive lowering speeds.

Two basically different types of auxiliary brakes have been developed, a hydraulic brake and an electro-magnetic brake. The hydraulic brake utilizes fluid friction to absorb some of the work done in

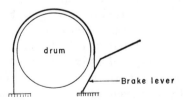

FIG. 11-23. Principle of the mechanical brake.

lowering equipment into the hole. The hydraulic brake is designed to impel fluid in a direction opposite to the rotation of the drum, thus tending to slow the drum rotation. The electro-magnetic brake is essentially two opposed magnetic fields, the magnitude of which are de-

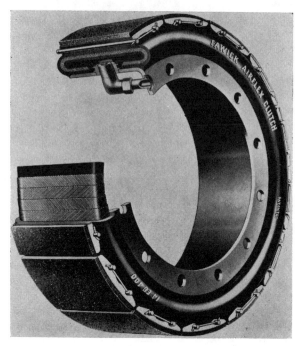

FIG. 11–24. Air clutch.

termined by the speed of rotation of the drum and the amount of external current supplied. In order for either the hydraulic or electro-magnetic brake to become effective, some movement of the drum is required. Therefore, neither of these types of brakes will bring the hoisting drum to a dead stop. It remains for the friction brake to effect complete stoppage. Some type of auxiliary brake is standard equipment on most medium- and deep-drilling rigs.

The draw-works clutch is used to engage mechanically the hoisting drum with the transmitted power. One of the earliest type of clutches was a mechanical or jaw clutch, so called because of the interlocking nature of the clutch assembly. This type of clutch was satisfactory for a steam power plant, because the steam power could be applied gradually. However, with the advent of internal combustion engine power plants, the jaw clutch was not satisfactory, since it caused shock loads

Fig. 11-25. Offshore rig in operation.

to be applied to the system. Pneumatic clutches were designed to overcome the disadvantages of the old mechanical clutch. The pneumatic or air-operated clutch is actually a friction clutch. It consists basically of an expanding element on the inside of which is attached a number of friction *shoes* which cause the clutch to become engaged as the pneumatic element is expanded. Figure 11-24 shows the operating principle of the air clutch.

CHAPTER 12

Drilling-Fluid Circulation

THE CIRCULATING SYSTEM

In usual rotary drilling operations, more horsepower hours of work are utilized in circulating drilling mud than in any other single operation. The pumps, flow lines, drill pipe, nozzles, and space areas through which mud flows deserve most careful engineering consideration. Although a vast amount of design work and experimental investigation have been performed on the various pieces of equipment in the circulating system, new ideas and concepts in drilling periodically require new and thorough engineering analyses.

In normal mud circulation, the mud is pumped down through the drill pipe, discharged through the bit, and returns to the surface in the annular space outside the drill pipe and inside the drilled hole and casing placed in the well. In reverse circulation, which can be used only where there is no tendency to lose mud into exposed rock formations, the drilling mud is pumped down the annulus and returns to the surface through the inside of the drill pipe. The various pieces of equipment through which the mud passes are described in the following paragraphs in the order of normal circulation.

Mud pits and tanks are required for holding an excess volume of mud at the surface. The volume so held in the circulating system is usually between 300 and 700 barrels. Mud pits scraped out in the earth with a bulldozer are usually two in number, each five to seven feet wide, about forty feet long, and three or four feet in depth. The long flow pattern provided by this shape permits maximum settling of sand and maximum release of entrained gas bubbles in the mud. Steel mud pits often have about the same dimensions, since such pits may be readily transported on the highways by trucks. Three such steel pits are often provided, and they are coupled together with one or more twelve- to eighteen-inch-diameter conduits. Four-inch-diameter steel pipe is often placed around the rims of such steel mud pits and furnished with connections where mud guns (jets) may be mounted for jetting down onto the surface of the mud. Jets are usually operated from an extra or auxiliary mud-mixing pump similar to but smaller than the circulating-mud pump. Steel mud pits and tanks have the advantage over pits dug in the earth that there is less chance of the mud's becoming contaminated with blowing sand or other materials. The steel pits and tanks also rest above ground, so that they may furnish some amount of pressure into the mud-pump suction. The drill cuttings and cavings and any mud discarded from the circulating system are generally run or jetted into the reserve pit. This is an area, adjacent to the mud pits, about one hundred feet square, surrounded by earthen banks about three feet high. Mud from the reserve pit has sometimes been reused in cases where mud has been lost into formations exposed in the well bore.

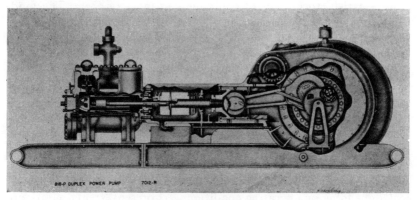

FIG. 12–1. Duplex power mud pump.

Mud pumps. Two types of mud pumps are in common use. These are the direct-connected, two-cylinder (duplex), double-acting steam pump and the two-cylinder, double-acting power pump in which the pistons are driven from a crankshaft. Both of these are piston-type pumps and they pump mud on both sides of the piston. On one side of the piston, the capacity per stroke is determined by the diameter of the cylinder and the length of the stroke, while on the other side, the volume of the piston rod passing in and out through the stuffing box must be subtracted to obtain the capacity per stroke. There is an inlet and an outlet valve on both ends of each cylinder.

The cylinders in mud pumps are made with replaceable liners, for mud contains abrasive sand and the cylindrical surfaces are sooner or later scored to such an extent that they must be replaced. The liners are simply smooth, hollow cylinders which fit inside of the mud pump, but the inside surfaces are ground smooth and highly polished and made of especially hardened steel by the several manufacturers. The sealing elements on the pistons usually consist of hard rubber cups which are pressed out against the inside of the liners by the pressure in the fluid which is being pumped. Since the pistons are double acting, each piston must have one set of cups facing one direction and another set facing the opposite direction. The rubber cups are also replaceable. The valves are actuated by the presssure in the fluid being pumped. These also are replaceable and usually consist of steel valves which seal against hard rubber seats or an equivalent arrangement. No other mechanical arrangement has been developed to handle the high pressures, large volumes, and abrasive fluid conditions encountered in circulating drilling fluids.

The direct-acting steam pumps are fundamentally and hydraulically a satisfactory piece of equipment in most respects. The mud-fluid discharge from these pumps is quite regular, with no large volume or pressure surges. With the exception of the valve areas, these pumps consist essentially of cylinders, pistons, and rods. As such, they are relatively light in weight for the work performed. However, these pumps operate slowly, at speeds of about fifty strokes (cycles) per minute. Steam pressure is seldom available in excess of 350 psi. If internal friction within the pump, both fluid and mechanical, were neglected, a pump equipped with seven-inch-diameter mud cylinders and fourteen-inch-diameter steam cylinders could deliver mud at 4×350 psi, or 1,400 psi, if it were furnished steam at 350 psi. The available delivery pressure might be from about 60 to 80 per cent of the theoretical, depending on operating conditions. At present, steam pumps are being made with steam cylinders of diameters greater than two times the diameter of the mud cylinders, such as an $18 \times 8 \times 20$-inch pump, sometimes designated as $18 \times 8 \times 20$-inch, which has an eighteen-inch-diameter steam cylinder, and eight-inch-diameter mud cylinder, and a twenty-inch stroke.

Power pumps, which are those in which the pistons are driven from a crankshaft, may be made with any number of cylinders, and they may be made either single or double acting. Single-acting pumps are those which pump mud on only one side of the piston, while those that are double acting pump mud on both sides of each piston. The only power pump in common use for drilling-mud circulation is the two-cylinder, double-acting pump. The large volumes and higher pressures required for faster drilling rates, particularly with jet-type bits, have imposed severe operating conditions on these pumps. The crankshaft of one of these pumps is usually gear driven with a speed reduction of the order of five to one from a drive shaft mounted on the top of the pump. The drive shaft is commonly driven by a multiple V-belt pulley from an internal combustion engine, an electric, or a steam drive. For use with electric rigs, electric motors are being mounted directly on the drive shaft of the pump.

Power pumps are far from being a perfect hydraulic machine. The momentum of the rotating parts of the pump tends to impart harmonic-type motion to the pistons, which means that the piston velocity is high near the center of the stroke and approaches zero at either end of the stroke. Four discharge strokes are superimposed on each other during each revolution. Although the discharge is continuous, it is not smooth. Only specially dampened pressure gauges can be used to measure the pressure in the mud as it is pumped into the top of the drill string. These inherent disadvantages have been overcome by developing extremely rugged machines. Gears, crankshaft bearings, and frames have been increased in size to the point where the mud pump is commonly the heaviest single piece of equipment that must be moved from one location to another. Rotary hoists, or draw works, are heavier than power pumps when assembled, but the larger hoists can be moved in sections while the power pumps cannot. The fluid-discharge pressure surges from the pump are transmitted to the drill string.

The mud suction line between the mud pit or mud tank should be at least one and one-half times the diameter of the mud-pump liners in order that the cylinders of the mud pump may fill completely during each pumping stroke. The use of centrifugal pumps for maintaining a positive pressure on the intake side of power pumps was introduced in the early 1950's. This has permitted mud pumps to be run at speeds about 50 per cent greater than the customary speed of 40 or 50 rpm.

The discharge line from the mud pumps is customarily four- or six-inch double extra-heavy pipe. It is customary to use a surge chamber which will retain a small volume of compressed air in its top near the pump discharge. The compressed air serves as a cushion for partially smoothing the discharge pressure surges from the pump. The effective size of the air cushion will be inversely proportional to the ratio of the pressures, according to Boyle's Law. Thus, a surge chamber with a volume of 4 cubic feet would offer, at a pump discharge pressure of 1,000 psi gauge, an effective cushion volume of only 15/1,015 (4), or about 0.05 cubic feet. The discharge line from the mud pump should always contain a pressure relief valve, and this should be equipped and guarded so that no injuries to personnel will occur when the valve pops open. Equipment should be chosen and installed for its operating ability.

The rotary hose conducts the mud from the upper end of the standpipe, the vertical steel pipe which extends about halfway up the derrick, to the swivel which supports the top of the drill string. Rotary hoses are customarily three inches or more in internal diameter in order that no appreciable fluid-flow pressure losses will occur in the hose. The flexible feature permits raising and lowering the drill string during drilling operations while mud is being pumped down through the drill string.

The swivel performs two functions. It permits rotation of the drill string and conducts mud into it. The swivel is supported by the traveling block in the derrick, and in turn the swivel supports the kelly stem, which is rotated by the rotary table. Within the swivel, two primary elements are involved. These are a bearing, usually a tapered roller bearing, which will support the rotating load, and a rotating pressure seal. About three-inch-diameter fluid passage is provided, and the member which provides for gently reversing the direction of mud flow from the suspended rotary hose is known as the gooseneck. Modern kellys are also usually provided with a bore of about three inches in order to conduct the required fluid flow without appreciable pressure losses.

The internal diameters of the drill pipe, tool joints, and drill collars, and also the total cross-sectional areas of the discharge jets in the bit, determine the major portion of the fluid-flow pressure losses, which in turn determine the pressure which must be supplied by the mud pump. This is largely as it should be—in the sense that fluid-flow pressure losses which increase the total mud pressure applied against rock formations exposed in the bore hole should be kept to a minimum. Obviously this applies, and can only apply, in the case of normal circulation where the drilling fluid is pumped down through the drill string and the flow losses in the annulus only are imposed against the exposed rocks. In the many areas where formations may break down and mud losses may occur, there is a definite obligation to maintain low fluid-circulating pressure losses in the annulus outside of the drill pipe. This is accomplished in two ways: (1) by maintaining a larger cross-sectional flow area in the annulus outside of the drill pipe, as compared to the flow area inside the drill pipe; and (2) by maintaining a drilling fluid of low effective viscosity and gel strengths. Avoiding the rapid lowering of drill pipe and rapid starting of circulation are also factors, to be discussed in Chapter 18. Torque strength requirements tend to provide a drill pipe of adequate, perhaps optimum, internal diameter. The internal diameters of the pipe couplings (tool joints) as well as the internal diameter of the drill collars must also be large enough to prevent excessive pressure losses during mud circulation.

The flow of drilling fluid returns to the surface in the annular space outside of the drill pipe. The greater diameter of the annular space in the lower section of the hole is generally determined by the drill bit size, and in the upper section by the diameter of the smallest size of casing which has been set and cemented in the well. The upward flow of mud passes through the blow-out preventer and then into the nearly horizontal mud-flow line under the derrick floor, which leads the mud back to the mud pits.

A shale shaker, which customarily sits on top of the first mud pit, receives the mud returning from the flow line. The shale shaker is essentially a screen that is used to separate drill cuttings and cavings from the mud. Two types are in common use. In one type the screen is in the form of a tapered cylinder which is rotated by the flow of drilling fluid. The other is a rubber-mounted sloping flat screen which is vibrated by the rapid rotation of an eccentric mass driven either by electric or hydraulic motor. In both, the drilling fluid falls by gravity through the screen and the drill cuttings and any cavings present pass over the end of the screen. The screen openings are generally rectangular.

The shale shaker is the favored position for adding water to the mud, for at that position the mud returning from the well is more fluid as compared to the slowly moving mud in the pits. A small stream of water distributed through a perforated pipe is often used to wash the mud materials off the drill cuttings before they pass over the end of the shale shaker.

Other pieces of equipment which are sometimes used include vacuum chambers, which remove gas from the mud, and centrifuges. Centrifuges are used to recover high specific gravity mud weighting material for maintaining high density muds; they are also used for rejecting sand and similar material from the mud, for the prupose of maintaining minimum density drilling muds.

Drilling-Fluid Circulation Requirements

The rate of drilling-fluid circulation is determined by the necessary upward flow velocity required for removing cavings and drill cuttings from the hole and by the jetting requirements of the bit. The inherent advantage of the rotary system of drilling is that a fluid is circulated for the purpose of removing the drill cuttings and maintaining a hole in such condition that the drill string can be withdrawn readily and returned to the bottom whenever necessary. Normally, the fluid is circulated down through the hollow drill string and returns to the surface through the annulus outside the drill pipe. Upward mud-flow velocities of 100 to 200 feet per minute have been found to be satisfactory under ordinary operating conditions. Rates of drilling-mud circulation which are in excess of those required for maintaining the hole are used to obtain faster drilling penetration rates. The additional costs in equipment and horsepower which are required for the additional mud circulation rates must be justified in terms of faster drilling rates. Information on the relations between drilling rates and the hydraulic horsepower furnished to the bit is contained in Chapter 13.

Flow Pressure Losses—Bernoulli's Theorem

When the principle of the conservation of energy is applied to the flow of fluids, the resulting equations are known as Bernoulli's Theorem. Where no changes in chemical energy are involved, such as would be caused in a combustion process, for example, it is customary to consider the changes in the potential energy, the kinetic energy, and the pressure volume energy of the fluid. Where two points, point 1 and point 2, in a flow system, are considered, the following equation may be written:

$$h_1 + \frac{U_1^2}{2g} + \frac{p_1}{\rho_1} - F + W = h_2 + \frac{U_2^2}{2g} + \frac{p_2}{\rho_2} \qquad (12\text{-}1)$$

where

h = height above a chosen reference elevation, ft.
U = flow velocity, ft./sec.
p = pressure of the fluid, lb./ft.^{2}
ρ = density of the fluid, lb./ft.^{3}
g = acceleration of gravity, 32.2 ft./sec.^{2}
F = sum of flowing pressure losses
W = sum of mechanical energy added

In the above equation, all of the terms are expressed in feet, or in foot-pounds of energy per pound of fluid. The theorem applies readily to systems in which a homogeneous fluid flows under such conditions that there is continuity of the fluid throughout the system.

In the case of circulating drilling mud, as in drilling a well, where the mud is pumped out of the mud pit and circulated back into the pit, the elevation terms h_1 and h_2 cancel each other. The velocity terms are readily calculated to be negligible. The pressure terms p_1 and p_2 cancel each other since both are atmospheric pressure and there are no changes in density. The equation then reduces to $W = F$. The term W represents the work, or hydraulic horsepower, which must be supplied by the mud pump. The term F represents the flow pressure losses, which consist of flow friction losses which occur throughout the flow passages and the pressure loss which occurs at the nozzles of the bit.

Bernoulli's Theorem may be written for the entire flow system as outlined above, or it may be written over a portion of the system. For example, it may be written for the discharge nozzles of the bit. This is the only place in the ordinary drill string where the velocity terms become large enough to contribute appreciably to the total pressure loss in mud circulation.

A common case where continuity is not maintained in pumping fluids occurs in cementing casing whenever the cement slurry is heavier than the mud. For example, where a 14 ppg cement slurry is pumped into the top of a string of casing, with both the casing and the annulus filled with 10 ppg mud, the heavier fluid inside the casing will flow downward because of its own weight. During this stage, there is a vacuum in the top part of the casing and the pump pressure required is only that required to overcome flow friction in the surface piping (which may often be of the order of 500 psi). The hydrostatic heads in the annulus and inside of the casing balance when the vertical feet of cement and of mud are equal in each. Continuity is restored when the casing becomes filled with liquid.

Types of Flow and Flow Patterns

The flow of liquids in channels such as pipes is customarily divided into two regions. In the first region, which occurs at slower rates of flow, the individual particles of the fluid move forward in straight or perhaps gently curving lines which conform to the confining walls of the flow channel.

For example, when such flow occurs inside a straight section of pipe, each individual particle of liquid moves forward in a line parallel to the axis of the pipe. Not all particles move forward with the same velocity, however. The particles of liquid that are in contact with the wall of the flow channel are stationary, or nearly so. This is particularly true in such cases as those of water or oil flowing through steel pipe, where the liquid wets the pipe wall. In other cases, such as mercury flowing through a glass tube, where the mercury does not wet the glass, there is evidence of slippage at the liquid-solid interface, and no stationary layer of fluid exists adjacent to the solid. In general, however, the velocity of flow increases from the confining walls of the flow channel and reaches a maximum velocity at the center of the channel, which is the point farthest away from the stationary confining walls.

The first or slower region of flow is divided generally into two categories according to the viscous nature of the flowing substance. For the case of true or Newtonian liquids, the region is referred to as *viscous*, or as *laminar*, or as *streamline* flow. Examples of such true liquids are water, gasoline, kerosene, and most fluid crude oils. Such liquids obey Newton's concept of viscosity, and the effective rate of shearing or sliding of adjacent fluid layers within the liquid is directly proportional to the pressure or force causing such flow. In the case of non-Newtonian liquids, the region is generally referred to as *plastic flow*. Drilling muds furnish excellent examples of plastic fluids. The general viscous nature of drilling muds is discussed in Chapter 7. In both viscous and plastic flow, the individual particles move forward in straight lines, conforming to the walls of the flow channel. However, in traversing from the wall of the flow channel to its center, the velocities and velocity distribution vary for different fluids according to the degree that they deviate from a true liquid. This in turn determines the amount of fluid which will flow for any given pressure drop imposed over the length of the flow channel.

Plug flow is an exaggerated type of plastic flow in which all of the shearing effort within the fluid takes place close to the pipe wall. The fluid occupying the central part of the cross-sectional area of the flow channel assumes sufficient body or structure so that it moves forward with no relative motion, very much as a solid plug of wood or rubber might be pumped through a pipe.

The second region or type of flow is referred to as *turbulent flow*. Most commercial and industrial transportation of fluids is carried out under conditions of turbulent flow, for generally too great an investment would be required for the large-diameter pipes which would be necessary to insure viscous-type flow. The rates of mud circulation required in rotary drilling insure turbulent flow conditions everywhere in the system except in the mud pits and possibly in enlarged sections of the open, uncased hole. In turbulent flow, the individual particles of liquid no longer move forward in straight lines or curved lines conforming to the channel walls. Instead, the fluid swirls and tumbles about inside the flow channel. The viewpoint may be taken that the forces causing flow become too great to be accommodated by the sliding motions of viscous flow, and a rolling type of motion is substituted in its place. This rolling type of motion involves the continuous transfer of momentum from one region in the fluid to another, hence the density of the fluid becomes a factor in the flow pressure drop. The viscosity of the fluid assumes a very minor function and disappears entirely as a factor of flow pressure drop at infinitely high rates of flow. In practical cases of turbulent flow, however, it is generally believed that there is a thin region in viscous flow next to the wall of the pipe or flow channel, while the main body of the fluid in the center of the pipe is tumbling about in turbulent flow. The velocity of flow, accordingly, increases rapidly away from the wall of the pipe, and then is fairly constant, on average, throughout the main body of the fluid. Both viscous and plastic fluids exhibit turbulent flow. Flow-friction pressure losses are calculated for both in the same manner.

The Reynolds Number

Experimental investigations have disclosed the various types of flow which may occur inside of conduits such as pipe. Analysis of the results of such experiments makes it possible to predict whether the flow will be of the viscous or turbulent nature. The criterion used for this purpose is known as the *Reynolds number*. This is a dimensionless number defined as follows:

$$Re = \frac{dU\rho}{\mu} \qquad (12\text{--}2)$$

where

Re = Reynolds number, dimensionless
d = pipe diameter, ft.
U = average flow velocity, ft./sec.
ρ = fluid density, lb./ft.^{3}
μ = fluid viscosity, lb./(ft.-sec.)

Since the Reynolds number is dimensionless, any consistent system of units may be used to obtain the same numerical value. In the metric system, the diameter would be expressed in centimeters, the velocity in centimeters per second, the density in grams per cubic centimeter, and the viscosity in poises. If the physical conditions were the same, the same numerical value would be obtained for the Reynolds number. It has been observed experimentally that flow changes from viscous (or plastic) to turbulent in the region of Reynolds number values of 2,000 to 3,000. At lower values the flow is streamline in nature; at higher values it is turbulent.

For flow channels whose shape is other than circular, the equivalent diameter is used. The equivalent diameter is defined as

$$\text{Equivalent Diameter} = \frac{(4)\,(\text{Cross-sectional Area})}{\text{Wetted Perimeter}} \qquad (12\text{--}3)$$

For a circular pipe, this expression reduces to the actual diameter. For the annulus between two circular pipes, it reduces to the difference between the inside diameter of the larger pipe and the outside diameter of the smaller pipe. In the latter case it may be noted that the wetted perimeter is determined by the inside surface of the larger pipe and the outside surface of the smaller pipe, since the flowing fluid contacts both of these surfaces.

The units which are in common use are often chosen for their convenient numerical size. It is seldom that consistent units are used throughout. For example, the diameter of the drill pipe is expressed in inches while its length is given in feet. The density of the mud is most conveniently expressed in pounds per gallon. The rate of flow is most conveniently expressed in barrels per minute (although gallons per minute is quite commonly used). The viscosity is most conveniently expressed in centipoises. Therefore, it is desirable to determine a conversion factor so that the Reynolds number may be found directly from these common units.

For the case of flow inside of a circular pipe, the following substitutions are made:

$$d = \frac{D}{12}$$

$$U = \frac{Q(5.6146)(144)(4)}{60\pi D^2}$$

$$= 17.157 \frac{Q}{D^2}$$

$$\rho = 7.4805G$$

$$\mu = 0.000672\eta$$

$$Re = \frac{(D/12)\left(17.157 \dfrac{Q}{D^2}\right)(7.4805G)}{0.000672\eta}$$

$$= 15{,}915 \frac{QG}{D\eta} \qquad (12\text{-}4)$$

where

Re = Reynolds number, dimensionless
Q = rate of flow, bbl./min.
   (NOTE: 1 bbl. = 42 U. S. gal.)
G = fluid density, lb./gal.
D = internal pipe diameter, inches
η = fluid viscosity, centipoises
d = internal pipe diameter, ft.
U = average flow velocity, ft./sec.

ρ = fluid density, lb./ft.³
μ = fluid viscosity, lb./(ft.-sec.)
5.6146 = ft.³/bbl.
7.4805 = gal./ft.³

For the case of flow in the annulus between two circular pipes, the following substitutions are made:

$$d = \frac{D_4 - D_3}{12}$$

$$U = \frac{Q(5.6146)(144)(4)}{60\pi(D_4^2 - D_3^2)}$$

$$= 17.157 \frac{Q}{(D_4^2 - D_3^2)}$$

$$= \frac{17.157Q}{(D_4 + D_3)(D_4 - D_3)}$$

$$\rho = 7.4805G$$

$$\mu = 0.000672\eta$$

$$Re = \frac{\dfrac{(D_4 - D_3)}{12} \dfrac{17.157Q}{(D_4 + D_3)(D_4 - D_3)} 7.4805G}{0.000672\eta}$$

$$= 15{,}915 \frac{QG}{(D_4 + D_3)\eta} \qquad (12\text{-}5)$$

where

D_4 = inside diameter of larger pipe, in.
D_3 = outside diameter of smaller pipe, in.

The Fanning Friction Factor

The *Fanning friction factor* is widely used in determining flow-friction pressure losses.[1] It is a dimensionless number defined as

$$f = \left(\frac{pd}{4L}\right) \Big/ \left(\frac{U^2\rho}{2g}\right)$$

$$= \frac{pdg}{2U^2\rho L} \qquad (12\text{-}6)$$

---

[1] An equivalent solution for flow friction losses utilizes the "Stanton" diagram, in which the friction factor has a numerical value four times the numerical value of the Fanning friction factor

where

f = Fanning friction factor, dimensionless
p = flow-friction pressure loss, lb./ft.^{2}
d = pipe diameter, ft.
L = length of pipe, ft.
U = average flow velocity, ft./sec.
ρ = fluid density, lb./ft.^{3}
g = acceleration of gravity, 32.174 ft./sec.^{2}

This number is the ratio of the shearing force in the fluid at the inside wall of the flow channel to the average kinetic energy of the flowing fluid. The above equation may be solved for the pressure to obtain

$$p = \frac{2fU^2\rho L}{dg} \qquad (12\text{--}7)$$

This relation may be expressed in more convenient units. For the case of flow inside of a circular pipe, the following substitutions are made:

$$p = 144P$$

$$U = 17.157 \frac{Q}{D^2}$$

$$\rho = 7.4805G$$

$$d = \frac{D}{12}$$

$$g = 32.174$$

$$144P = \frac{2f\left(17.157 \dfrac{Q}{D^2}\right)^2 7.4805GL}{32.174 \dfrac{D}{12}}$$

$$P = 11.406 \frac{fQ^2GL}{D^5} \qquad (12\text{--}8)$$

where

P = flow-friction pressure loss, psi
Q = rate of flow, bbl./min.
L = length of pipe, ft.
G = fluid density, ppg
D = internal pipe diameter, in.

For the case of flow in the annulus between two circular pipes, the following substitutions are made:

$$p = 144P$$

$$U = 17.157 \frac{Q}{(D_4^2 - D_3^2)}$$

$$\rho = 7.4805G$$

$$d = \frac{D_4 - D_3}{12}$$

$$g = 32.174$$

$$144P = \frac{2f\left[17.157\dfrac{Q}{D_4^2 - D_3^2}\right]^2 7.4805GL}{32.174 \dfrac{(D_4 - D_3)}{12}}$$

$$P = 11.406 \frac{fQ^2GL}{(D_4 - D_3)^3(D_4 + D_3)^2} \tag{12-9}$$

A plot of the Fanning friction factor *vs.* the Reynolds number is shown in Fig. 12-2. Inspection of the flow conditions will show that one factor which must be included in some manner is the roughness factor, which might be defined as the ratio of the height of the average rough protrusions in the surface to the diameter of the flow channel. This is usually taken care of by showing different correlation lines in

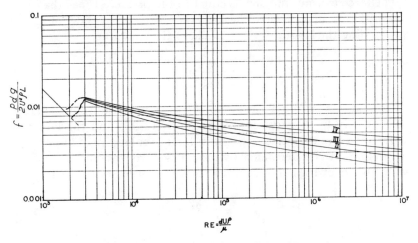

FIG. 12-2. Fanning friction factor *vs.* Reynolds number.

the turbulent region for various kinds of materials and pipe diameters. Figure 12–2 may be used directly in solving for flow-friction pressure losses. Line I is for brass pipe, line II for 1- to 4-inch steel pipe, line III for casing annulus, line IV for rough annulus. The numerical value of the Reynolds number can be determined from Equations (12–2), (12–4), or (12–5). The corresponding value of f determined from Fig. 12–2 can then be used in Equations (12–7), (12–8), or (12–9) to calculate the flow-friction pressure loss

THE CHARACTERISTICS OF VISCOUS AND PLASTIC FLOW

The equation which describes the viscous or streamline flow of true liquids in circular pipes is known as Poiseuille's Law. It is derived directly from Newton's equation of viscosity,

$$\frac{F}{A} = -\eta \frac{dv}{dr} \qquad (12\text{–}10)$$

where

$\dfrac{F}{A}$ = unit shear force causing flow

η = the viscosity

$\dfrac{dv}{dr}$ = the velocity gradient

The force causing the flow is equal to the pressure times the cross-sectional area, and the area sheared is the area of a cylinder of radius, r, extending the length of the flow channel. Accordingly,

$$\frac{F}{A} = \frac{p\pi r^2}{2\pi r L} = \frac{pr}{2L} \qquad (12\text{–}11)$$

The expression for shearing force per unit area from Equation (12–11) may be substituted in Equation (12–10) and the latter integrated to give the velocity gradient for a circular pipe. The constant of integration is determined by assuming zero velocity at the pipe wall. The following expression results for the velocity gradient:

$$v = \frac{p}{4\eta L}(R^2 - r^2)$$

where R is the actual radius of the pipe and r is a variable radius, inside the pipe, for purposes of analysis. This equation was used to calculate the velocity distribution shown in Fig. 12-3.

The total flow is calculated from the basic relation

$$dq = v\, dA$$

where q is the volume rate of flow and A the cross-sectional flow area. For a circular pipe, dA is replaced by $2\pi r dr$. Integration gives the familiar expression for total flow,

$$q = \frac{\pi p R^4}{8\eta L} \qquad (12\text{-}12)$$

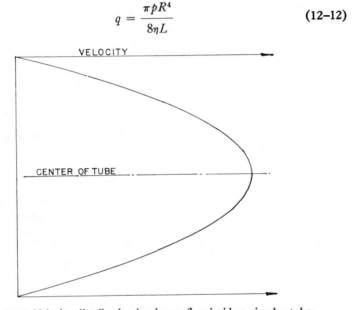

FIG. 12-3. Velocity distribution in viscous flow inside a circular tube.

Equation (12-12) is Poiseuille's Law in its usual form. It is understood however, that the pressure is expressed in absolute units, such as dynes per square centimeter, because of the definition of the poise. In order to use common pressure units, the gravitational constant, g, must be introduced. Also, in order to be consistent in English units, it is preferable to replace the viscosity term, η, with the foot-pound-second equivalent

$$q = \frac{\pi g p R^4}{8\mu L}$$

$$q = \frac{\pi g p d^4}{128 \mu L} \qquad (12\text{-}13)$$

This equation coincides with the viscous flow line on the Fanning friction factor *vs.* the Reynolds number diagram. Since q may be expressed in terms of the average velocity as $(\pi/4)Ud^2$, Equation (12–13) may be arranged algebraically as

$$\frac{pdg}{2U^2L} = 16\frac{\mu}{dU}$$

Dividing both sides by the density ρ gives both the Fanning friction factor and the Reynolds number:

$$\frac{pdg}{2U^2\rho L} = 16\frac{\mu}{dU\rho}$$

$$f = \frac{16}{Re} \qquad (12\text{–}14)$$

Equation (12–14) is the equation of the viscous flow line on the Reynolds number chart. On the log-log plot, this is a straight line with a negative slope of unity.

Further insight into the mechanics of both viscous and plastic flow is given by considering the regions within the flow channel which contribute the most to the total flow. Since the resistance to flow originates at the wall of the flow channel, it follows that any movement of fluid near the channel wall carries with it all of the fluid which occupies the more central portion of the flow channel. This process is progressive from the wall to the center of the channel. Equation (12–12) gives the total flow for the case of a true or Newtonian liquid flowing through a circular pipe of radius R. If r is considered a variable radius inside the channel, for purposes of analysis, the contribution to total flow which is made by fluid shear inside of radius r would be given by

$$q_r = \frac{\pi p r^4}{8\eta L}$$

The fraction of the total flow which is contributed by fluid shear outside of radius r, near the wall of the flow channel, is given by

$$\frac{q_T - q_r}{q_T} = 1 - \left(\frac{r}{R}\right)^4 \qquad (12\text{–}15)$$

Equation (12–15) was used in constructing Fig. 12–4, which shows for example that the fluid shearing which occurs between the pipe wall and the distance halfway to the center of the pipe accounts for about

94 per cent of the total flow, for true liquids. For such plastic fluids as drilling muds, the major portion of the total flow would be accommodated by the fluid shear at distances even closer to the pipe wall. The significance of this is that any viscosity relationships which are used in plastic flow calculations for such fluids as drilling muds must accurately describe the viscous nature of the fluid for flow conditions near the wall of the channel. The flow conditions near the center of the channel have little influence on the total flow.

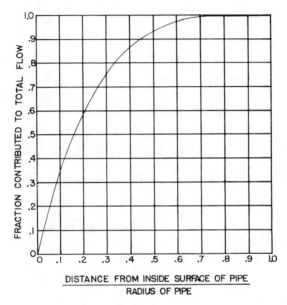

FIG. 12–4. Contribution to total flow *vs.* distance from pipe wall for viscous flow inside a circular pipe.

Two equations have been proposed for describing the viscous nature of such plastic fluids as drilling muds. The first of these, proposed by Bingham, is

$$T - T_B = \eta_p \frac{dv}{dr} \qquad (12\text{–}16)$$

where

$$T = \frac{F}{A} = \frac{pr}{2L}$$

= unit shearing force causing flow
T_B = Bingham yield point

η_p = plastic viscosity

$\dfrac{dv}{dr}$ = velocity gradient

This relation integrates into the Bingham-Buckingham Equation:

$$q = \frac{\pi R^4}{8\eta_p L}\left[p - \frac{4}{3}p_0\left(1 + \frac{p_0^3}{4p^3}\right)\right] \qquad (12\text{-}17)$$

where

$$p_0 = \frac{2LT_B}{R}$$

This equation recognizes the existence of plug flow in the center of the pipe, which might occur at extremely slow rates of flow. Where p is large compared to p_0, the equation reduces to

$$q = \frac{\pi R^4}{8\eta_p L}\left(p - \frac{4}{3}p_0\right) \qquad (12\text{-}18)$$

Good results have been reported from using Equation (12–18), although the straight-line relation proposed by Bingham, Equation (12–16), was observed not to apply exactly to the muds investigated.[2]

The second equation which has been used to describe the viscous nature of plastic fluids was originally proposed by Ostwald:

$$T = K\left(-\frac{dv}{dr}\right)^m \qquad (12\text{-}19)$$

This equation plots as a straight line on log-log paper, whereas the Bingham Equation plots as a straight line on Cartesian co-ordinates. Consequently, the Ostwald Equation is more versatile and can accommodate curvature in the T vs. dv/dr data. If T is again set equal to $pr/2L$, the equation integrates to

$$q = \frac{\pi}{4}\left(\frac{p}{2K'L}\right)^{1/m} R^{(3m+1)/m} \qquad (12\text{-}20)$$

where

$$K' = K\left(\frac{3m+1}{4m}\right)^m$$

[2] R. W. Beck, W. F. Nuss, and T. H. Dunn, "The Flow Properties of Drilling Muds," *API Drilling and Production Practice* (1947), 9–22; J. C. Melrose and W. B. Lilienthal, "Plastic Flow Properties of Drilling Fluids—Measurement and Application," *Trans. A.I.M.E. Petroleum Division*, Vol. 192 (1951), 159–64.

Equation (12–20) can be rearranged to fit the general equation $f = 16/Re$ which applies to Newtonian liquids and the following expression obtained for the Reynolds number:

$$Re = \frac{d^m U^{2-m} \rho}{g K' 8^{m-1}} \qquad (12\text{–}21)$$

The preceding relations have been developed by Metzner and Reed[3] and Saunders and Melton[4] to obtain excellent correlations between measured fluid properties and rates of flow through circular pipes. They explicitly state that the values of K and m used must correspond to the shear stress, $pd/4L$, which exists at the pipe wall. However, m and K have been found to be constant over 10-fold ranges of flow rates. Methods of obtaining basic measurements from viscometer data are discussed by these authors and also by Cardwell.[5]

None of viscosity relations which have been developed apply to highly thixotropic fluid. Fluids which develop a gel structure during quiescent standing may develop higher viscosities during prolonged flow at slow rates. Pigott[6] observed that muds flowing in the viscous range often exhibit viscosities which are inversely proportional to the rate of flow. That is, as rates of flow are reduced in the plastic flow region, the pressures required do not drop accordingly.

For true liquids, such as water or oil flowing in the viscous flow region, Equation (12–13) may be expressed as follows:

For circular pipes

$$P = 0.0115 \frac{Q \eta L}{D^4} \qquad (12\text{–}22)$$

For an annulus

$$P = 0.0115 \frac{Q \eta L}{(D_4 - D_3)^3 (D_4 + D_3)} \qquad (12\text{–}23)$$

For a plastic fluid flowing inside a circular pipe, Equation (12–18) may be written as

---

[3] A. B. Metzner and J. C. Reed, "Flow of Non-Newtonian Fluids—Correlation of the Laminar, Transition, and Turbulent-Flow Regions," AI Ch E *Journal*, Vol. I, No. 4 (1955), 434.

[4] C. D. Saunders and L. L. Melton, "Rheological Measurements of Non-Newtonian Fluids," Paper 716-G, A.I.M.E. Petroleum Branch Fall Meeting, October 14–17, 1956, Los Angeles, Calif.

[5] W. T. Cardwell, Jr., "Drilling—Fluid Viscosimetry," *API Drilling and Production Practice* (1941), 104–12.

[6] R. J. S. Pigott, "Mud Flow in Drilling," *API Drilling and Production Practice* (1941), 91–103.

$$Q = 87 \frac{D^4}{\eta_P L}\left[P - \frac{LY_P}{225D}\right] \qquad (12\text{-}24)$$

where

P = flow friction pressure loss, psi
Q = rate of flow, bbl./min.
η = viscosity, centipoises
L = length of pipe, ft.
D = diameter, in.
D_4 = large diameter of annulus, in.
D_3 = small diameter of annulus, in.
η_P = plastic viscosity, centipoises
Y_P = Bingham yield point, lb./hundred ft.^{2}

Example: Use Equations (12–24) and (12–20) to calculate the rate of mud flow which will be obtained for the following conditions. A pressure drop of 120 psi is imposed on a one-inch I.D. pipe of 500 feet length. The mud density is 10.5 ppg. The viscosity was measured with a Fann viscometer. The scale reading at 300 rpm was 58 and the reading was 98 at 600 rpm.

Solution: According to the characteristics of the Fann viscometer,

$$\eta_p = R_{600} - R_{300}$$
$$= 98 - 58$$
$$= 40 \text{ centipoises, plastic viscosity}$$
$$Y_p = R_{300} - \text{plastic viscosity}$$
$$= 58 - 40$$
$$= 18 \text{ lb./hundred ft.}^2, \text{ yield point}$$

Before proceeding further, the shearing force at the pipe wall will be investigated. This shearing force is determined as

$$\text{Shear force, wall} = \frac{pd}{4L}$$
$$= \frac{(120 \text{ lb./in.}^2)(454 \text{ gr./lb.})(980 \text{ dynes/gr.})(1 \text{ in.})}{(6.45 \text{ cm.}^2/\text{in.}^2)(12 \text{ in./ft.})(4)(500 \text{ ft.})}$$
$$= 346 \text{ dynes/cm.}^2$$

The shearing forces in the viscometer can also be calculated from the basic concept of viscosity. The Fann viscometer is calibrated to read directly the apparent viscosity at 300 rpm. and the apparent viscosity at 600 rpm is accordingly half of the reading at 600 rpm. From the definition of apparent viscosity, unit shearing force equals apparent viscosity multiplied by the rate of shear. Savins and Roper[7] give the average rate of shear of the Fann viscometer as 479 sec.$^{-1}$ at 300 rpm and 958 sec.$^{-1}$ at 600 rpm. To obtain the force in dynes/cm.^{2}, the viscosity must be in poises.

At 300 rpm, for the mud described,

$$\text{Unit shearing force} = (0.58 \text{ poises})(479 \text{ sec.}^{-1})$$
$$= 278 \text{ dynes/cm.}^2$$

At 600 rpm, for the mud described,

$$\text{Unit shearing force} = \left(\frac{R_{600}}{2}\right)\left(\frac{1}{100}\right)(958 \text{ sec.}^{-1})$$
$$= (0.46 \text{ poises})(958 \text{ sec.}^{-1})$$
$$= 441 \text{ dynes/cm.}^2$$

From the foregoing analysis, it may be seen that the conditions in the problem are such that the unit shearing force on the inside of the pipe wall is about midway between the lower and higher measurement made with the viscometer. In this special case, the mud flow calculated from Bingham-type relations should be about the same as that from calculations based on the Ostwald viscosity relation.

For the Bingham-type calculation, Equation (12–24) may be used as follows:

$$Q = 87 \frac{D^4}{\eta_p L}\left(P - \frac{LY_p}{225D}\right)$$

$$= \frac{(87)(1 \text{ in.})^4}{(40 \text{ cp.})(500 \text{ ft.})}\left(120 \text{ psi} - \frac{(500 \text{ ft.})(18 \text{ lb./hundred ft.}^2)}{(225)(1 \text{ in.})}\right)$$

$$= \frac{(87)(1)}{(40)(500)}(120 - 40)$$

$$= 0.348 \text{ bbl./min.}$$

The Ostwald-type calculation requires first the determination of the constants K and m. If Equations (12–19) and (12–20) are considered as being in fundamental c.g.s. units, the procedure is as follows:

---

[7] J. G. Savins and W. F. Roper, "A Direct-indicating Viscometer for Drilling Fluids," *API Drilling and Production Practice* (1954), 7–22.

$$m = \frac{\log\left(\dfrac{\text{shear force}_2}{\text{shear force}_1}\right)}{\log\left(\dfrac{\text{shear rate}_2}{\text{shear rate}_1}\right)}$$

$$= \frac{\log\left(\dfrac{441 \text{ dynes/cm.}^2}{278 \text{ dynes/cm.}^2}\right)}{\log\left(\dfrac{958 \text{ sec.}^{-1}}{479 \text{ sec.}^{-1}}\right)}$$

$$= 0.665$$

$$K = \frac{(\text{shear force}_1)}{(\text{shear rate}_1)^m} = \frac{(278 \text{ dynes/cm.}^2)}{(479 \text{ sec.}^{-1})^{0.665}}$$

$$= 4.60$$

Also,

$$K' = K\left(\frac{3m+1}{4m}\right)^m$$

$$= 4.60\left(\frac{(3)(0.665)+1}{(4)(0.665)}\right)^{0.665}$$

$$= 4.975$$

The substitutions made in Equation (12–20) are made by considering all quantities to be in fundamental c.g.s. units. In this system, pressure is in dynes/cm.^{2}, volume flow rate is in c.c./sec., and the length, diameter, and radius are in cm. The substitutions will be made in terms of common symbols:

$$q = \frac{\pi}{4}\left(\frac{p}{2K'L_{cm}}\right)^{1/m} R^{(3m+1)/m}$$

$$\frac{(Q \text{ bbl.min.})(5.615)(30.48)^3}{60} = 0.7854\left[\frac{(P)(454)(980)}{(6.45)(2)K'30.48L}\right]^{1/m}\left[\frac{2.54D}{2}\right]^{(3m+1)/m}$$

$$\frac{Q(5.615)(30.48)^3}{60} = \frac{\pi}{4}\left[\frac{P(454)(980)}{(6.45)(2)K'30.48L}\right]^{1/m}\left[\frac{2.54D}{2}\right]^{(3m+1)/m}$$

$$Q = 0.000297 \left[\frac{1{,}130P}{K'L}\right]^{1/m} [1.27D]^{(3m+1)/m}$$

$$= 0.000297 \left[\frac{(1{,}130)(120 \text{ psi})}{(4.975)(500 \text{ ft.})}\right]^{1/0.665} [(1.27)(1 \text{ in.})]^{4.5}$$

$$= (0.000297)(54.6)^{1.503}(1.27)^{4.5}$$

$$= 0.356 \text{ bbl./min.}$$

It may be noted that Equations (12–20) and (12–24) give practically identical results in the above example. However, the conditions in the example were chosen with this result in mind. That is, the shear force at the pipe wall was midway between the two shearing forces used in the viscometer when measuring the viscosity constants. The conclusions to be drawn are that the fluid shearing close to the pipe wall largely controls the total flow and that the viscosity constants used in plastic flow calculations should correspond to flow conditions at the pipe wall.

Example: Calculate the rate of flow of an oil of 40 centipoises viscosity through 500 feet of one-inch ID pipe, where the flow pressure loss is 120 psi. The specific gravity of the oil is 0.85.

Solution: It will first be assumed that the flow is in the viscous region and a corresponding solution obtained. Subsequently the Reynolds number will be determined to ascertain that flow is in the viscous region.

Equation (12–22) may be used as follows:

$$Q = \frac{PD^4}{0.0115\eta L}$$

$$= \frac{(120 \text{ psi})(1 \text{ in.})^4}{(0.0115)(40 \text{ cp.})(500 \text{ ft.})}$$

$$= 0.521 \text{ bbl./min.}$$

Equation (12–4) may be used to determine the Reynolds number:

$$Re = 15{,}915 \frac{QG}{D\eta}$$

$$= \frac{(15{,}915)(0.521 \text{ bbl./min.})(0.85)(8.33 \text{ lb./gal.})}{(1 \text{ in.})(40 \text{ cp.})}$$

$$= 1{,}470$$

The value of the Reynolds number (less than 2,000) indicates that flow is in the viscous region and that the solution is valid.

Turbulent-Flow Viscosity

The calculation of flow-friction pressure losses in the important region of turbulent flow is accomplished through the use of the Fanning friction factor–Reynolds number correlation, or equivalent mathematical treatment.

In the case of true liquids, all quantities necessary in the calculation can be evaluated readily by direct measurements. For the purpose of similar calculations in the case of plastic drilling muds, all necessary quantities can be evaluated by direct measurement, except the viscosity of the fluid. Fortunately, in the region of practical interest, pressure losses vary approximately with the one-fifth power of the viscosity, so that precise values of viscosity are not required. Nevertheless, the effects of viscosity are significant and cannot be neglected in calculating flow-pressure losses.

The turbulent viscosity of drilling muds has been evaluated indirectly by calculation. The flow-pressure relations of a piping system have been measured precisely by flowing water through the system. Since the viscosity of the water is a known quantity, a friction factor–Reynolds number plot can be made for the particular piping system. Subsequent flow of drilling muds through the system permitted the calculation of the viscosity exhibited by the mud while in turbulent flow. The correlation which was obtained by Beck, Nuss, and Dunn between the turbulent viscosity and the plastic viscosity as measured with a rotating-cup viscometer is given by the equation

$$\eta_T = \frac{\eta_P}{3.2} \qquad (12\text{--}25)$$

where

η_T = effective turbulent viscosity
η_P = plastic viscosity (corresponding to Bingham-type plastic fluid)

A different correlation was obtained by Havenaar[8] as shown in Fig. 12–5. The best straight line through the data points was given as

$$\frac{\eta_T}{\eta_w} = 46.5\phi \qquad (12\text{--}26)$$

where

η_T = turbulent viscosity
η_w = viscosity of the liquid phase (water)
ϕ = volume fraction of solids suspended in the drilling mud

[8] I. Havenaar, "The Pumpability of Clay-Water Drilling Fluids," *Trans. A.I.M.E. Petroleum Division*, Vol. 201 (1954), 287–93.

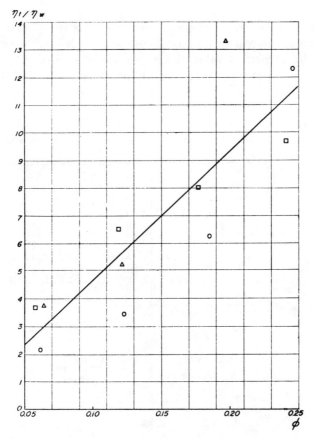

FIG. 12-5. The relation between η_T/η_W and the volume fraction of dispersed phase ϕ; clay muds, O; bentonite barite muds, □; field muds, △.

It was found experimentally that Equation (12-26) could be used to calculate the effects of temperature on the turbulent viscosity, where the viscosity of the water would be known at both temperatures. It is also significant that, within the turbulent range, the flow velocity had no effects on the viscosity of the muds tested.

Equation (12-26) is very similar to a relation derived mathematically by Einstein for the viscosity of dilute suspensions of spherical particles. For the derived equation, it was specified that the solution be diluted to the extent that the rigid spherical particles would not interfere with each other. The usual form for Einstein's equation is

$$\frac{\eta_s}{\eta_w} = 1 + 2.5\phi \qquad (12\text{-}27)$$

where η_s is the viscosity of the suspension, η_w is the viscosity of the liquid phase, and ϕ is the fractional volume occupied by the suspended solids. The numerical value of the coefficient 46.5 compared to 2.5 incorporates an average shape factor for the particles but it reflects particularly the interference between particles in average clay-water-barite muds.

Equation (12–26) was used to calculate the probable relationships between mud density and turbulent viscosity. Where a mud is com-

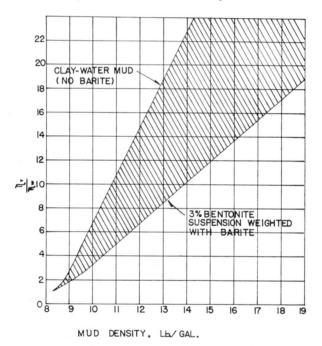

Fig. 12–6. Relation between ratio of turbulent viscosity to water viscosity and density of water-base mud.

posed of clays and water, the volume fraction of solids will be greater than if barite weighting material is used to obtain the same density of mud. The probable limiting cases are shown in Fig. 12–6. One line represents a mud containing only water and clay with no barite. The lower line represents a mud containing a minimum amount of clay for suspending barite weighting material. Data are not available at the present time to distinguish the relative effects of different kinds of clays and weighting materials except on the basis of their specific gravity.

Flow-Friction Pressure Losses—Turbulent Flow

Flow-friction pressure losses in either the viscous or turbulent region can be calculated by direct use of the Fanning friction factor–Reynolds number plot. However, it is often convenient to use an equation which is directly applicable to the turbulent-flow region. This is particularly true when it is desirable to analyze the effect of a particular factor, such as the pipe diameter, on the pressure or rate of

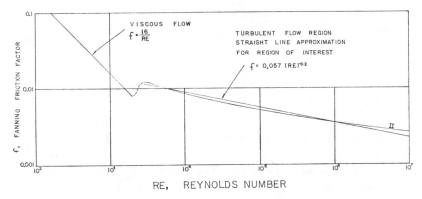

FIG. 12-7. Turbulent-flow equation.

flow. For this purpose, it is customary to draw a straight line on the friction-factor chart which will approximate the curved relations on the chart throughout the field of interest. Such a line has been drawn on Fig. 12-7, and the equation of the line on the chart is

$$f = 0.057(Re)^{-0.2} \tag{12-28}$$

Substituting the symbols defining f and Re gives

$$\frac{pdg}{2U^2\rho L} = 0.057\left(\frac{\mu}{dU\rho}\right)^{0.2}$$

Solved for p, the following is obtained:

$$p = \frac{(2)(0.057)U^{1.8}\rho^{0.8}\mu^{0.2}L}{gd^{1.8}}$$

The substitutions in terms of commonly used units were next carried out as outlined under discussions of the Reynolds number and Fanning friction factor. The following equations result:

For a circular pipe,

$$P = 0.094 \frac{Q^{1.8}G^{0.8}\eta_T{}^{0.2}L}{D^{4.8}} \tag{12-29}$$

For an annulus,

$$P = 0.094 \frac{Q^{1.8} G^{0.8} \eta_T^{0.2} L}{(D_4 - D_3)^3 (D_4 + D_3)^{1.8}} \qquad (12\text{--}30)$$

where

P = flow-friction pressure loss, psi
Q = rate of flow, bbl./min.
G = fluid density, lb./gal.
η_T = turbulent viscosity, centipoises
L = length of pipe, ft.
D = pipe diameter, in.
D_4 = large diameter of annulus, in.
D_3 = small diameter of annulus, in.

These equations approximate the line usually given for commercial iron or steel pipe of 1 to 4 inches diameter from Reynolds numbers of 3,000 to 1,000,000.

The accuracy of flow-friction pressure-loss calculations is many times not better than within about 10 per cent of measured values. The calculation is extremely sensitive to the value used for the diameter, which enters in about to the fifth power. Accordingly, an error of 2 per cent in the diameter would cause an error of about 10 per cent in the calculated pressure.

Tables 12–1 and 12–2 contain values of pressure drops which were calculated from Equations (12–29) and (12–30). The hydrostatic pressure of the mud was calculated from the relation

$$P = 0.052 G h$$

where h is the depth in feet. The calculation of the total pressure in the mud at any point is facilitated by remembering that the hydrostatic heads inside and outside of the drill string balance each other. Flow pressure losses are additive in vertical flow the same as they are in horizontal flow. For downward flow, the pressure at the bottom of a particular flow section equals the pressure at the top plus the difference in hydrostatic head minus the flow pressure loss. For upward flow, the pressure at the bottom of a particular flow section equals the pressure at the top plus the difference in hydrostatic head plus the flow pressure loss. The results of these calculations are presented graphically in Fig. 12–9. The table and graphs illustrate that considerably less pressure is exerted against exposed rock formations when using normal circulation, as compared to reverse circulation.

TABLE 12–1
Values of Pressure Drops

Flow Section	Length, ft. L	Diameter, in. D		Re	Velocity, ft./sec. U	$D^{4.8}$	$L/D^{4.8}$	Pressure Drop through Section, psi
(Pressure gauge)								(1,500)
Standpipe	50	3.152		46,200	13.8	245	0.20	9
Rotary Hose	50	3		48,600	15.3	194	0.26	12
Swivel	—	—		—	—	—	—	—
Kelly	41	4		36,400	8.6	770	0.05	2
Drill pipe	9,600							814
Pipe	8,956	3.826		38,000	9.4	630	14.21	(660)
Tool joints	644	3		48,600	15.3	194	3.32	(154)
Drill Collars	400	2.25		64,800	27.2	49.0	8.16	379
Bit								127
(Annulus)		D_1	D_2					
Drill-collar hole	400	7.875	6.25	10,300	6.0	505	0.79	37
Drill-pipe hole	6,600							104
Pipe hole	6,246	7.875	4.5	11,800	3.3	3,530	1.77	(82)
Tool-joint hole	354	7.875	6	10,500	5.3	750	0.47	(22)
Drill-pipe casing	3,000							16
Pipe casing	2,839	8.921	4.5	10,800	2.3	9,290	0.30	(14)
Tool-joint casing	161	8.921	6	9,800	3.2	3,250	0.05	(2)
Totals							29.58	1,500

Pressure Drop Through Bit Nozzles

Drilling mud is discharged at the bottom of the hole through watercourses or nozzles in the bit. In order to aid in drilling, the nozzles have a relatively small cross-sectional area so that the mud is discharged at high velocities. The kinetic energy of the mud stream flowing out of the jets is dissipated in fluid friction at the bottom of

TABLE 12–2
Values of Pressure Drops

Flow Section	Flow-Pressure Drop, psi	Depth to Lower End of Section, ft.	Hydrostatic Mud Head, psi	Normal Circulation			Reverse Circulation		
				Cumulative Flow-Pressure Loss to Lower End, psi	Total Pressure in Mud, psi	Mud Density Equivalent to Total Pressure, ppg	Cumulative Flow-Pressure Loss to Lower End, psi	Total Pressure in Mud, psi	Mud Density Equivalent to Total Pressure, ppg
(Pressure gauge)					1,500			1,500	
Standpipe	9								
Rotary hose	12								
Kelly	2			23	1,477		1,477	23	
Drill pipe	814	9,600	5,980	837	6,643		663	6,817	
Drill collars	379	10,000	6,240	1,216	6,524		284	7,456	
Bit	127	10,000	6,240	1,343	6,397		157	7,583	
(Annulus)									
Drill-collar hole	37	10,000	6,240	1,343	6,397	12.3	157	7,583	14.6
Drill-pipe hole	104	9,600	5,980	1,380	6,100	12.2	120	7,360	14.7
Drill-pipe casing	16	3,000	1,870	1,484	1,886	12.1	16	3,354	21.5

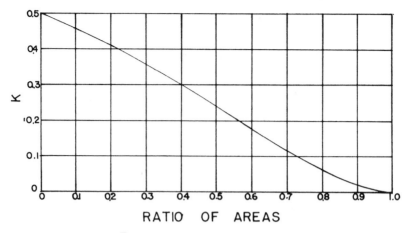

Fig. 12-8. Entrance-loss coefficient.

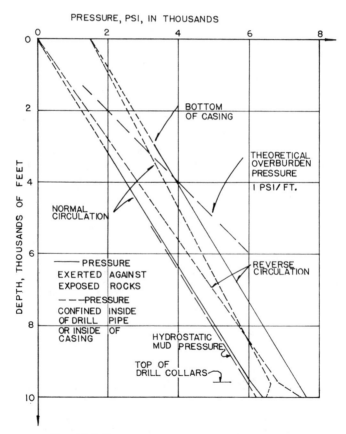

Fig. 12-9. Pressure *vs.* depth in normal and reverse circulation (calculated example).

DRILLING-FLUID CIRCULATION

the hole and converted to heat. In addition to the exit loss of kinetic energy, there is also an entrance loss as the fluid flows into the restricted area. The amount of this loss may be controlled to some extent by the design of the entrance into the nozzles. In the general case, the entrance loss is expressed as a fraction k of the kinetic energy of the stream inside the nozzles. The total pressure loss across the bit can be found by

$$P_B = 0.1465(1 + k)\frac{GQ^2}{J^2} \qquad (12\text{-}31)$$

where

$G =$ mud density, lb./gal.
$Q =$ rate of flow, bbl./min.
$J =$ total cross-sectional flow area of all jets, sq. in.

Example: Calculate the pump pressure required to circulate 8 bbl./min. of 12 ppg water-base mud down through $4\frac{1}{2}$-inch 16.6-lb./ft. drill pipe and 400 feet of drill collars and return to the surface. The bit has three watercourses, each of which is 11/16 inch in diameter. The drill collar ID is $2\frac{1}{4}$ inches and the OD is $6\frac{1}{4}$ inches. The ID of the tool joints is 3 inches and their OD is 6 inches. Bit size is $7\frac{7}{8}$ inches. The intermediate casing string consists of 3,000 ft. of $9\frac{5}{8}$-inch 36-lb./ft. casing. Depth of the well is 10,000 ft.

Solution: The lengths of the flow sections, inside the drill pipe, are computed according to the recommendations in "Hydraulics in Rotary Drilling,"[9] which recommends using a length of 67 feet of tool joint bore and 933 feet of drill pipe proper per 1,000 feet total length of drill pipe. (Exceptions to this recommendation are in the cases of API regular and Acme regular tool joints.) In the annulus, the lengths of the sections are computed by using a length of 20 inches, 1,667 feet, for the made-up tool joints, and assuming an average length of 31 feet per joint of pipe in the made-up drill string. This amounts to (1,000 ft./31 ft.) (1.667 ft.) or 53.7 feet of tool joint OD per thousand feet of made-up drill pipe. Since the tool joints, located between the individual lengths of drill pipe, have different cross-sectional flow areas, they should be considered separately in the flow-pressure loss calculation. Unless there is drastic reduction in flow area, the contraction and enlargement losses at the individual tool joints are negligible. For convenience in calculation, the lengths contributed by the individual tool joints are combined. The following illustrates the calculation of the length of the tool-joint OD section in the drill-pipe hole annulus:

[9] Publication of Hughes Tool Company (*Bulletin No. 1-A*, revised, April, 1954).

Length of drill-pipe hole annulus

$= 10{,}000$ ft. T.D. $-$ 400 ft. drill collars $-$ 3,000 ft. casing

$= 6{,}600$ ft.

Length of tool joint O.D. section $= (53.7$ ft./1,000 ft.$)(6{,}600$ ft.$)$

$= 354$ ft.

Length of drill pipe O.D. section $= 6{,}600$ ft. $-$ 354 ft.

$= 6{,}246$ ft.

The Reynolds number for each flow section was calculated from Equations (12–4) and (12–5). The results are listed in Table 12–1. Since the numerical values of the Reynolds number are above 2,100, all sections are in turbulent flow. Accordingly, Equations (12–29) and (12–30) are applicable and were used in calculating the flow-friction pressure losses.

The turbulent viscosity of the mud was taken from Fig. 12–6. The upper density limit is about 12 ppg for a clay-water mud which contains no barite. Without more specific information available, it was assumed that some barite had been used in the mud and the turbulent viscosity was taken as 10.5 centipoises, which is about midway between the probable upper and lower limits.

For each flow section inside of the drill string, Equation 12–29 is arranged as follows:

$$P = 0.094 Q^{1.8} G^{0.8} \eta_T^{0.2} \frac{L}{D^{4.8}}$$

$$= (0.094)(8 \text{ bbl./min.})^{1.8}(12 \text{ ppg})^{0.8}(10.5 \text{ cp.})^{0.2} \frac{L}{D^{4.8}}$$

$$= 46.4 \frac{L}{D^{4.8}}$$

The values of Q, G and η_T substituted in Equation (12–30) are the same as above. Accordingly, the flow-friction pressure loss for any section in the annulus is

$$P = 46.4 \frac{L}{(D_4 + D_3)^{1.8}(D_4 - D_3)^3}$$

The values of the $L/D^{4.8}$ terms, and the corresponding terms for the annulus, are listed in Table 12–1. These were converted to pressure losses as indicated above.

In order to calculate the pressure drop across the bit watercourses, the total cross-sectional area of the three watercourses was calculated to be 1.113 square inches. In determining the entrance-loss coefficient k, the cross-sectional flow area inside the drill collars was calculated to be 3.97 square inches. The ratio of areas is 1.113/3.97 or 0.281. A value of $K=0.40$ was used for the nozzle-entrance loss. The pressure drop across the bit was then calculated as follows:

$$P_B = 0.1465(1 + k) \frac{GQ^2}{A^2}$$

$$= (0.1465)(1.40) \frac{(12 \text{ ppg})(8 \text{ bbl./min.})^2}{(1.113 \text{ sq. in.})^2}$$

$$= 127 \text{ psi}$$

Hydraulic Horsepower

The hydraulic horsepower delivered by a mud pump depends upon the quantity of fluid delivered by the pump and the pressure increase effected by the pump. One horsepower is defined as equal to 33,000 foot-pounds per minute or 550 foot-pounds per second. Accordingly, hydraulic horsepower may be calculated as follows:

$$\text{Hydraulic horsepower} = \frac{pq}{550} \tag{12-32}$$

where

p = pressure, lb./ft.^{2}
q = rate of flow, ft.^{3}/sec.
550 = ft-lb./(sec.-hp)

The relation may be expressed in more common units as follows:

$$\text{Hydraulic horsepower} = \frac{(144P)(5.615Q)}{33,000}$$

$$= \frac{PQ}{40.8} \tag{12-33}$$

where

P = delivered pressure increase, psi
Q = rate of flow, bbl./min.

By way of example, if a mud pump were delivering 8 bbl./min. at a gauge pressure of 950 psig, and where the intake is at atmospheric pressure, the hydraulic horsepower would be

$$\text{Hydraulic horsepower} = \frac{(950 \text{ psi})(8 \text{ bbl./min.})}{40.8}$$

$$= 186 \text{ hp}$$

If the over-all mechanical efficiency of the mud pump were 80 per cent, the horsepower supplied to the pump would be

$$\text{Input horsepower} = \frac{186 \text{ horsepower}}{0.80 \text{ mechanical efficiency}}$$

$$= 233 \text{ hp}$$

Lifting of Drill Cuttings and Cavings

The equations governing the settling of rock particles have been given by Pigott as follows. In the viscous region Stokes' law applies:

$$v = \frac{2gd^2(\rho_1 - \rho_2)}{36\mu} \tag{12-34}$$

In turbulent flow, Rittinger's formula applies

$$v = 9\sqrt{\frac{d(\rho_1 - \rho_2)}{\rho_2}} \tag{12-35}$$

where

v = slip velocity of spherical particle, ft./sec.
ρ_1 = density of particle, lb./ft.^{3}
ρ_2 = density of mud, lb./ft.^{3}
d = diameter of spherical particle, ft.
μ = mud viscosity, lb./ft.-sec.
 = centipoises $\times 0.000672$

Since most drill cuttings and cavings tend to be flat in shape, Pigott suggests that the probable settling velocity will be about 40 per cent of that calculated by the above equations. It has also been suggested that the slippage of drill cuttings or cavings cannot be governed by Stokes' law when the mud is in turbulent flow. Ordinarily no difficulty is encountered in raising drill cuttings or the usual cavings (if any) in good rotary drilling practice. Difficulty is occasionally encountered when a light rig with a small mud pump is used in remedial work. The usual remedy in such cases is to add sufficient bentonite to the mud so that the effective viscosity becomes sufficient to raise the large cavings.

Required Circulation Rates for Deep Air-Gas Drilling

The amount of gas or air which must flow upward through the annulus while drilling is proceeding is determined primarily by the requirements for lifting drill cuttings. Air drilling at shallow depths, including quarrying operations, has demonstrated that an upward air velocity of about 3,000 ft./min. at atmospheric conditions is satisfactory for lifting drill cuttings. The method used in calculating required upward flow rates consists of calculating an upward flow velocity at the bottom of the hole which will have a lifting power equal to an air velocity of 3,000 ft./sec.[10] The density of the gas at the bottom of the well is influenced by the discharge pressure at the top of the hole, the flow-friction pressure loss between the bottom and top of the hole, and the hydrostatic head of the gas bearing its load of cuttings. Accordingly, the rate of drilling itself influences the required circulation rate, since it influences the hydrostatic head of the flowing column.

For the purpose of determining the upward flow velocity at the bottom of the hole which will be equivalent to an atmospheric air-flow velocity of v_A, Rittenger's formula for slip velocity may be written for atmospheric air-flow conditions, denoted by subscript A, and for bottom-hole conditions, denoted by subscript B:

$$v_{SA} = C \sqrt{\frac{\rho_S - \rho_A}{\rho_A}}$$

$$v_{SB} = C \sqrt{\frac{\rho_S - \rho_B}{\rho_B}}$$

where v_{SA} is the cutting slip (settling) velocity in air, v_{SB} is the slip velocity in gas at bottom-hole conditions, ρ_S is the density of the cuttings, ρ_A is the density of atmospheric air, and ρ_B is the density of gas at bottom-hole conditions.

If a constant-percentage slip velocity is to be permitted, then the following relation holds:

$$\frac{v_{SA}}{v_A} = \frac{v_{SB}}{v_B}$$

or

---

[10] R. R. Angel, "Volume Requirements for Air or Gas Drilling," Paper No. 873-G, Annual Fall Meeting of the Society of Petroleum Engineers of A.I.M.E., October 6–9, 1957, Dallas, Texas.

$$\frac{v_B}{v_A} = \frac{v_{SB}}{v_{SA}}$$

where v_A is the flow velocity of air at standard (surface) conditions and v_B is the gas flow velocity of bottom-hole conditions. Introducing the preceding equations for slip velocities gives

$$\frac{v_B}{v_A} = \sqrt{\frac{\rho_A(\rho_S - \rho_B)}{\rho_B(\rho_S - \rho_A)}}$$

Unless a high back-pressure would be held on the top of the well, the density of the gas, as well as the density of air, is negligible, compared to the density of the drill cuttings, and the relation reduces to

$$\frac{v_B}{v_A} = \sqrt{\frac{\rho_A}{\rho_B}}$$

or

$$v_B^2 \rho_B = v_A^2 \rho_A$$

A more useful relation is obtained by making the following substitutions:

$$v_B = \frac{Q_A P_A T_B Z_B}{A_B P_B T_A}$$

$$\rho_B = \frac{144 P_B M_G}{Z_B R T_B}$$

$$\rho_A = \frac{144 P_A M_A}{R T_A}$$

Using the preceding substitutions and simplifying gives

$$Q_A^2 = v_A^2 A_B^2 \frac{M_A}{M_G} \frac{P_B T_A}{P_A T_B Z_B}$$

or

$$P_B^2 = \left(\frac{Q_A}{v_A A_B}\right)^4 \left(\frac{s P_A T_B Z_B}{T_A}\right)^2 \qquad (12\text{--}36)$$

where Q_A is the gas flow in ft.^{3}/min. measured at P_A and T_A, v_A is the equivalent air velocity referred to conditions P_A and T_A in ft./min. (for example, 3,000 ft./min.), A_B is the cross-sectional flow area at the bottom of the hole in ft.^{2}, P_B is the bottom-hole pressure in psia, T_B is the bottom-hole temperature in °R and Z_B is the compressibility factor of the gas at bottom-hole conditions, and s is the specific gravity of the gas.

The above equations give the required quantity of gas, as measured at standard conditions P_A and T_A which must be circulated to lift the drill cuttings, assuming that constant per cent slippage of the cutting is permissible. In order to solve the equation, the bottom-hole pressure term must be eliminated, and Equation (12–36) is aranged for subsequent substitution into the vertical gas-flow equation.

The vertical gas-flow equation is derived from Bernoulli's Theorem written in differential form:

$$\frac{dp}{\rho} + \frac{UdU}{g} + dF + dh = 0 \qquad (12\text{--}37)$$

where p is pressure in lb./ft.^{2}, U is velocity in ft./sec., h is height in feet, and F is the flow-friction loss in feet. The equation is developed by substituting values for density and velocity which are derived from the gas laws. The substitution for flow friction loss is derived from Fanning's equation. The height is expressed in terms of an angle measured from the horizontal, so that slant as well as vertical holes may be considered. The substitutions used are

$$\rho = \frac{pM}{ZRT}(1+w)$$

where

$$w = \frac{\text{weight of rock fragments}}{\text{weight of flowing gas}}$$

$$U = \frac{qTp_0Z}{AT_0p}$$

where

$q = $ ft.^{3}/sec. gas measured at p_0 and T_0
$A = $ cross-sectional flow area, ft.^{2}

$$dU = -\frac{qTp_0Z}{AT_0p^2}dp$$

$$dF = \frac{2fU^2}{gd}dL$$

$$= \frac{2f}{gd}\left(\frac{qTp_0Z}{AT_0p}\right)^2 dL$$

$$dh = \sin\theta dL$$

Following the substitutions, the equation is arranged algebraically as follows for integration:

$$\int_{p_T}^{p_B} \frac{p\,dp}{\dfrac{2fM(1+w)TZ}{gdR}\left(\dfrac{qp_0}{AT_0}\right)^2 + \dfrac{M(1+w)\sin\theta}{ZRT}}$$

$$- \int_{p_T}^{p_B} \frac{dp}{\left[\dfrac{2f}{d} + \left(\dfrac{AT_0}{qTp_0Z}\right)^2 gp^2 \sin\theta\right]p} = \int_0^L dL$$

This is of the form

$$\int_{p_T}^{p_B} \frac{p\,dp}{a'' + bp^2} - \int_{p_T}^{p_B} \frac{dp}{(c + kp^2)p} = \int_0^L dL$$

The integral is

$$\frac{1}{2b} \log\left[\frac{bp_B^2 + a''}{bp_T^2 + a''}\right] - \frac{1}{2c} \log \frac{p_B^2(c + kp_T^2)}{p_T^2(c + kp_B^2)} = L$$

This may be arranged algebraically to

$$\frac{bp_B^2 + a''}{bp_T^2 + a''} = e^{2bL} \left[\frac{p_B^2(c + kp_T^2)}{p_T^2(c + kp_B^2)}\right]^{b/c}$$

The bracket on the right-hand side has a value of unity when $q=0$, and at Reynolds number of 5×10^7 its value is about 1.002. It may therefore be dropped from the equation to give

$$p_B^2 = e^{2bL} p_T^2 + (e^{2bL} - 1)\frac{a''}{b}$$

Letting $2bL = m$ gives

$$p_B^2 = e^m p_T^2 + (e^m - 1)\frac{a''}{b} \qquad (12\text{-}38)$$

It may be noted that all terms are in units of (lb./ft.)². If the pressure in the a'' term is changed to psi, the equation may be written as

$$P_B^2 = e^m P_T^2 + (e^m - 1)\frac{a}{b} \qquad (12\text{-}39)$$

Equation (12-39) above is for upward flow with q considered positive in the upward direction. If q is taken as a positive quantity in downward flow, the equation may be written, for downward flow, as

$$P_B^2 = e^m P_T^2 - (e^m - 1)\frac{a}{b} \qquad (12\text{--}40)$$

where

P_B = bottom-hole pressure, psi
P_T = top-hole pressure, psi
e = 2.718
$m = 2bL$

$$= \frac{(2)(29.0)\,sh\,(1+w)}{1{,}544ZT}$$

$$= \frac{sh\,(1+w)}{26.62ZT} \qquad (12\text{--}41)$$

s = specific gravity of gas, air = 1

$$\frac{a}{b} = \frac{2fL}{gdh}\left(\frac{qP_0ZT}{AT_0}\right)^2$$

If the diameter is expressed in inches and the rate of flow is expressed in standard-condition cubic feet per minute, the following results:

$$\frac{a}{b} = \frac{2fL}{300gDh}\left(\frac{QP_0ZT}{AT_0}\right)^2$$

where

f = Fanning friction factor, dimensionless
L = length of well, ft.
g = acceleration of gravity, 32.174 ft./sec.^{2}
D = diameter, or equivalent diameter, in.
h = vertical depth of well, ft.
Q = gas flow referred to standard conditions, ft.^{3}/min.
P_0 = standard pressure for gas volume measurement, psia
T_0 = standard temperature for gas volume measurement, °R
Z = average gas compressibility factor, for average flow conditions
T = average gas flow temperature, °R
A = cross-sectional flow area, ft.^{2}

It is mathemically possible to treat the temperature gradient in the well as a linear function of depth. However, it has been shown by

Angel that the difference in the calculated bottom-hole pressure is small for air-gas drilling conditions.

The value of w may be readily evaluated in terms of the drilling rate, if the slip velocity of the cuttings is assumed to be negligible with respect to the upward gas-flow velocity. In the treatment, the volume of the drill cuttings is assumed to be very small compared to the volume of the flowing gas.

$$w = \frac{\text{weight of rock}}{\text{weight of gas}}$$

$$= \frac{(\text{ft.}^3 \text{ rock drilled/min.})(\text{density of rock, lb./ft.}^3)}{(\text{ft.}^3 \text{ gas flowing/min.})(\text{density of gas, lb./ft.}^3)}$$

$$= \frac{(0.7854)D_H{}^2(62.4)S_R(1{,}544)(540)}{(144)t(14.7)(144)(29.0)sQ_A}$$

$$= 4.6224 \frac{S_R D_H{}^2}{stQ_A}$$

If S_R, the specific gravity of the rock, is taken as 2.70, then,

$$w = \frac{12.5 D_H{}^2}{stQ_A} \qquad (12\text{--}42)$$

where

$D_H =$ = drilled-hole diameter, inches
$t =$ drilling rate expressed as minutes per foot

If it were established that the average upward velocity of the drill cuttings is three-fourths of the gas velocity and the rock was drilling at 4 minutes per foot, a value of $t = 3$ min./ft. should be used for calculating w and m.

The equation for upward flow of gas may now be written as

$$P_B{}^2 = e^m P_T{}^2 + (e^m - 1)\frac{2fL}{300gDh}\left(\frac{TZP_A Q_A}{AT_A}\right)^2 \qquad (12\text{--}43)$$

The symbols P_A and T_A are used to designate the standard pressure and temperature for measuring Q_A, in order to correspond with Equation (12–36). The value of $P_B{}^2$ as determined by lifting requirements and given in Equation (12–36) may be substituted in Equation (12–43) and arranged in the quadratic form as

$$\left(\frac{sP_A Z_B T_B}{T_A}\right)^2 \frac{Q_A{}^4}{(V_A A)^4} - \frac{(e^m - 1)2fL}{300gDh}\left(\frac{TZP_0}{AT_0}\right)^2 Q_A{}^2 - e^m P_T{}^2 = 0$$

By the quadratic solution,

$$Q_A{}^2 = \frac{-b' \pm \sqrt{(b')^2 - 4a'c'}}{2a'}$$

$$Q_A = \sqrt{-\frac{b'}{2a'}}\left[1 \pm \sqrt{1 - \frac{4a'c'}{(b')^2}}\right]^{1/2}$$

Choosing the positive solution, the above equation may be written as

$$Q_A = x[1 + \sqrt{1 + y}]^{1/2} \tag{12-44}$$

where

$$x = \sqrt{-\frac{b'}{ab'}}$$

$$= \sqrt{\frac{(e^m - 1)fL}{300gDh} \frac{ZTAV_A{}^2}{sZ_BT_B}} \tag{12-45}$$

$$y = -\frac{4a'c'}{(b')^2}$$

$$y = \frac{e^m}{(e^m - 1)^2}\left(\frac{300\sigma Dhs}{fLV_A{}^2}\right)^2\left(\frac{Z_B{}^2}{Z^4}\right)\left(\frac{T_A{}^2T_B{}^2}{T^4}\right)\left(\frac{P_T}{P_A}\right)^2 \tag{12-46}$$

Equation (12–44) can be solved for Q, the required amount of air or gas needed for lifting the drill cuttings. The values of x and y are determined from Equations (12–45) and (12–46). However, a trial-by-error solution must be used because the value of m, obtained from Equations (12–41) and (12–42), contains Q_A. It is therefore necessary to assume successive values of Q in the m term until the resulting solution agrees with the assumed value. Such calculations are performed very rapidly on electronic computers.

Some results obtained from Equation (12–44) are shown in Figs. 12–10 and 12–11. In the calculations, a bottom-hole temperature gradient of 1 degree F. per 100 feet was assumed, and the average flowing temperature was assumed to be midway between the bottom-hole temperature and the surface temperature of 80 degrees F. The gas compressibility factors were taken from the Natural Gasoline Supply Men's Association Technical Manual. For the rough wall of the annulus, a value of 0.012 was assumed for the Fanning friction factor. The discharge pressure at the top of the hole was assumed to

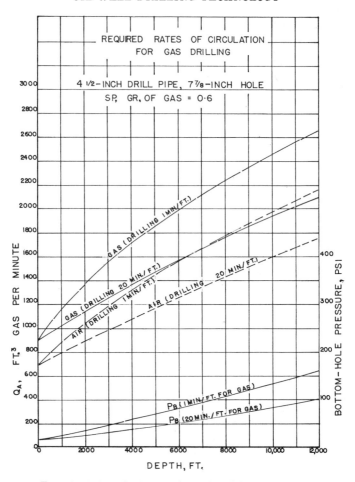

Fig. 12-10. Required rates of circulation for gas drilling.

be 14.7 psia. The gas-volume measurement is referred to a pressure base of 14.7 psia and a temperature base of 80 degrees F. It may be observed that the required rate of air or gas circulation depends both upon the depth and the rate of drilling.

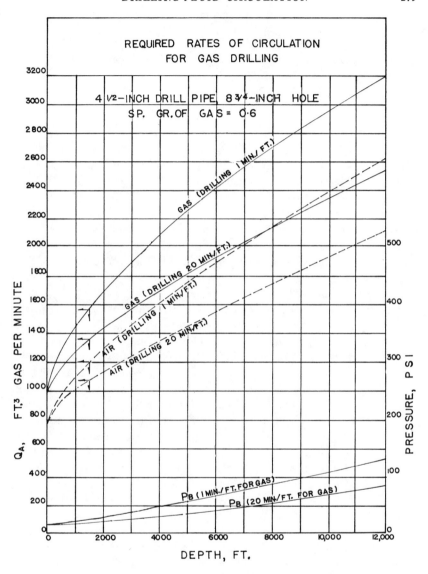

FIG. 12–11. Required rates of circulation for gas drilling.

CHAPTER 13

Drilling Practices

Rotary Drilling

Choice of Bit

The size of the hole to be drilled is generally specified in the casing-bit-size program in the master plan of the well or in the contract, as discussed in Chapter 8. The diameter of the bit must not exceed the drift diameter of the casing through which it must pass, particularly in the case of such bits as the diamond bit, which has a full circumference and is absolutely rigid. Considerations of pressure surges discussed in Chapter 18 make it desirable that the bit have more than minimum clearance and that sufficient mud-flow area be provided in the annulus outside of the drill collars, drill pipe, and tool joints that rapid pipe movements will not cause mud losses or swabbing-in of producing formations.

The type of bit selected for drilling a certain depth interval in the hole should be governed by the rock to be drilled. The design principles and operating characteristics of the rolling cutter bits used in soft formations and those used in hard formations have been discussed in Chapter 10. Formations are generally classified as "soft" or "hard" according to the compressive or crushing strength of the rock and vary from soft shales and sandstones to dense limestones, quartzites, and granites. It has been pointed out that the life of a bit is normally terminated by one or more of the following circumstances: (1) the teeth become worn to the extent they will not effectively chip or scrape; (2) the gauge (peripheral) surfaces become so worn that further drilling would result in an undersized hole; (3) bearing races and parts become so worn, chiefly as a result of abrasive substances in the mud that bearing parts or cones might be lost in the hole; and (4) bearing races and pins (rollers) spall and break under bottom-hole operating conditions so that the cutter wheel is locked and worn flat.[1] This source also stresses the importance of metallurgy and the quality in the steel used in the bit, as well as the proper use of hard-surfacing techniques and materials. It is pointed out that case-hardening of the rolling member after milling is desirable to produce hard teeth and bearing surfaces. Hard-surfacing materials are preferably welded to the gauge surfaces only of hard-formation bits since the drilling

[1] R. A. Bobo, R. S. Hock, and G. S. Ormsby, "Keys to Successful Competitive Drilling," *World Oil*, Vol. CXXXIX, No. 6 (November, 1954), 172–76.

weights used would result in such material's spalling off of the bit teeth. The sharpness of the tooth angle on such bits is of minor importance provided adequate drilling weight is carried on the bit. On the other hand, hard-surfacing materials are used on one or more surfaces of the teeth of soft-formation bits so that they will wear sharp during drilling operations. Bits with short, closely spaced teeth permit the manufacturers to build the bearings bigger and stronger so that more drilling weight can be carried on the bits. Conversely, longer and more widely spaced teeth drill faster and more effectively in softer formations, and intermediate designs are best adapted for drilling formations which are intermediate between the very soft and the very hard.

The experience gained in drilling any particular rock formation within a certain area permits the selection of the bit type best suited for drilling the particular interval. More specifically, the most suitable bit can be selected from among those used under the drilling conditions practiced at the time trial comparison were made, usually in the early life of a particular field or area. Table 13–1 was prepared by H. G. Bentson to aid in choosing proper bits.

Drilling Weights and Rotary Speeds

The weight which is applied to the bit, the speed at which the bit is rotated, and the weight, size, and strength of the drill collar are closely interrelated in rotary drilling. The drill collar, the lower portion of the drill string, should be considered as the tool which holds the bit and guides and forces it against the rock formations. As such, the weight, stiffness, mass distribution, and vibration characteristics of the drill collar influence the action of the bit and the degree of stability with which it is held on the bottom. The importance of the drill collar increases with the amount of drilling weight applied to the bit. Weight on the bit and rotary speed are sometimes treated as separate variables, perhaps without conscious thought of the drill string and drill collars. Introspection will reveal, however, that when weight on the bit and rotary speed are being considered, other specific operating factors are simultaneously assumed as they apply to a certain area and depth interval.

In drilling the harder rocks, a uniform weight may not be held on the bit throughout its run. When a new bit is first rotated, the driller may apply a comparatively light weight in order that the full-length teeth may wear their own drilling pattern into the bottom of the hole. The pattern of ridges and valleys formed by the bit on the bottom of the hole varies among different types of bits and the amount of wear

TABLE 13-1
Bit-selection Table*

Condition of dull bit	Possible causes	Possible remedies
Excessive bearing wear	Excessive rotary speed Excessive rotating time Excessive weight on bit Excessive sand in circulating fluid Unstabilized drill collars Improper type of bit	Slower rotary speed Reduced rotating hours Lighter weight on bit Removal of sand from circulating fluid Stabilize drill collars Use harder formation bit type having larger bearing structure
Excessive broken teeth	Improper type of bit Improper "break-in" procedure used for new bit Excessive weight on bit for type used	Use harder formation bit type having greater number of teeth Proper "break-in" procedure for new bit Lighter weight on bit
Unbalanced tooth wear	Improper type of bit Improper "break-in" procedure used for new bit, resulting in broken teeth, which appear to be worn when bit is pulled	Use different type of bit based upon the rows of teeth which are excessively worn on the dull bit Proper "break-in" procedure for new bit
Excessive tooth wear	Excessive rotary speed Improper type of bit Use of non-hard-faced type	Slower rotary speed Use harder formation bit having greater number of teeth Use type of bit having hard-faced teeth
Excessive cupping of tooth crests	Double hard-faced teeth Insufficient weight for adequate penetration Improper formation for double hard-faced teeth	Use bit with single face hard-facing or single face and "tipped" hard-facing Heavier weight on bit
Bradding of teeth	Excessive bit weight on dull bit Formation too hard for type of bit used	Replace bits after less rotating time Use harder formation bit having greater number of teeth
Fluid cut teeth and cone	Excessive circulation rate of fluid Excessive sand in circulating fluid	Reduction in circulation fluid rate Removal of sand from circulating fluid Use jet circulation bit
Excessively undergage	Improper bit type Excessive rotating time	Use type of bit having greater gage protection Reduced rotating hours
Skidded due to balling	Excessive weight on bit Improper type of bit Insufficient fluid circulation rate	Lighter weight on bit Use of softer formation bit type having teeth more widely spaced Increased fluid circulation rate

* From H. G. Bentson, "Rock Bit Design, Selection and Evaluation," *Oil and Gas Journal*, Vol. LV, No. 15 (April, 1957), 143.

of the particular bit. Such patterns of ridges and valleys in effect rotate as drilling progresses into the rock. A break-in period during which lighter weight is carried on the bit permits the bit teeth and the bottom of the hole to adjust to each other and may prevent breakage of the teeth and yield a greater footage drilled per bit. As the drilling progresses, the weight may be gradually increased to the desired amount. For example, under such practice, about 70 per cent of the footage drilled will be under conditions of fairly constant bit weight. Near the end of the run, as the bit approaches a worn-out condition, the weight may be increased in order to maintain the desired drilling rate.

The rotation between the weight on the bit and the rate of penetration for both drag-type and rolling cutter rock bits have been studied both in the laboratory and in the field. The relationship appears to be linear. In some tests, the relationship appears to be linear after some threshold weight, or force, has been exceeded. This would seem to imply that some initial force is required in such cases in order to overcome the initial crushing resistance of the particular rock, and further applications of force on the bit result in proportional increases in the rate of rock failure. Since rock formations differ in their susceptibility to drilling, or the rate at which bits will penetrate through the rock differs, the results of such drilling-rate tests are often reported on a percentage basis. Such data permit an estimation of the percentage of increase in drilling rate which might be expected from any given increase in the weight on the bit. During tests run for the purpose of determining the effect of the weight on the bit on drilling rate, it is necessary to hold constant all other factors, such as rotary speed and hydraulic factors, in order to isolate the one variable. Of all the variables which affect the rate of penetration of the bit, the influence of the weight force applied to the bit has been most satisfactorily isolated and defined.

Figures 13–1 and 13–2 illustrate the influence of bit weight on the rate of penetration of drag-type bits. The upper limit of bit weight appears to depend upon the rate at which cuttings are removed by mud-fluid circulation and the weight at which the bit is forced directly into the formation so that the rotary speed and torque capacities are exceeded.

Figures 13–3 to 13–5 illustrate the effect of bit weight on the penetration rate of toothed-wheel and carbide-studded bit. These graphs were plotted from data taken during an investigation sponsored by the American Association of Oil Well Drilling Contractors in the Dora Roberts Field near Odessa, Texas, in the depth interval be-

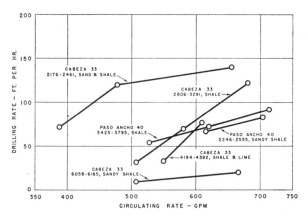

Fig. 13–1. Combined effect of circulation rate, nozzle-fluid velocity, and bit weight at constant speed of rotation on rate of drilling with two-way drag-type bits (mud properties constant, uniform shale—South Texas area).

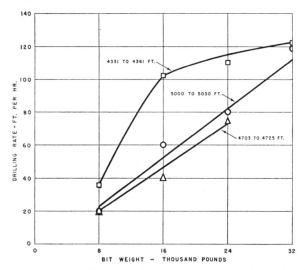

Fig. 13–2. Variation of drilling rate of two-way drag-type bits with bit weight at constant rate of circulation and speed of rotation (uniform shale; rate of circulation, 820 gal./min.; nozzle-fluid velocity, 107 ft./sec.—South Texas area).

tween 10,190 feet and 13,000 feet.[2] The results of the investigation were reported at the seventeenth annual meeting of the AAODC in Tulsa, October 13–15, 1957, by H. B. Woods and E. M. Galle (Hughes Tool Company, Houston, Texas). They say that in analyzing the data it became readily apparent that formation changes occurred many

[2] H. B. Woods and E. M. Galle, "Effects of Weight on Penetration," *The Drilling Contractor*, Vol. XIII, No. 6 (October, 1957), 74–79.

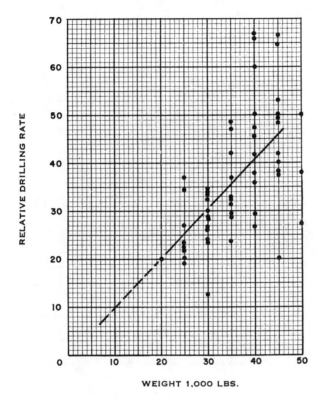

FIG. 13-3. Effect of weight on penetration rate.

times during individual tests and that formation "drillability" changed greatly from one test to another. These changes in formation drillability prevented any direct averaging of all test results and made it expedient to construct combined plots of the relative drilling rates obtained under different weights during individual tests. Figure 13-3 illustrates the typical spread of individual data points. Figure 13-4 is an averaged composite curve for all tests made with toothed-wheel rock bits during the investigation. Figure 13-5 is a similar plot of data obtained with carbide-studded button bits. The most important relationship is the slope of the curves in the operating range. The data indicate that in the range investigated a 50 per cent weight increase will produce approximately 65 per cent increase in drilling rate.

AAODC PENETRATION RATE TESTS ROY PARKS B-23 AND B-25
TOOTHED BITS

TESTS: (B-23) 1, 3-8, 43-46

(B-25) 1-7, 9, 13-15, 17, 18, 22-24, 26-28, 32, 45, 46, 49, 51-53, 55

COMBINED DATA

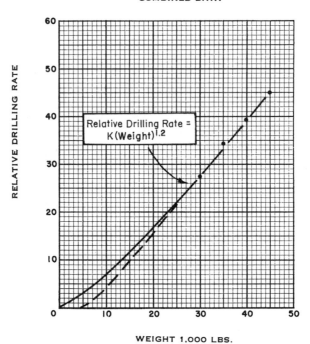

FIG. 13-4. Effect of weight on penetration rate.

The amount of weight used on the bits drilling into the harder formations has been continuously increased, particularly since about 1940. Drilling practices in the Permian Basin were summarized in 1955 by R. J. Bromell,[3] who reported that bit weights were considerably increased during the previous two-year period. Prior to about 1955, it was common practice to run 30,000 to 40,000 pounds on $12\frac{1}{4}$-inch bits (2,500 to 3,300 lb./in. diameter) and 30,000 to 45,000 pounds on $8\frac{3}{4}$-inch bits (3,400 to 5,100 lb./in.). By 1955, it was common practice to run 70,000 to 110,000 pounds on $12\frac{1}{4}$-inch bits (5,700 to 9,000 lb./in.) and 50,000 to 80,000 pounds on $8\frac{3}{4}$-inch bits (5,700 to

[3] R. J. Bromell, "Drilling Practices in the Permian Basin," *API Drilling and Production Practice* (1955), 93–102.

9,200 lb./in.). The drilling weight used with $7\frac{7}{8}$-inch bits had increased, and commonly 35,000 to 50,000 pounds (4,500 to 6,400 lb./in. were run on these bits, and as much as 70,000 pounds (8,900 lb./in.) was occasionally used. Similarly, the weight run on $6\frac{3}{4}$-inch bits had increased to 30,000 to 35,00 pounds (4,500 to 5,200 lb./in.), with as much as 50,000 pounds (7,400 lb./in.) occasionally used. This source stated that increased weights on the $12\frac{1}{4}$-inch bits were associated with jet bits, and without their aid much of the benefit derived from using increased weights in the softer formations would have been impossible. Some operators also reported faster penetration rates and greater footage drilled per bit with larger drill collars, such as $7\frac{3}{4}$-inch drill colars in an $8\frac{3}{4}$-inch hole, than in offset (adjacent) wells using smaller drill collars, even when the indicated weight on the bit was the same for both operations. The faster penetration and greater foot-

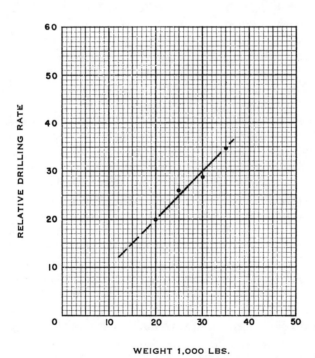

FIG. 13–5. Effect of weight on penetration rate.

age were attributed to the fact that there was less buckling of the drill collars, so that wall friction was lessened and more of the indicated drilling weight actually applied to the bit, and further, to the fact that the bit was held more squarely on the bottom by the decrease in buckling. This emphasizes again that it is important to consider the tool for holding or applying the bit as well as the bit weight and rotating speed.

The effect of rotary speed on the rate of penetration of toothed-wheel bits is much more difficult to evaluate than the effect of bit weight. Apparent inconsistencies sometimes appear in the data. The rotation of the drill string and bit, when the bit is drilling on bottom, is associated with vibrations which originate at the bit. In many instances, the action of the bit on the bottom of the hole can be interpreted from the feel of the vibrations in the derrick floor and in the area immediately adjacent to the drilling rig. Such vibrations probably influence the action of the bit on the rock and may be instrumental in rock failure and so aid the drilling process. Unfortunately, excessive vibration causes materials to fail—bit teeth and bearings for example—and can cause fatigue failures in the drill collars and drill pipe. In order to avoid such unfavorable results, drilling practice in the United States has tended toward slower rotary speeds with greater bit weights to obtain faster rates of penetration.

It would appear that a first assumption might be made that drilling rate is proportional to rotary speed, since drilling occurs by virtue of the contracts between the bit teeth and the rock formation, and that the number of such contacts should be porportional to the rotary speed. For this assumption to be true, however, the individual contacts between the bit teeth and the formation would have to be equally effective in disintegrating rock at both high and low rotational speeds. Such linear assumptions are not substantiated either by laboratory investigations or by controlled field drilling tests. All tests seem to indicate that the rate of penetration is less than linear with the rotary speed. Figure 13–6 illustrates laboratory data taken with small rolling cutter (micro) bits.[4] Figures 13–7 through 13–10 represent results obtained in the investigation conducted by the Weight-Speed-Penetration Study Committee of the AAODC and reported at the annual meeting in October, 1957.[5] Figures 13–7 and 13–8 illustrate the effects of local formation changes and general formation changes which had to be mathematically eliminated in analyzing the data in

[4] W. R. Wardroup and G. E. Cannon, "Some Factors Contributing to Increased Drilling Rates," *API Drilling and Production Practice* (1956), 274–82.

[5] J. E. Eckel and D. S. Rowley, "Effects of Speed on Penetration Rates," *The Drilling Contractor*, Vol. XIV, No. 1 (December, 1957), 50–53.

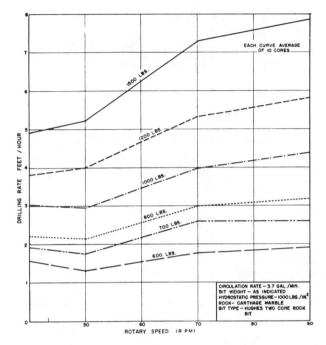

Fig. 13–6. Drilling rate *vs.* rotary speed (laboratory data).

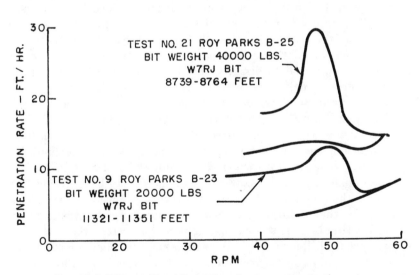

Fig. 13–7. Effects of local formation changes on penetration rate.

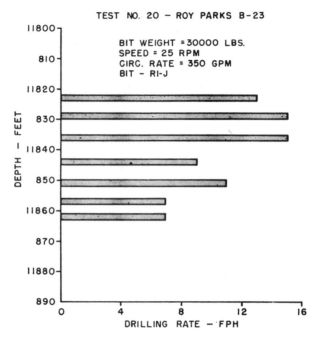

Fig. 13–8. Variations in drilling rate for constant weight, constant speed, and constant circulation rate.

order to obtain relationships involving only the factors of weight and rotary speed. Figure 13–9 shows the relation obtained between relative drilling rate and rotary speed for carbide-studded bits. Figure 13–10 shows the relation obtained between relative drilling rate and rotary speed for toothed-wheel bits. On this last figure, the open circles indicate the slight disagreement between this analysis and the above-described analysis of the effects of weight on drilling rate.

On the number of occasions it has been observed that slowing the rotary speed, from 80 rpm to 60 rpm for example, has resulted in an increase in the drilling rate. Thorough investigations of the relationship of drilling rate and rotary speed reported above do not indicate any such anomalies. Two explanations might be offered to explain these anomalies: first, the slower rate of speed may have altered the nature of the contact of the drill string with the wall of the hole in such a way that more of the apparent drilling weight was actually applied to the bit; or, second, ir might be postulated that, at some rotary speeds, drill-string vibrations are so effectively dampened that lower penetration rates are obtained than would be obtained at speeds either above or below the particular rotary speed. The oppo-

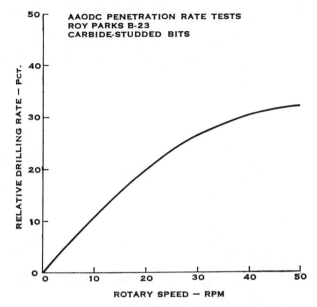

Fig. 13–9. Effect of rotary speed on penetration rate.

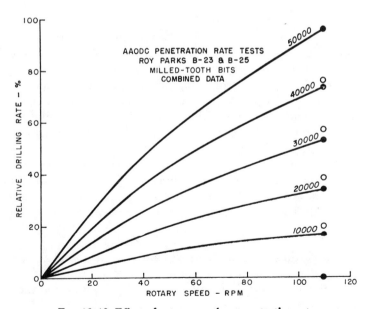

Fig. 13–10. Effect of rotary speed on penetration rate.

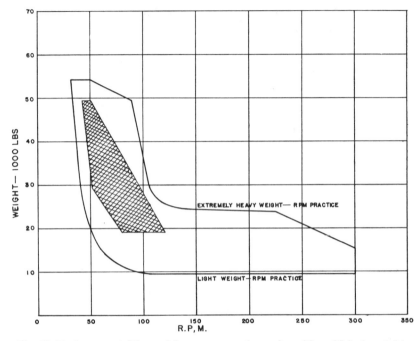

Fig. 13–11. Average drilling weight–rotary speed practice—7⅝- to 7⅞-inch rock bits.

site condition has been recognized in that natural vibration frequencies exist for each individual drill string and that rotary speeds in resonance with this frequency may result in severe vibrations which lead to drill-string failures. Such vibrations tend to be dampened by the contact between the drill string and the wall of the hole.

The severity of service imposed by operating conditions on the bit and the drill string increases with both the total weight applied on the bit and the rotary speed. Consequently, as the drilling weight on the bit is increased, it is common practice to decrease the rotary speed. Figures 13–11 through 13–14 illustrate the limits of common practice for bit weight–rotary speed combinations. These figures represent averages compiled from field observations and are reproduced from *The Tool Pusher's Manual.*[6]

Hydraulic Effects

The composition and characteristics of the drilling fluid and its bottom-hole pressure, rate of circulation, and velocity influence the rate of penetration of the bit. In many cases the drilling fluid and the man-

[6] Page A-4, 1 and 2. Published by the American Association of Oilwell Drilling Contractors, Dallas.

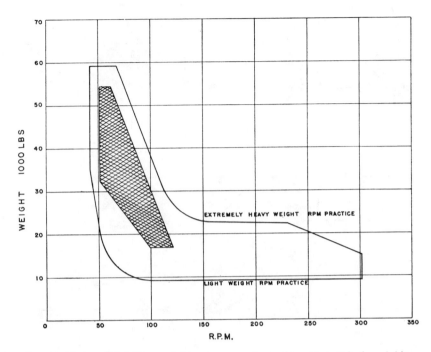

Fig. 13–12. Average drilling weight–rotary speed practice—$8\frac{1}{2}$- to 9-inch rock bits.

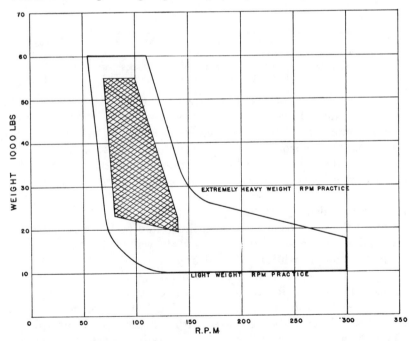

Fig. 13–13. Average drilling weight–rotary speed practice—$9\frac{5}{8}$- to $9\frac{7}{8}$-inch rock bits.

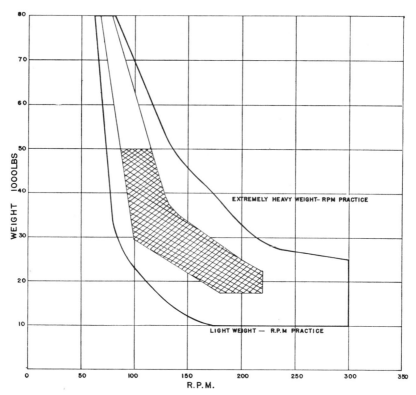

Fig. 13-14. Average drilling weight–rotary speed practice—
$10\frac{5}{8}$-, 11-, and $12\frac{1}{4}$-inch rock bits.

ner in which it is circulated are the controlling or limiting factors that govern the rate of drilling. The dominant specific factors connected with the mud which exert the greatest influence on drilling rate unquestionably vary from one region to another because of differences in formations drilled and in drilling practices. However, in all cases the circulatory system is an integral part of the rotary-drilling process, and as such it must be used to the best advantage in order to obtain the optimum drilling rate and progress.

One of the functions which the drilling fluid performs is to provide lubrication between the drill string and the wall of the hole. The mud itself may act as a lubricant, and any filter cake deposited by the mud may also serve as a lubricant. Generally, torque requirements for rotating the drill string present no particular problem, although in some cases the torque becomes excessive. One cause of such high torque is believed to be associated with curvature in the drilled hole where the weight of the drill pipe bears heavily against one side of the

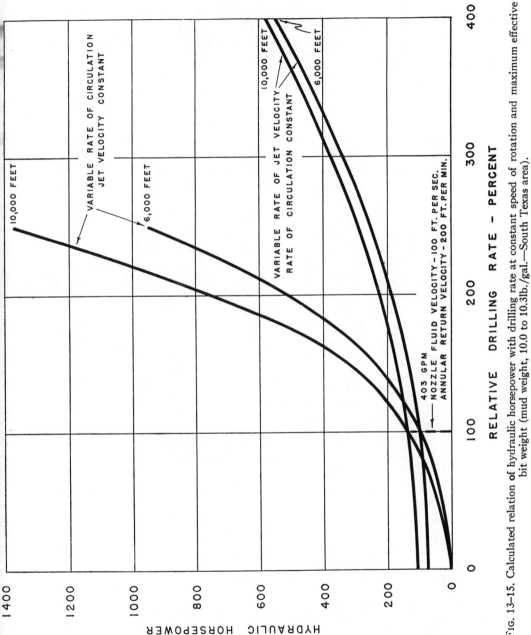

Fig. 13-15. Calculated relation of hydraulic horsepower with drilling rate at constant speed of rotation and maximum effective bit weight (mud weight, 10.0 to 10.3 lb./gal.—South Texas area).

hole. Reduction of drill-pipe torque may permit better control of the drilling weight applied to the bit. Both the use of oil-emulsion muds and the addition of powdered graphite to the mud have resulted in very marked reductions in drill-string torque.

The wetting characteristics of the drilling mud influence drilling operations in some instances. When oil is used in the mud, the oil generally wets the steel drill pipe, tool joints, drill collars, and bit teeth. When the steel is wet with oil, shale drill cuttings do not adhere to the metal. Balling around the top of the drill collars and tool joints, in which masses of shale drill cuttings or cavings become packed together and quite firmly attached to the metal, is minimized by using oil-base or oil-emulsion muds. They allow greater freedom of movement of the drill pipe and permit free circulation of the mud. Round trips are more easily made, and mud losses into formations are minimized. The wetting characteristics of oil may be partly responsible for increased rates of penetration in drilling shales with oil-emulsion muds. The presence of dissolved salts and organic compounds has been shown to affect the strength of the surface layer of mineral crystals in contact with the solution. Ordinary muds, of course, contain a number of such substances in solution.

In drilling with drag-type bits, the hydraulic effects on the rate of penetration appear to be associated chiefly with the total kinetic energy, or hydraulic horsepower, delivered through the bit jets to the bottom of the hole. It has been demonstrated that both the volume rate of circulation and the jet velocity affect the rate of penetration. In the softer formations, some material is unquestionably removed by the jetting action of the drilling mud. However, a more important factor appears to be that more weight can be carried on the bit when more hydraulic energy is present to remove loosened material. Typical relationships obtained from experiments with fish-tail bits in the Gulf Coast area are illustrated in Fig. 13–15.[7] An analysis of the data has been made to aid in selecting optimum-size bit nozzles, and Figs. 13–16 and 13–17 illustrate typical results obtained in the analysis.[8] Also an analysis has been made of the effects of nozzle design.[9]

With rocks bits of the rolling cutter type, the rate of penetration in

[7] J. P. Nolley, G. E. Cannon, and D. Ragland, "The Relation of Nozzle Fluid Velocity to Rate of Penetration with Drag-type Rotary Bits," *API Drilling and Production Practice* (1948), 22–42.

[8] J. R. Eckel and J. P. Nolley, "An Analysis of Hydraulic Factors Affecting the Rate of Penetration of Drag-type Rotary Bits," *API Drilling and Production Practice* (1949), 9–37.

[9] J. R. Eckel and W. J. Bielstein, "Nozzle Design and Its Effect on Drilling Rate and Pump Operation," *API Drilling and Production Practice* (1951), 28–46.

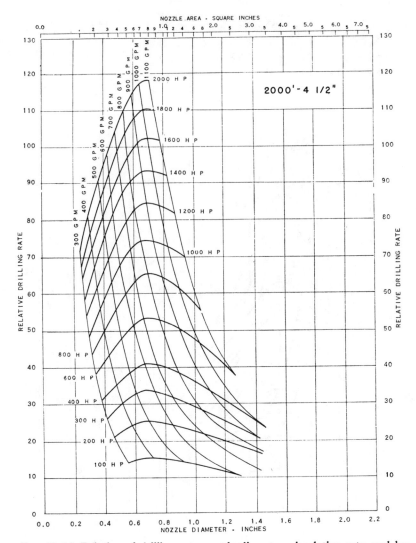

Fig. 13–16. Relation of drilling rate, nozzle diameter, circulating rate, and hydraulic horsepower at 2,000-ft. depth in two-way drag-type bits (4½-in. O.D. 16.60-lb. I.F. drill pipe; 10.0 to 10.3 lb./gal. mud).

drilling the softer formations is strongly affected by the hydraulic horsepower developed in the bit nozzles. Figures 13–18 and 13–19 illustrate relationships which have been obtained in drilling soft formations.[10] Additional data in Fig. 13–20 show results obtained from ap-

[10] W. J. Bielstein and G. E. Cannon, "Factors Affecting the Rate of Penetration of Rock Bits," *API Drilling and Production Practice* (1950), 61–78.

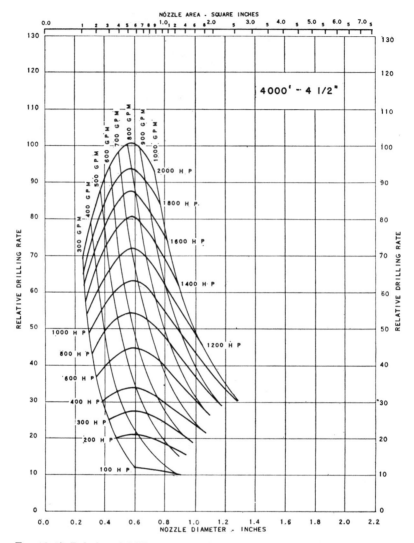

FIG. 13–17. Relation of drilling rate, nozzle diameter, circulating rate, and hydraulic horsepower at 4,000-ft. depth in two-way drag-type bits (4½ O.D. 16.60-lb. I.F. drill pipe; 10.0 to 10.3 lb./gal. mud).

plication of increased hydraulic horsepower at the bit.[11] The last figure includes the effects of carrying additional weight on the bit made possible by increased ability to clean the bit and bottom of the

[11] E. C. Hellums, "The Effect of Pump Horsepower on the Rate of Penetration," *API Drilling and Production Practice* (1952), 83–93.

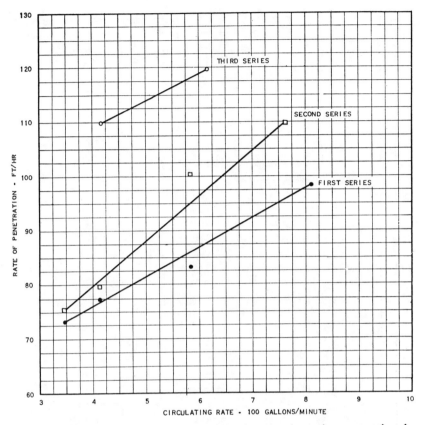

FIG. 13–18. Relation of circulating rate to rate of penetration, using two-cutting-element jet rock bits at constant rotary speed, bit weight, and nozzle-fluid velocity.

hole as a result of additional hydraulic horsepower. The advantages of using additional hydraulic horsepower at the bit are generally restricted to jet-type bits which jet the mud directly on the bottom of the hole. Although increased hydraulic horsepower apparently gives some advantage in drilling the harder formations, the advantage appears to become less as the rock drilled becomes harder and denser.

The composition of the drilling fluid exerts a controlling influence on the rate of penetration in drilling with toothed-wheel bits. Drilling is fastest with air or gas, with water and oil ranking next and muds following in varying degrees. Comparative drilling data on the 5,000-

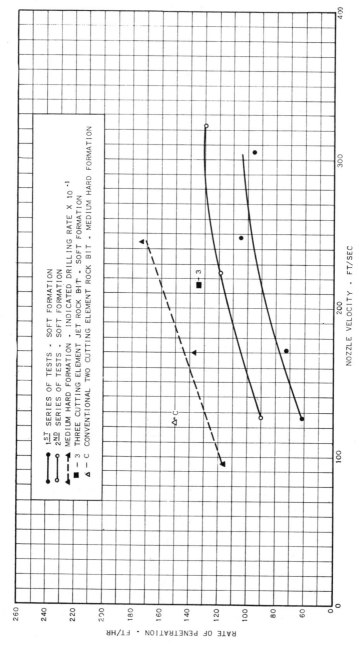

Fig. 13-19. Relation of nozzle-fluid velocity to rate of penetration, using two-cutting-element jet rock bits at constant rotary speed, bit weight, and circulation rate.

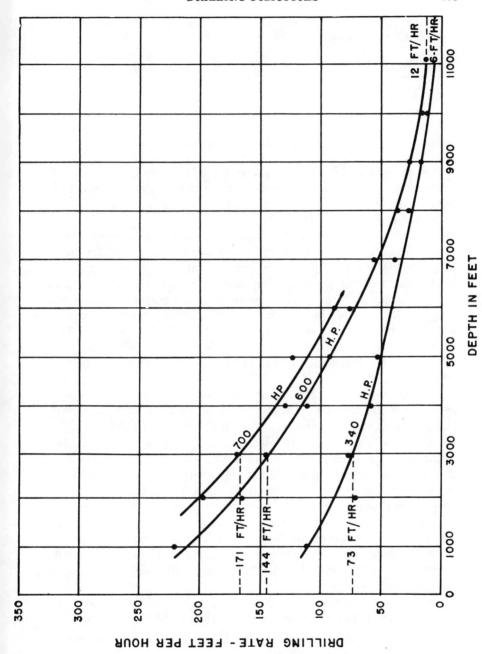

Fig. 13–20. Hydraulic horsepower study of wells drilled in South Louisiana.

to 10,000-foot interval (hard rock section) in the Permian Basin is contained in the following table.[12]

Well	A	B	C	D
Drilling fluid	Air-gas	Water-oil emulsion	Water	Gel mud*
Trip and rotating hours	250	520	650	940
Bits used	12	39	58	82

*The term gel mud refers to a water-bentonite-base mud.

Part of the effect is simply a pressure effect. There is less pressure at the bottom of the hole when drilling with air or gas. The effect of the density of drilling mud on the rate of drilling is included in the data shown in Fig. 13–21.[13]

Other mud properties which influence drilling rate are viscosity and filtration (water loss). There are probably several ways in which these properties influence drilling rate, and the filtration quality may be interrelated with the pressure effect noted above since it influences the pressure gradient through the rock face on the bottom of the hole. Less viscous mud is able to remove cuttings more readily from the bottom of the hole. The nature of the filter cake is also important because it influences the friction between the bit teeth and the bottom of the hole. The effects of mud on drilling rate are further discussed in Chapters 7 and 15.

In ordinary clay-water mud, the clays impart high viscosity as well as reducing the filter loss of the mud. In such mud systems, the clay fraction is sometimes kept to a minimum in order to obtain faster drilling rates. These muds are referred to as "minimum solids" muds. Centrifuges are used for this purpose and also for maintaining low density (light-weight) muds. Muds are sometimes maintained thin, by additions of water or minimum additions of bentonite, for hole-making purposes, even though some caving results that could be stopped by further additions of bentonite. A slight amount of caving can often be tolerated by an alert drilling crew. The particular nature of the shales exposed in the bore hole and the depth and investment involved in the hole already drilled must be considered in connection with this practice.

The exact relationship between drilling rates and mud properties under bottom-hole conditions are not known. Laboratory data, such

[12] From H. E. Mallory, "Improving the Mud Program," *The Drilling Contractor*, Vol. XIII, No. 3 (April, 1957), 56–57.

[13] John R. Eckel, "Effects of Mud Properties on Drilling Rate," *API Drilling and Production Practice* (1954), 119–25.

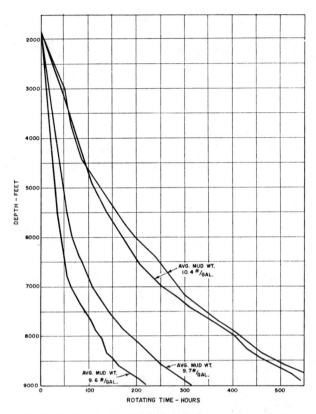

FIG. 13–21. Field data showing effect of mud weight on drilling rate (South Mississippi).

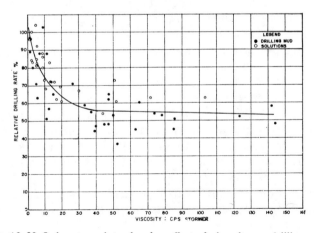

FIG. 13–22. Laboratory data showing effect of viscosity on drilling rate.

as that shown in Fig. 13–22, give relationships found under the test conditions used in the particular laboratory experiments. Both in the laboratory and in the field the greatest reductions in drilling rate appear to occur with the first additions of colloidal solids to water. While all solids present in the mud tend to reduce drilling rate, the colloids generally have a greater effect than the inert solids. Further additions of such colloids as bentonite, starch, and others continue to give further reductions in drilling rates. The following table shows comparative data on two wells drilled in New Mexico.[14] In well B, a definite effort was made to maintain a light-weight low-solids mud.

	Well A Gel-chemical mud	Well B Low solids oil- emulsion mud
Footage drilled	4,959	5,464
Bits used	24	19
Rotating hours	258	192
Ft. per bit (av.)	206	266
Ft. per hr. (av. on bottom)	19	26
Per cent increase, ft. per bit	—	29
Per cent increase, penetration rate	—	37

Air-Gas Drilling

Rotary drilling in which air or natural gas is circulated for the purpose of removing drill cuttings from the hole is used for drilling competent rocks where the control of pressures does not present a serious problem. The circulation rates required for removal of the cuttings are discussed in Chapter 12. In some areas natural gas is taken from near-by wells, circulated through the drilling well, and burned as it leaves the flow line. Where natural gas under pressure is not available, air compressors are used to force air down through the drill pipe. After-coolers are used with compressors in order to protect the rotary hose and other rubber parts from excess heat. Usually such compressors are auxiliary units driven by independent motors, although compressors have been driven by the main motors in some cases, particularly in Pennsylvania. Both the volume and the delivery pressure required are largely dependent on depth, and are further influenced by the rate of penetration, size of pipe and hole, and well conditions. Although it is unusual, pressures up to 1,500 psi have sometimes been used.

A large portion of the drill cuttings are blown out of the flow line in the form of dust. Where such dust must be controlled, it is often

---

[14] From *ibid*.

wetted by a spray of water in the flow line, or an equivalent method of wetting with water is employed. Filters have also been used.

Most of the development of air-gas drilling has occurred in the Permian Basin and Four Corners areas.[15] The practice has been to reduce weight on the bit of 50 to 75 per cent of that used when drilling with mud. Rotary speeds are also reduced to 50 to 60 per cent of ordinary values. Under these conditions the drill cuttings can be removed from the hole about as fast as they are formed. This is an important factor, for circulation must be periodically interrupted in order to add a joint of drill pipe to the drill string. Also, under these conditions, a longer bit life is obtained and more feet are drilled per bit. A single bit is usually run from eighteen to thirty-six hours. Total feet drilled per bit, as well as the rate of penetration, is an important factor in drilling costs, particularly at greater depths where more rig time is required in order to make a round trip to replace a worn bit. The penetration rates are from 50 to 500 or more per cent greater than when drilling with mud.

Water entering the hole from permeable formations or "weeping shales" presents one of the more serious problems in air-gas drilling. Small amounts of water cause the small dustlike drill cuttings to adhere to each other, to the drill pipe, and to the wall of the hole. The damp cuttings may ball up and stick the drill string. Sufficient additional water, fifteen to twenty gallons per minute, is added to keep the drill cuttings in a mobile condition. Foaming agents are also used at rates of one-half to one gallon per hour to aid in keeping the cuttings and water dispersed. The addition of foaming agents is reported to cause some reduction in drilling rate.

The maximum amount of water which has been handled economically is seventy-five to one hundred barrels per hour. If more water than this is entering the hole, it must be shut off, or water or mud must be resorted to for further drilling. Water which flows into the hole from permeable sandstones can be shut off by pumping into the pores a liquid plastic that will subsequently harden to a solid. Fractured limestones and shales present more difficult problems in shutting off water. In many instances locating the source of the water is the most difficult problem of all.

When air is used, there is some danger of fire and explosions. Some down-hole explosions have occurred. It is believed that these have been associated with plugging of the annulus so that pressure was in-

[15] Information from joint API-AAODC Air-Gas Drilling Committee, presented in an address by W. L. Brantly before local API meeting, Oklahoma City, February 11, 1958.

creasing in the system at the time the explosion occurred. In normal operations, pressures continuously decrease under circulating conditions. Serious corrosion of the drill string has also occurred, and saturated lime water has been added as a countermeasure.

Straight-Hole Drilling

Wells are drilled for the purpose of penetrating the rock at some definite subsurface location. In most instances this is accomplished by drilling a reasonably straight and reasonably vertical hole. When the surface location is inaccessible or where several wells are drilled from the same drilling platform, the wells are surveyed and purposely slanted by means of special tools and techniques in order to reach the desired subsurface location. An opposite technique has been used in areas containing inclined beds which cause the drilled hole to drift consistently from the vertical. In such case the surface location has been chosen so that the well would naturally drift to the desired subsurface location, thus permitting easier and more rapid drilling.

There are three concepts regarding the directional progress of a hole that are distinguished from one another. A vertical hole is one that is maintained within a few degrees, usually within three to five degrees, of the vertical, and usually no particular attention is paid to the horizontal direction or directions of the deviation. It would be possible for a vertical hole, as above defined, to contain a number of sufficiently sharp bends, or dog-legs, that would cause trouble in the subsequent drilling and producing of the well. A straight hole is one that contains no sharp curvature, and the term is usually restricted to either vertical holes or holes which maintain a constant deviation from the vertical. A directionally drilled well is one which is deviated from the vertical and whose direction is controlled so that it follows a predetermined course.

Periodic checks are made on the directional progress of the hole during routine drilling operations. In the ordinary case where it is desired to drill a vertical well, such checks consist only of plumb readings in order to determine if the bottom section of the hole is within the usual permissible three- to five-degree deviation from the vertical. The survey instrument is placed inside the top of the drill pipe and allowed to fall freely to bottom, sometimes with the aid of drilling-mud circulation. The instrument lodges inside the drill collars just above the bit, and it is preset at the surface to record after a definite time interval. Modern instruments are self-checking. One way in which this is accomplished is to set the instrument so that it makes two recordings several seconds apart. If the two readings are the same, it may be presumed that the instrument was at bottom and stationary at both times of recording. Usually the instruments are run in the drill pipe just before the pipe is pulled to replace a worn bit. Where this practice gives sufficient information on direction, it eliminates using a wire line in order to recover the instrument.

Two recognized causes of deviation of the hole from the vertical are inclined rock strata and the bending of the drill string above the bit. An area containing inclined rock strata which cause consistent deviation is referred to as "crooked-hole country." As a general rule, a hole drilled into hard strata tends to drift upstructure (because of drill string bending), whereas a hole drilled into soft formations tends to drift downstructure or along the contour.

The techniques used for drilling a straight vertical hole include using less weight on the bit, heavier drill collars, more rigid drill collars, larger drill collars with relatively small clearance between the drill collars and the hole, and the placing of stabilizers (roller reamers or their equivalent) at various positions in the drill-collar string in order to control any bending of the drill collars. Where other methods fail, directional drilling techniques may be employed.

The technique of drilling a straight hole and the relationship between drill-string bending and hole deviations were placed on a mathematical basis in work done by Arthur Lubinski and his collaborators in a series of four papers on the subject.[16] The position ordinarily assumed by the drill string is shown in Fig. 13–23. It may be noted that the drill string lies on the lower side of the hole except where it approaches the bit, which ordinarily is centered in the bore hole. In the unsupported portion between the bit and the point of tangency, where it comes in contact with the wall of the hole, the drill string follows an elastic curve. The weight of the lower, unsupported portion of the drill string and the bit tends to bring the hole toward the vertical. When longitudinal force, or weight on the bit, is applied through the drill string, there is a tendency to direct the hole away from the vertical. In Fig. 13–23, the resulting force is shown deviating from the vertical by angle ϕ. The deviation of the hole from the vertical is angle α, while r is the "apparent radius" or one-half of the diametral clearance between hole and drill collars. It may be noted that where

[16] A. Lubinski, "A Study of the Buckling of Rotary Drill Strings," *API Drilling and Production Practice* (1950), 178–214; G. C. MacDonald and A. Lubinski, "Straight-hole Drilling in Crooked-hole Country," *API Drilling and Production Practice* (1951), 80–90; A. Lubinski and H. B. Woods, "Factors Affecting the Angle of Inclination and Dog-Legging in Rotary Bore Holes," *API Drilling and Production Practice* (1953), 222–50; A. Lubinski and H. B. Woods, "Practical Charts for Solving Problems on Hole Deviation," *API Drilling and Production Practice* (1954), 56–71.

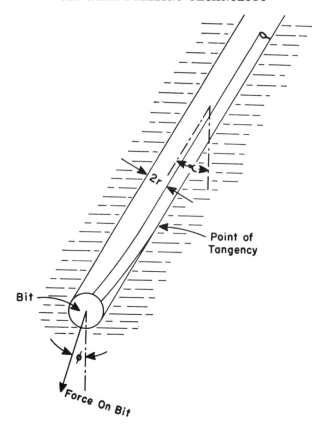

Fig. 13-23. Position of bit and drill collars while drilling.

conditions are such that $\phi = \alpha$, or $\phi/\alpha = 1$, the hole deviation tends to remain constant.

The equation of the elastic curve of the unsupported portion of the drill string was obtained as follows. It should be noted that angle α, the deviation from the vertical, is assumed to be small in the following derivation. However, modifications of the resulting relationships have been shown to be sufficiently accurate for large values of α, up to 90 degrees.[17] The elastic curve of the unsupported section is

$$M = EI \frac{d^2Y}{dX^2} \tag{13-1}$$

---

[17] Lubinski and Woods, "Practical Charts for Solving Problems on Hole Deviation," *loc. cit.*

where

M = bending moment
E = Young's modulus
I = moment of inertia of the section
X = vertical axis
Y = horizontal axis

Taking the derivative of the bending moment gives the shearing force.

$$A = EI \frac{d^3 Y}{dX^3} \qquad (13\text{-}2)$$

where

A = shearing force

Referring to Fig. 13–24, the shearing force in the unsupported portion between B and A can be expressed as

$$A = \text{(effective vertical compression)} \sin \bar{\alpha} - F_2 \cos \bar{\alpha}$$
$$= [W_2 - p(X_2 - X)] \sin \bar{\alpha} - F_2 \cos \bar{\alpha} \qquad (13\text{-}3)$$

where

W_2 = weight on the bit
F_2 = horizontal reaction at the bit
$X_2 - X$ = unsupported length below any point X
p = effective weight per foot of drill collars, immersed in mud

$$= (\text{wt./ft. in air}) \left(\frac{\text{density of steel} - \text{density of mud}}{\text{density of steel}} \right)$$

If the origin is chosen so that $W_2 = pX_2$, Equation (13–3) reduces to

$$A = pX \sin \bar{\alpha} - F_2 \cos \bar{\alpha} \qquad (13\text{-}4)$$

Where $\bar{\alpha}$ is small, $\cos \bar{\alpha}$ is sensibly equal to unity and $\sin \bar{\alpha}$ equals $\tan \bar{\alpha}$, so that Equation (13–4) may be expressed as

$$A = pX \tan \bar{\alpha} - F_2 \qquad (13\text{-}5)$$

Since $\tan \bar{\alpha} = -dY/dX$, Equation (13–5) becomes

$$A = -pX \frac{dY}{dX} - F_2 \qquad (13\text{-}6)$$

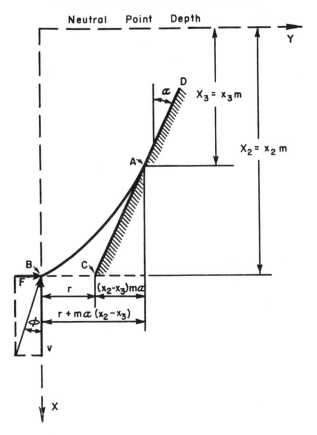

Fig. 13-24. Analysis of drill collar bending.

Substituting the expression for A in Equation (13-6) in Equation (13-2) gives

$$EI \frac{d^3Y}{dX^3} + pX \frac{dY}{dX} + F_2 = 0 \qquad (13\text{-}7)$$

The solution of this equation must satisfy the following conditions: (1) the horizontal displacement between points A and B equals $r + (X_2 - X_3)\alpha$, where α is expressed in radians; (2) the bending moment is zero at point A and also at point B; and (3) the drill string is tangent to the hole at point A.

The solution of equation, obtained by the method of iteration,[18] is

[18] Lubinski and Woods, "Factors Affecting the Angle of Inclination and Dog-Legging in Rotary Bore Holes," *loc. cit.*

$$\frac{r}{\alpha m} = \frac{x_1}{\dfrac{24}{x_1{}^3} - \left(\dfrac{24}{\pi^2}\right)\left(\dfrac{x_2}{x_1}\right) + \left(\dfrac{28}{\pi^2} - \dfrac{144}{\pi^4} - 1\right)}$$

$$\frac{\phi}{\alpha} = \frac{r}{\alpha m}\left[\frac{1}{x_1} - \frac{1}{x_2}\left(\frac{1}{2} - \frac{2}{\pi^2}\right)\right] + 1 - \frac{x_1}{2x_2}$$

NOTE: x is length in dimensionless units, so that $x = X/m$.

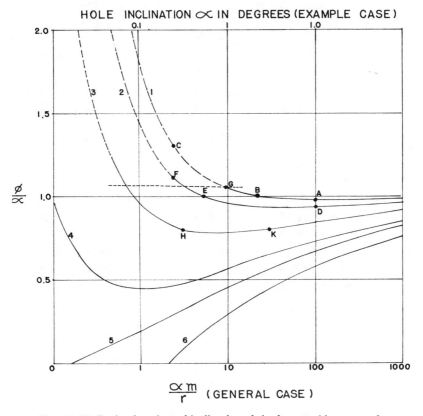

FIG. 13–25. Ratio of angle ϕ of inclination of the force on bit over angle α of hole inclination.

	Curve					
	1	2	3	4	5	6
Example case: 6½-in. drill collars, 8¾-in. hole, 10 lb./gal. mud. Weight in pounds	36,000	18,000	9,000	5,400	4,500	3,600
General case: Weight in dimensionless units	8	4	2	1.2	1	0.8

If the simplifying assumptions used for the case of nearly vertical holes are omitted, the more general solutions are[19]

$$\frac{r}{m \sin \alpha} = \frac{x_1}{\frac{24}{x_1^3} - \left(\frac{24}{\pi^2}\right)\left(\frac{x_2}{x_1}\right) + \cos\alpha \left(\frac{28}{\pi^2} - \frac{144}{\pi^4} - 1\right)}$$

$$\frac{\sin\alpha - \tan(\alpha - \phi)}{\sin\alpha} = \frac{r}{m \sin\alpha}\left[\frac{1}{x_1} - \frac{\cos\alpha}{x_2}\left(\frac{1}{2} - \frac{2}{\pi^2}\right)\right]$$

$$+ 1 - \frac{x_1}{2x_2} \cos\alpha$$

The solution of the equation is presented graphically in Figs. 13–25 and 13–26. The lower abscissa on Fig. 13–25 shows values of $\alpha m/r$. The quantity m is the length of a dimensionless unit defined as follows:

$$m = \sqrt[3]{\frac{EI}{p}}$$

The significance of the chart is shown by taking the case where drill collars of 6.25 inches OD and 2.25 inches ID are being used to drill a hole of 8.75 inches diameter with 36,000 pounds drilling weight on the bit. Consistent foot-pound units will be used in the calculation. A mud density of 10 ppg is assumed.

$E = (30 \times 10^6 \text{ psi})(144 \text{ in.}^2/\text{ft.}^2)$

$ = 4.32 \times 10^9 \text{ lb./ft.}^2$

$I = \dfrac{\pi}{64}(D_2^4 - D_1^4)$

$ = \dfrac{\pi}{64}[(6.25)^4 - (2.25)^4]\left(\dfrac{1}{12}\right)^4$

$ = 0.00355 \text{ ft.}^4$

$p = (0.7854)\left[\dfrac{(6.25)^2 - (2.25)^2}{144}\right](488)\left[\dfrac{488 - (7.48)(10)}{488}\right]$

$ = (90.5)(0.847)$

$ = 76.63 \text{ lb./ft., immersed in 10 ppg mud}$

[19] Ibid.

$$m = \sqrt[3]{\frac{(4.32 \times 10^9 \text{ lb./ft.})(0.00355 \text{ ft.}^4)}{76.63 \text{ lb./ft.}}}$$

$$= 58.50 \text{ ft.}$$

The weight of one dimensionless unit is

$$mp = (58.50 \text{ ft.})(76.63 \text{ lb./ft.})$$

$$= 4483 \text{ lb.}$$

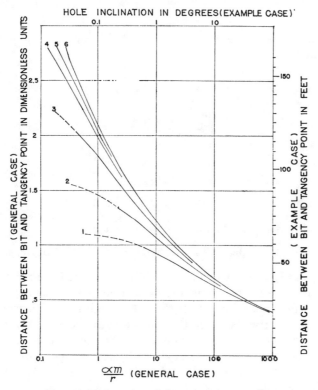

FIG. 13-26. Location of the point of tangency.

	Curve					
	1	2	3	4	5	6
Example case: 6¼-in. drill collars, 8¾-in. hole, 10 lb./gal. mud.						
Weight in pounds	36,000	18,000	9,000	5,400	4,500	3,600
General case:						
Weight in dimensionless units	8	4	2	1.2	1	0.8

The dimensionless weight on the bit is

$$\frac{36{,}000 \text{ lb.}}{4{,}483 \text{ lb.}} = 8.03 \text{ dimensionless units}$$

This weight corresponds to curve 1 on Fig. 13–25, which is for eight dimensionless units. Curve 1 shows that when $\phi/\alpha = 1$, $\alpha m/r = 22.5$. The value of r is

$$r = \left(\frac{1}{2}\right)\left(\frac{8.75 \text{ in.} - 6.25 \text{ in.}}{12 \text{ in./ft.}}\right)$$

$$= 0.1042 \text{ ft.}$$

The equilibrium deviation angle for the drilling conditions may be determined:

$$\frac{\alpha m}{r} = 22.5$$

$$\alpha = 22.5 \frac{r}{m}$$

$$= \frac{(22.5)(0.1042 \text{ ft.})}{58.50 \text{ ft.}}$$

$$= 0.0400 \text{ radians}$$

Expressed in degrees, the angle of deviation is

$$(0.0400 \text{ radians})(57.3 \text{ radians/degree}) = 2.3 \text{ degrees}$$

This shows that for the drilling conditions specified the normal angle of inclination from the vertical is 2.3 degrees, which is in good agreement with field observations. If the hole should have either greater or less deviation, it will change to coincide with the equilibrium deviation from the vertical. It may be noted that the values given in degrees on the upper abscissa coincide with the example solution. Figure 13–26 shows that the point of tangency would be 48 feet from the bottom of the hole.

In the foregoing example it was assumed that the formation was isotropic and would drill as readily in one direction as another. In order to explain such phenomena as the tendency of a hole to drill, or deviate, upstructure in hard rock, it has appeared necessary to assume that formations are generally anisotropic. That is, the rock has more resistance to drilling in a direction parallel to the bedding planes than it has to drilling directly through the bedding planes. This was handled mathematically[20] by the use of an anisotropic index, h, and the values of the ordinate on Fig. 13-25 were modified as follows:

$$\frac{\phi}{\alpha} = \frac{1}{1-h}$$

The value of h is zero for an isotropic formation, and values between zero and 0.075 were found to explain field observations. It may be observed that in the case of horizontal formations anisotropic formations tend to produce a more vertical hole than do isotropic formations. For example, where h has a value of 0.05, the equilibrium value of ϕ/α becomes 1.0526, shown on Fig. 13-25 as a short dash line. Line 1 intersects this value at an equilibrium deviation of 1 degree from the vertical, as compared to 2.3 degrees for the isotropic formation. If the formations are inclined, then the condition of anisotropy tends to cause the bit to penetrate in a direction normal to the bedding planes. This causes wells to drift upstructure.

Two causes of dog-legs, or sudden bends, in drilled holes are illustrated in Figs. 13-27 and 13-28. In the first case, the direction in which conditions of anisotropy tend to deviate the hole is reversed when passing through the unconformity. The use of a reamer (stabilizer) a short distance above the bit would decrease the severity of the dog-leg. Figure 13-28 illustrates a dog-leg caused by a sudden drop in the drilling weight, which tends to bring the hole rapidly back to the vertical. A gradual reduction in drilling weight, over a distance of fifteen to thirty feet, would cause a gradual bending back toward the vertical and thus avoid forming a severe dog-leg.

Directional Drilling

Directional drilling is used to straighten crooked holes and return them to the vertical, to sidetrack lost tools or obstructions, and to direct the course of the hole in a predetermined path to a predetermined bottom-hole location.[21] Wells are directed to underground locations not under the derrick floor because the surface locations are inaccessible or economically prohibitive. Wells are drilled from the shore to locations under water, and multiple wells are drilled from a single marine drilling platform. Wells are directed under overhanging salt domes in order to avoid drilling through the salt.

Directional drilling requires accurate underground surveys that

[20] *Ibid.*
[21] G. L. Kothny, "Underground Well Surveying, Directed Drilling, Side-wall Sampling, and Polar Core Orientation," *API Drilling and Production Practice* (1941), 76-90.

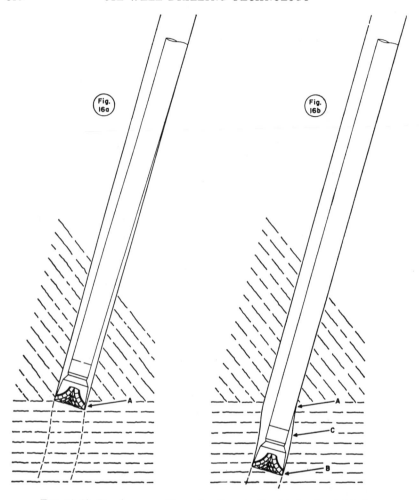

Fig. 13–27. Dog-leg caused by going from inclined to horizontal beds.

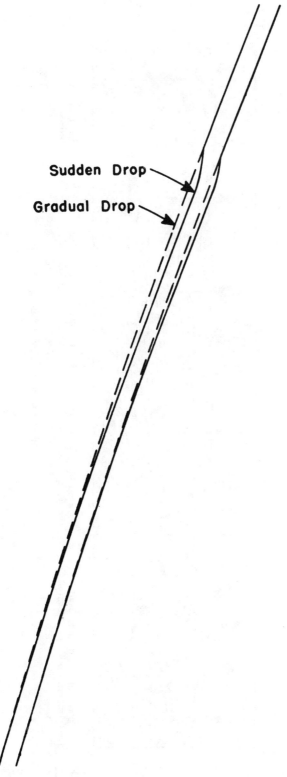

Fig. 13-28. Effect of drop of weight.

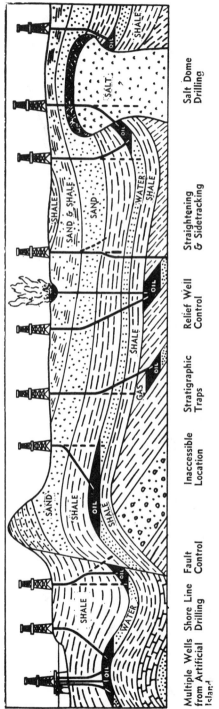

Fig. 13-29. Applications of directional drilling.

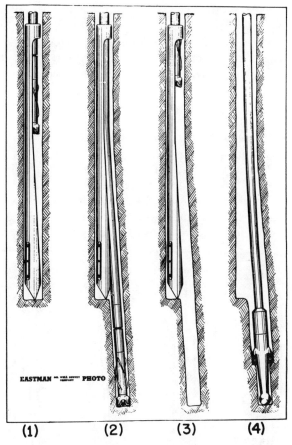

FIG. 13–30. Operation of whipstock: (1) on bottom in oriented position before pin is sheared; (2) drilling assembly in rathole; (3) whipstock in pick-up position; (4) reaming rathole to full gauge with hole opener.

give the amount and direction of hole deviation from the vertical throughout the depth of the well. Pendulums, or plumb bobs, are used to determine the amount of deviation from the vertical where such deviation is less than about 10 degrees. Spherical level boxes are used in cases of greater vertical deviation. The horizontal direction of the deviation is usually indicated by a magnetic compass. The readings of both the pendulum and the compass needle are simultaneously photographed or otherwise recorded at successive depth points in the well. When the magnetic compass is used to determine horizontal direction, the surveying instrument is usually positioned inside of a nonmagnetic stainless-steel drill collar. An earlier practice consisted of using a bit with a hole through the center which permitted the surveying instrument to protrude through the lower end of the drill string. Compass readings should be corrected for the declination from true north which occurs at the location. The compass cannot be used inside sections of a hole which have been cased with steel. In this case a gyroscope is available for indicating direction, the instrument having a slightly larger OD than the instruments using magnetic compasses.

Deflecting tools include the whipstock, knuckle joint, spudding bit and bit with unbalanced jets.

The *whipstock* is essentially a wedge. Permanent whipstocks are cemented or otherwise secured to the rock, but they often cause trouble later on by falling over into the hole when they are placed in shale, which softens with time during subsequent drilling. Retrievable whipstocks commonly have a chisel point which is spudded into

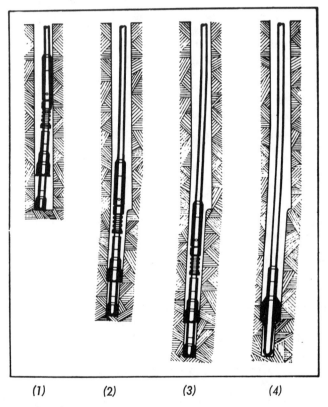

(1)   *(2)*   *(3)*   *(4)*

Fig. 13–31. Operation of knuckle joint: (1) on bottom, in oriented position; (2) starting new hole; (3) completion of knuckle joint run; (4) enlarging hole to full gauge with hole opener.

shale, and the subsequent application of additional weight shears an inch-diameter steel pin so that the bit can drill ahead for a few feet while the whipstock remains in position. A heavy cylindrical collar at the top of the whipstock provides for its withdrawal when the drill string is raised.

The *knuckle joint* consists of a short section equipped with bit and reamer which is attached to the drill string by a ball-and-socket type of joint. The short section is held at a predetermined angle with the axis of the drill string by the force of a spring. The small bit is spudded into one side of the bottom of the hole, and during subsequent rotation of the drill pipe, the ball-and-socket joint permits the short section to drill ahead in the direction in which it was first oriented. When regular drilling is resumed with a full-size bit, the full-size bit tends to follow and ream out the smaller hole drilled by the knuckle-joint bit.

The *spudding bit*, as its name implies, is spudded into one side of the bottom of the hole and drills ahead into soft shale without the aid of a flexible joint.

Unbalanced *jets* on a tricone bit, where one jet is larger than the other two, obtain directional deviation by jetting the bottom of the hole. In soft formations, more hole is jetted in the direction of the largest jet. Normal drilling is then resumed with the same bit and regular drill string. In all except the softer shales, hole deviation may require a diamond bit and whipstock.

The principles of drill-string bending which were discussed under "Straight-Hole Drilling" are also used in connection with directional drilling. The following table has been presented as generalized information which may serve as a guide to illustrate principles.[22] The data apply generally for drilling conditions of 70 rpm with 16,000 pounds drilling weight for $4\frac{1}{2}$-inch drill pipe in $10\frac{5}{8}$- to $11\frac{1}{2}$-inch-diameter holes.

Increase in angle per 100 ft. of hole, degrees	Drill collars with reamer below, ft.
0.0–0.5	80–120
0.5–1.0	40
1.0–2.0	30
2.0–3.0	20
3.0–5.0	8–12
5.0–7.0	0
7+	No drill collars, 60 ft. of $3\frac{1}{2}$-in. drill pipe above bit.

[22] From D. K. Weaver, "Practical Aspects of Directional Drilling," *API Drilling and Production Practice* (1946), 9–17.

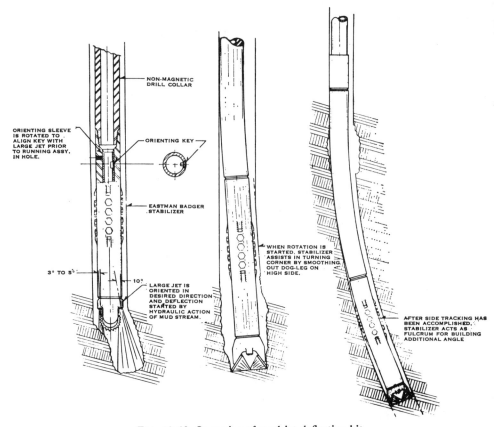

Fig. 13-32. Operation of mud jet deflection bit.

It may be observed in analyzing the table that drill-string bending similar to that shown in Fig. 13–24 will explain the results obtained. The drill string lies on the lower side of the hole except where the bit supports the lower end in the center of the hole. Consequently the longitudinal force in the drill string produces a tendency to increase the angle of inclination from the vertical. It should be noted that if the lower section of the hole is already curved in some direction, the longitudinal force on the drill string will cause the drill string to bend and come in contact with the hole at some point on the side opposite the direction of curvature. Accordingly, an increase in curvature will be caused in the same plane as the original curvature. Larger-diameter drill collars, of course, tend to straighten the hole. The angle of drift has been decreased by drilling ahead a short distance with a smaller-diameter bit, which naturally tends to drill into the lower side of the end of the hole. When the smaller hole is subsequently reamed with a larger bit, the full-size bit tends to follow the path of the smaller hole. By using such techniques, wells have been directionally drilled with only one or two or, in some cases, no whipstock settings.

The orientation of such deflecting tools as whipstocks has been controlled by oriented drill pipe and by nonmagnetic drill collar sections. Where drill pipe is oriented into the hole, clamps are temporarily fastened to each stand of pipe and transits mounted on the clamps in order to compute any rotation of the pipe while it is being lowered. Where nonmagnetic drill collar sections are used, they contain two small magnets and a landing seat for the surveying instrument. The angular displacement between the direction of the magnets and the direction of the deflecting tool can be determined at the surface before lowering the tool into the hole. When the tool reaches bottom, the surveying instrument can be seated in the nonmagnetic section. A photographic record of the positions of two compass needles, one responding to the magnetic field of the earth and the other responding to the two small magnets, gives the direction of the tool. Rotation of the top of the drill string combined with raising and lowering the pipe in order to relieve wall-friction effects permits accurate orientation of the deflecting tool. For deeper operations, the second method of orientation is more rapid and more accurate, since directions are measured at the bottom of the hole.

Diamond Coring

The proper operating conditions for diamond core bits are concerned primarily with the rotary speed, weight on the bit, and rate of fluid circulation. Most core barrels are capable of handling a forty-foot-long core. Since the core barrels are relatively thin-walled tubes, they should be stabilized. Diamond bits drill by a shearing action.[23] Loads build up rapidly on individual diamonds to the point where the rock under the diamond fractures, at which time the shearing load on the particular diamond drops to zero. Thus individual diamonds are subject to a form of impact loading during drilling. In order that such impact load is not too severe, the linear cutting speed of the diamonds on the major diameter of the bit should not be greater than eighty or ninety feet per minute.[24] The operating range for weight on the bit is between 250 and 500 pounds per square inch of projected cutting area. The desirable drilling-fluid pressure loss across the face

[23] E. B. Williams, Jr., "How to Get Most Out of Diamond Bits," *World Oil*, Vol. CXL, No. 5 (April, 1955), 198–201.
[24] *Ibid.*

of the diamond bit, in order that sufficient flow will occur between diamonds to remove cutting, is given in the following table.[25]

	Pressure drop, psi
Soft Formations	
Small stone bits	125–175
Medium stone bits	100–150
Large stone bits	75–125
Hard Formations	
Small stone bits	100–150
Medium stone bits	75–125

According to the above data, an $8\frac{3}{4}$-inch diamond bit which is cutting a $4\frac{3}{8}$-inch-diameter core should be rotated at about 40 rpm and carry 9,000 to 18,000 pounds of weight on the bit, depending upon the condition of the bit and the hardness of the rock. The amount of drilling fluid which can be circulated without raising the bit off bottom or unduly eroding the matrix holding the diamonds depends on the total area of the watercourses. However, common values are given in the following table:[26]

Bit diameter, inches	Volume of circulating fluid, gal./min.	
	Practical maximum	Minimum desired
3	75	16
4	140	25
5	210	37
6	300	50
7	360	66
8	440	85
9	520	120
10	575	145
11	620	175
12	665	210

Cable-Tool Operations

Cable-tool drilling methods have been employed continuously in the petroleum industry since the first oil well, the Drake well, was drilled near Titusville, Pennsylvania, in 1859. In this system, all tools used in drilling operations are lowered into the well on cables. The main surface machinery consists of hoists for withdrawing the various cables from the hole and some means of imparting reciprocal motion to the drilling cable. No fluid is circulated in the hole. Consequently,

[25] *Ibid.*
[26] From H. M. Stanier, "Review of Diamond Coring in California," *API Drilling and Production Practice* (1952), 243–56.

cable tools are often used for the drilling in of sensitive oil formations which might be damaged by rotary drilling mud, even though rotary methods were used to drill down to the top of the formation. The drilling operations are intermittent in nature and tend to give slower rates of penetration than rotary methods. However, the investment

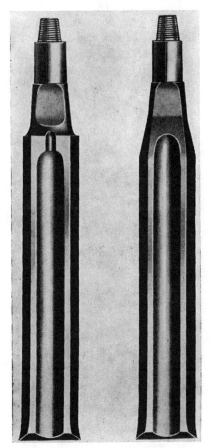

FIG. 13-33. Cable-tool drilling bits.

and operating costs are lower. Although many cable-tool wells have been drilled to depths of 8,000 or more feet, the best operating total depth limits are probably less than 2,000 feet.

The general drilling procedure consists of drilling about five to eight feet of hole, withdrawing the drilling tools, then removing the drilled rock material from the hole and again proceeding with actual drilling. The total amount of hoisting involved is proportional to the depth of the hole. The drilled rock material is mixed with water, and about one barrel of water must be present in the bottom of the hole. The reciprocal motion of the drilling tools causes a corresponding displacement of the fluid in the bottom of the hole, and the resulting up-

FIG. 13–34. Cable-tool jar.

ward and downward flow of fluid is quite effective in mixing such rock as shale into a thin suspension resembling rotary drilling mud. The grooved sides of the bit accommodate such fluid motion and are referred to as watercourses.

The mixed-up rock suspension is removed by means of tubes equipped with suitable valves. The diameter of such tubes is approximately three-fourths of the diameter of the hole, and their length is commonly between twenty and forty feet. They are lowered on a $\frac{3}{8}$- to $\frac{5}{8}$-inch-diameter steel line, referred to as the *sand line*. The *simple bailer* is such a tube equipped with a ball-type valve on its lower end.

FIG. 13–35. Sand pump.

However, the ball can be raised off the seat by means of a dart that is firmly attached to the ball and protrudes from the lower end of the bailer. The *sand pump* is such a tube equipped with a piston, and the steel sand line is attached to the piston. Therefore, when the sand pump is set on the bottom of the hole, the first upward motion of the sand line will raise the piston only, so that fluid is drawn into the lower end of the tube while it is on the bottom of the hole. The sand pump is used for picking up sand grains which settle to the bottom in the absence of sufficient clay for producing a thick suspension.

Steel lines of $\frac{3}{4}$-inch to about 1-inch diameter are commonly used for drilling cables. Tapered cables having larger diameter sections at the top of the hole and smaller diameter sections at the bottom of the hole are also used for better performance in deeper drilling. Manila lines approximately two inches in diameter were formerly widely used for drilling cables. Such cables are attached at their lower end to a rope socket, which is a part of the drill string. Swivel sockets that permit independent rotation of the lower end of the cable have been found to be advantageous when using steel cables.

The simplest drill string consists of a bit, drill stem, and rope socket. The various parts of the drill string are joined together by tapered tool joints. The stem is commonly about five inches in diameter and thirty to forty feet in length, depending somewhat upon the diameter to the hole to be drilled. The drill stem furnishes weight for increasing the force of the impact delivered by the bit, and its length and stiffness provide for drilling a straight hole. Where caving conditions exist that might cause the bit to become stuck, it is sometimes necessary to add a set of jars to the drill string. The working parts of jars are equivalent to two links of a steel chain and permit longitudinal movement. The part of the drill string above the jars is equivalent to a bottom-hole hammer which can be used to pound upward or downward on the lower part of the drill string. Sometimes a short stem, referred to as a *sinker bar*, is used above the jars to increase the force of such hammmer blows at the bottom of the hole.

The lower end of a common cable-tool bit resembles a blunt chisel. In drilling, the bit rotates sufficiently to strike successive different positions in order to make a round hole. Rotating tendencies, which are operative during the moment of contact, are imparted as the bit strikes the rough bottom of the hole when drilling in hard rocks. When drilling in soft rocks, the bit is rotated chiefly by the twist of the cable which develops from the varying tension in the cable during each stroke.

A comparison of drilling rates in holes of different diameters can be made as follows: For drilling any particular rock, the bit loadings expressed in pounds per inch of hole diameter should be about the same, since the bit makes linear contact with the bottom of the hole. The number of blows required for crushing the rock on the bottom face of the hole then depends upon the diameter of the hole. Consequently, for any given number of strokes per minute, the drilling rate becomes inversely proportional to the diameter. That is to say, it would take twice as long to drill a twelve-inch hole as it would to drill a six-inch hole.[27]

Successful cable-tool drilling depends upon imparting a satisfactory drilling motion to the drilling tools. Except near the surface, the tools are suspended at the end of an elastic cable. The tools, the bit, stem and rope socket, and the cable itself all have weight or mass that is in all instances supported by the cable above. The hardest blows are delivered in most rapid succession and with the smoothest action when the cable is under tension when the bit strikes the formation. This is referred to as "drilling on the stretch of the cable," and if the drilling motion is stopped at the lowest position, the tools will be suspended off bottom. In his analysis of cable-tool drilling, Bonham assumed that the instant the bit came in contact with the bottom of the hole, the cable should be under sufficient tension to give an upward acceleration of g (32.2 ft./sec.^{2}) to the tools.[28] Under such conditions, any rebound energy from the blow on bottom is immediately absorbed into the upward motion of the drill string. The upward stroke of lifting the cable must be sufficiently slow that the cable can lift the tools and contract throughout its length, which point is indicated by a vibration wave, or jar, at the top of the hole. The downward stroke follows and may be divided roughly into two principal parts. During the first part, the tools and cable fall freely. During the second part, the tools are decelerated and the cable stretched out to the required tension corresponding to the instant of impact on the bottom of the hole. The total length of the stroke is then governed by two factors, the stretch necessary to produce the required tension in the cable and the required free-fall distance. The required free-fall distance must be long enough to allow the tools to gain sufficient momentum to stretch the cable and have sufficient momentum remaining to strike an effective blow on the bottom of the hole.

The above description applies to a rather ideal cycle. The oscillating system represented by the cable and drilling tools is not free to vibrate as in air. Wells drilled by cable tools, as well as those drilled by rotary tools, tend to deviate slightly from the vertical, and for similar reasons. Therefore, the action of the cable is dampened by frictional contact with the wall of the hole. The action of the bit and drill stem

[27] C. F. Bonham, "Engineering Analysis of Cable Tool Drilling," *The Petroleum Engineer*, Vol. XXVII, No. 13 (December, 1955), pp. B93–B100.
[28] *Ibid.*

is dampened because of the fluid in the bottom part of the hole. When water enters the hole from water-bearing sands so that all or an appreciable part of the hole is filled with water, the action of the entire cable is further dampened by the viscous drag through the water. Under such conditions, the drilling motion is appreciably slowed down, and pressure effects on the rock at the bottom of the hole further reduce the tendency to rock fracture, so that penetration rates are greatly reduced.

CHAPTER 14

Fishing Tools and Practices

Whenever equipment is mechanically lowered into small holes within the earth, especially when large stresses are being placed on the equipment, the probability exists that sooner or later a mechanical failure will occur, and some part of the subsurface equipment will be left in the bore hole. Another common source of trouble encountered in drilling operations is the sticking of equipment in the hole on account of caving of formations, excessive mud filter-cake thickness, accumulation of drill cuttings around the bottom of equipment, or other similar causes which prevent the equipment in the hole from being removed without creating excessive stresses on the drilling rig. The technique of removing equipment which has become disengaged from the drilling tools or stuck is called *fishing*. The development of the fishing-tool industry has paralleled the growth of the drilling industry because of the very close basic relationship between the two industries. Fishing for equipment lost in a bore hole is an art which requires a thorough knowledge of new equipment and techniques and the stresses which can be applied to the equipment in the hole. Many of the tools used to recover equipment are specially designed for the particular job. However, on account of the similarity of equipment used in most drilling operations, certain more or less standard fishing tools have been developed and are available from fishing-tool or supply companies.

When equipment has become lost or stuck in the hole, it must be removed as rapidly as possible. In general, the longer this equipment remains in the hole the more difficult it is to recover it. The actual cost of the fishing tools is normally small compared to the cost of rig time and the investment in the well bore. If the equipment cannot be removed from the hole, it may be necessary to sidetrack (directionally drill) around the lost equipment, or else drill another hole, either of which would be a costly operation.

A broad classification of fishing tools divides them into two groups: (1) those used to recover tubular products and (2) those used to recover miscellaneous equipment. There are three basic types of fishing tools used in recovering tubular products (drill pipe, casing, and tubing): inside fishing tools, outside fishing tools, and hydraulic and impact tools.

Common causes of trouble requiring fishing operations are: fatigue failures, caused by excessive stresses in the drilling string, as when the rotary table continues to turn when the lower portion of the drill string becomes stuck; failure of down-the-hole equipment on account of corrosion or erosion by drilling fluids; parting of the drilling string because of excessive pull when attempting to free equipment which has become stuck; mechanical failure of parts of the drilling bit, causing some part of the bit to become lost; accidental dropping of tools or other undrillable objects into the hole; and sticking of drill pipe or casing. Drill pipe or casing can be stuck by caving formations, excessive mud filter cake build-up, accumulation of drill cuttings around the lower end of the pipe, and key-seating of the drill pipe.

If a fishing job becomes necessary, all factors should be carefully analyzed, including the size and shape of the top of the fish, the amount of fish in the hole, the character of the formations, the condition of the drilling mud, and the condition of the drilling rig, with particular emphasis on the condition of the derrick or mast and the drilling line. Relatively large stresses may be placed on these last two items of equipment, and caution must be exercised to insure that failure in above-ground equipment does not occur to further complicate the fishing operations. After a consideration of all factors, the proper fishing tool should be selected and run in the hole. It is important that the fish be recovered on the first attempt, if it is at all possible. If it is not recovered on the first attempt, recovery problems increase. Drill cuttings and cavings will settle around the fish, and the first fishing attempt may have moved the fish into a position which makes recovery even more difficult.

Fishing for Tubular Products
Inside Fishing Tools

Most fishing tools designed to catch the tubular fish from the inside are variations of the *spear* or *tap*. The basic elements of the *spear* type of fishing tool consist of a tapered interior along which one or more slips are free to move. A slip, shown diagrammatically in Fig. 14–1, is one of the commonest devices found in oil field equipment. It is comprised essentially of two parts: part A, the slip segment proper, which is a tapered section with serrated edges for engaging purposes; and part B, a tapered seat on which part A is free to move. The serrated edges, or teeth, of slip segment should be as hard as or harder than the equipment engaged. Equipment is free to move in one direction through the slips, but when movement is in the direction of increasing thickness of part B, the slip segment will move downward until the serrated edges grip the equipment, preventing further movement. Additional downward pull will only increase the effectiveness of the "grip" of the slips. The tool can move freely inside the fish, but when the tool is removed from the fish, the slips move downward on the tapered interior, coming into contract with the inner wall of the fish. The thickness of the tapered part of the tool is such that the slips will not allow the tool to be removed from the fish with continued upward pull.

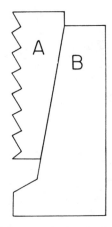

Fig. 14–1. Diagrammatic sketch of a slip segment.

One of the first spears to be used in recovering stuck pipe is shown in Fig. 14–2. When this type of spear has once engaged the fish, it cannot be released. The use of this equipment is somewhat limited, principally because of nonreleasing feature. Its principal advantage is its low cost, although this is of very little advantage in most rotary-drilling operations, where the principal requirement is fast, efficient removal of the fish. A later development of the original spear is shown in Fig. 14–3. A releasing mechanism is included in this tool, thus overcoming one of the basic disadvantages of the original spear.

A *tap*, shown in Fig. 14–4, is essentially a tapered tool with case-hardened threads. It is used primarily to engage a fish when the uppermost part of the fish consists of an inside-threaded element. The taper of the tool permits easier entry into the fish and also permits a positive engagement with the threads of the fish. The principal disadvantage of this tool is that it, like the original spear, is nonreleasable. In the event the tap becomes attached to the fish and the fish cannot be recovered, then additional problems of major importance are created unless methods are available to release the tap from the pipe being used in the recovery operations. Safety joints, which will be discussed later in this chapter, are one means of achieving this objective.

Fig. 14-2. Nonreleasing spear.   Fig. 14-3. Spear with releasing mechanism.

Outside Fishing Tools

Outside fishing tools must pass over the outside of the fish before attaching themselves. A very simple outside fishing tool, a *die collar*, is shown in Fig. 14–5. This tool is basically a short length of tubular material on the inside of which threads have been cut. The die collar is used to recover tubular material which has male threads on the uppermost portion. Once the die collar has engaged the fish, it cannot be released.

The *overshot* is another early style of fishing tool. It consists of a tapered bowl on which slips are free to move up and down, and the entire assembly is designed to fit over the upper segment of the fish.

FISHING TOOLS AND PRACTICES

Fig. 14-4. Rotary taper tap.

Fig. 14-5. Die collar.

The inside taper of the bowl is designed to permit the overshot to drop down over the fish, but as the overshot is retracted, the slips move on the tapered bowl to engage the exterior of the fish. A cross section of a simple overshot is shown in Fig. 14-6. The design of the original overshots were such that they were also nonreleasable, but a releasing method has been developed whereby the overshot can be released by rotation, which causes retraction of the tapered bowl with subsequent freeing of the slips. An additional feature that increases the effectiveness of an overshot is provision for circulating. A releasing, circulating overshot is one of the tools most commonly used in the recovery of tubular products, especially drill pipe.

Hydraulic and Impact Tools

When additional pull is desired which exceeds the capacity of the derrick or drilling line, a hydraulic pulling tool can be used. The *hydraulic pulling tool* is essentially a hydraulic jack with means for attaching to the fish and slips to engage the casing. By using a series of power pistons, force advantages as large as 100 to 1 are possible. This force is transmitted to the casing instead of the derrick, and, in general, much greater forces can be applied to the stuck pipe. The hydraulic pulling tool has good application in deep drilling, where much

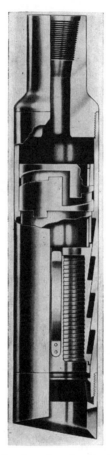

Fig. 14–6. Cross section of a simple overshot.

of the surface pull may be dissipated by the drag caused by crooked holes.

A *jar*, shown in Fig. 14–8, is an expansible tool composed of two parts which are free to move vertically a short distance in relation to each other. This tool is normally placed immediately above the fishing tool, and in the closed position the jar allows a short, free pull of the drill pipe; however, when the jar has been expanded completely, a relatively large impact force, or jar, is applied to all equipment located below the jar. This impact force, which momentarily is much greater than could be applied at the surface, is useful in many operations where equipment has become stuck in the hole.

A *hydraulic jar* utilizes a liquid power section to impart the impact

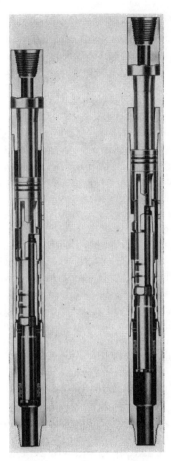

Fig. 14–7. Rotary jars.   Fig. 14–8. Cable-tool fishing jar.

blow. Most hydraulic jars obtain their impact force by the resistance of the power liquid as it is transferred to another chamber through a restricted opening. The pressure on the liquid is obtained by surface pull or lowering of weight, depending on whether the jar is used to strike an upward blow or a downward blow. As the liquid is being transferred under relatively large pressures, an exhaust port is opened, relieving the pressure immediately and resulting in a blow being struck by the tool. The tool may also be operated against a gas compression chamber, and when the balancing liquid pressure is released, the gas pressure drives a free-moving hammer which strikes the top of the tool.

Special Equipment for Recovering Tubular Products

Knuckle joints are special tools which have a swivel arrangement on the lower end to facilitate reaching to the side of the hole or into cavities to recover the pipe. Knuckle joints may be provided on both inside and outside fishing tools.

In many instances the stuck pipe cannot be recovered by simply grappling the pipe and pulling. In some areas, once the pipe has become stuck and circulation ceases, incompetent formations will continue to cave in around the pipe, making recovery by simple pulling impossible. In these cases it is standard practice to use *washover equipment* to recover the pipe. Typical washover equipment includes a washover shoe and sufficient washover pipe to wash over the stuck pipe the desired distance. The washover shoe is actually a type of drilling bit in that it has a cutting edge that can increase the rate of penetration through the cavings or settlings outside the stuck pipe. In many cases the washover shoe is simply a short length of pipe with teeth cut on one end. The washover shoe and wash pipe must have a larger internal diameter than the maximum outside diameter of the fish. The outside diameter of the wash pipe and rotary shoe should be less than the diameter of the original hole, however, or the rate of penetration will be much less because of the necessity of drilling additional formation. In addition, the possibilities of having the wash pipe stick are much greater if a larger hole is necessary, as the outside diameter of the washover shoe is not much larger than the outside diameter of the wash pipe.

In washing over stuck pipe, the washover shoe is run in the hole with the required length of wash pipe, using drill pipe or tubing for the remainder of the string. The length of wash pipe used will depend on the length of stuck pipe, the deviation of the hole, and the competency of the formations. If the deviation of the hole is large and the formations are incompetent, relatively short sections of wash pipe will have to be used. On the other hand, if the hole is straight and the formations are competent, several hundred feet of wash pipe can safely be run. After the maximum amount of stuck pipe has been washed over, this pipe is removed from the remaining stuck pipe by either mechanically cutting the pipe, chemically cutting the pipe, using a small, controlled explosion to sever the pipe at the desired interval (jet cutters), or using a small, controlled explosion to cause the stuck pipe to become separated at a collar (back-off shot).

Mechanical cutters can be designed to begin cutting either from the outside or from the inside. Chemical cutters are designed to cause a high-pressure stream of a strongly reactive chemical to be impinged on the internal surface of the pipe, so that a very clean cut through the pipe results.

The application of the shaped-charge principle to oil well perforating has also been extended to oil well fishing operations. In pipe-cutting operations the *shaped charge* is designed to cut the pipe without damaging any surrounding casing. This tool is lowered into the hole to the desired depth on an electrical cable, similar to that used in perforating or logging, and the shaped charge is detonated electrically from the surface. This procedure consumes very little rig time and produces a very clean cut.

The so-called *back-off shot* is a tool which utilizes a length of primacord as an explosive to cause pipe to become separated at a threaded connection. After this tool has been placed in the desired position, the pipe is placed in tension and a small amount of left-hand torque is applied. When the primacord is detonated, the combination of left-hand torque, tension, and explosive force will cause the pipe to become disengaged at the coupling. This tool is lowered into the hole and operated with an electrical cable. It has all the advantages of the shaped-charge cutter and the additional advantage that the pipe is not normally damaged.

Determining the actual point at which cavings or other material is causing pipe to be stuck is an important consideration in fishing operations. A useful method has been developed by Hayward[1] for determining the freeze point in drill pipe, casing, or other tubular goods which have become stuck. This method, which is described more completely in Chapter 19, requires a knowledge of the yield strength of the stuck pipe. The stretch in the pipe is measured after applying a known pull on the pipe at the surface. With this data it is possible to calculate the length of free pipe, and therefore the point at which the pipe is stuck is determined. This tool is used extensively in fishing operations, as it eliminates much of the guesswork in determining the depth at which the pipe is stuck.

Combinations of several of these fishing techniques have resulted in decreasing the time required for many fishing operations. The use of washover equipment in conjunction with the free-point indicator and back-off shot is standard in many areas. Using this combination of equipment, the stuck pipe is washed over for the maximum length, the free-point indicator is run to determine the length of free pipe, and the back-off shot is used to separate the pipe at the desired depth. In many cases several hundred additional feet of pipe can be recovered

[1] J. T. Hayward, "Methods of Determining How Much of a Frozen or Cemented Column of Pipe Is Free," *API Drilling and Production Practice* (1935).

below the pipe that was washed over, because an up-the-hole bridge of cavings may have been the principal cause of sticking. After this bridge has been cleared out by the washover pipe, the stuck pipe may be free for several hundred additional feet. The free-point indicator will reveal this information. Equipment has been designed to permit use of washover equipment, free-point indicator, back-off shot, and grappling equipment in one operation.

Miscellaneous Fishing Equipment

In addition to tubular material, other equipment, such as parts of drilling bits, pieces of tools, or equipment accidentally dropped in the hole, may necessitate fishing operations. Special equipment is also needed to recover these miscellaneous objects. Like the tools used to recover tubular material, the fishing equipment used for these miscellaneous objects is quite varied, and much of it is specially designed for a particular fishing operation. A few of the more or less standard tools are described below to illustrate the principles involved in this type of fishing operation.

The *milling tool* is one of the common fishing tools. It is used to grind the fish into small pieces which can be circulated out of the hole with the drilling fluid or removed in a junk basket. A milling tool may be an ordinary drilling bit, a special type of drilling bit, or a solid-head tool with a grinding surface. The use of carbide as a cutting surface is quite common. The carbide, in the form of small splinters, is interspersed in a specially blended matrix, and this mixture is applied to the head of the milling tool. This design results in a self-sharpening tool with a very long cutting life.

A *junk basket* is a tool run either separately or in conjunction with a milling device. It is designed to recover the smaller fragments which can be lifted by the circulating action of the drilling fluid, but may be too large to be circulated to the surface.

Since much of the material that may become lost in the hole is comprised principally of steel, the use of a magnet to recover small objects has been quite successful. *Magnetic fishing tools* may be either of the permanent-magnet type or the electromagnetic type. In the permanent-magnet type of tool, a powerful, permanent magnet is located inside a nonmagnetic material. The lower end of the tool is the engaging face, and the magnetic force is concentrated at this lower face. This tool is usually run on drill pipe or tubing. A circulating hole is provided on most magnetic fishing tools to permit normal drilling-mud circulation during the course of the fishing operation. The electromagnetic type of fishing tool is run on an electrical cable, and once the tool is in the hole over the fish, an electrical current is passed through the tool to energize the magnet. As much as several hundred pounds of lost metallic material can be recovered in one operation by using these magnetic fishing tools.

Another utilization of the shaped-charge explosive principle is the *shaped-charge fragmentizer*, which is used in the fragmentizing of small objects so that they can subsequently be recovered with a junk basket. This process may even permit drilling operations to proceed without recovery of the fragments. The operation consists of lowering the tool in the hole on an electrical cable until the tool is in contact with the object to be destroyed. The explosive energy of the tool is directed downward, and when the charge is detonated, this force is concentrated on the object, and its fragmentization results. It can be used to fragmentize bits, bit parts, slips, and other relatively small objects.

A necessary item in most fishing-tool strings is a *safety joint*, a device which permits release from the fish if it cannot be pulled and the releasing mechanism of the fishing tools become inoperative. In most fishing operations the safety joint is located immediately above the fishing tool. Most safety joints are released with left-hand torque. When it becomes necessary to utilize the features of the safety joint, the upper part of the tool is retrieved with the tubing or drill pipe, and the lower part remains with the tools in the hole.

Because of their very nature, fishing operations may become quite expensive. Before a fishing operation is initiated, therefore, the operator should figure the investment in the present hole and lost equipment, and then in the light of subsequent fishing operations, he can determine intelligently whether it is economically feasible to continue fishing. Although not the normal procedure, in some instances it may be more economical to leave the equipment in the hole, skid the rig, and begin drilling a new hole. In many cases, expensive fishing operations have continued for months, and in the end, a new hole had to be drilled. It should be realized, of course, that another few days of fishing could result in recovery of the fish; nevertheless, the financial considerations should be continually reviewed to determine the most economical procedure.

Proper and frequent inspection of equipment will reduce the frequency of fishing jobs, although it will be impossible to eliminate them completely because of the nature of drilling operations. Mechanical failures are the cause of most lost or stuck tools, and if undue operating strain is not imposed on the equipment, frequent inspections made, and poor equipment removed, then the number of fishing operations can certainly be reduced to a minimum.

Cable-Tool Fishing Devices

Recovering stuck or lost tools in a cable-tool operation is essentially no different from recovering equipment in a rotary-drilled hole in that the equipment to be recovered must be carefully studied and a tool selected that will satisfactorily perform the job. The principal difference in rotary and cable-tool operations is that in the latter, fishing tools are lowered into the hole on a wire line rather than drill pipe.

Fishing jars are widely used in cable-tool fishing operations, because in most cases the drilling line and rig preclude the use of large pulling strains, and impact blows exerted by the jars must be relied upon to recover equipment that has become stuck.

CHAPTER 15

Development of Drilling Systems

Purpose of Drilling

The ultimate purpose of drilling an oil or gas well is to provide a conduit, from the reservoir to the surface, which will permit the commercial withdrawal of fluids from the reservoir. All wells drilled should yield geological information for purposes of reservoir control and evaluation and discovery of resources. The diameter of the well and the equipment used in drilling and producing the well represent an economic balance. Certain factors which affect drilling rate, and therefore affect drilling costs, are discussed in following sections. Some of the attempts that have been made to modify existing drilling practice or to develop new systems of drilling are also described. The governing principles are economic considerations and the extreme ruggedness required in equipment capable of boring a deep hole through hard rock.

Effects of Rock Properties on Bit Performance

The effects of rock properties on bit performance are discussed generally in terms of the operation of a toothed-wheel, or rolling cutter, type of bit, unless otherwise noted.

Hardness. Rock hardness governs the abrasive action of the rock on the bit. Other conditions being equal, the rock hardness is a factor determining the rate of wear on the teeth of the bit, and any particles of the drilled rock which may enter the bearings affect the rate of wear of the bearings. The hardness of the rock being drilled and the hardness of the small rock particles in the mud, together with the weight which must be carried on the bit, affect the length of time which the bit may be rotated on bottom. Bits may be run until the teeth are worn smooth in drilling such rocks as hard limestones, and the rate of drilling may continue to be satisfactory when the bit is in this condition if sufficient weight is carried on it. However, the bit must be pulled and replaced before bearing failure occurs, for such failure would leave parts of the cutters and bearings in the hole so that a fishing job would be necessary.

The hardness scale which is most often referred to in describing rocks is that proposed by Mohs. In this scale, the harder rock will scratch the softer rock when the two are rubbed together. The steel used in rock bits has a hardness between six and seven on the scale. Bit manufacturers cover parts of the teeth on their bits with a coating of harder materials, such as tungsten carbide, in order to boost their wearing quality.

MOH'S SCALE OF HARDNESS

1. Talc	6. Orthoclose feldspar
2. Gypsum	7. Quartz
3. Calcite	8. Topaz
4. Fluorite	9. Corundum
5. Apatite	10. Diamond

The effect of hardness on the rate of drilling is usually controlled or at least strongly affected by other properties, so that no direct correlations have been obtained between hardness and drilling rate. Rocks are seldom composed of a single mineral, and this fact in itself would make such a correlation difficult even if no other factors were involved.

Mechanical structure-fracture characteristics. Rocks, as well as materials in general, may be described as brittle or as fibrous and tough, or according to varying degrees of these qualities. As an example, pure chert, which is a hard mineral, often drills readily at such rates as fifteen minutes per foot. Chert exhibits a conchoidal fracture which facilitates drilling when sufficient weight is applied to the bit. However, the drilling of chert-bearing limestone may be much slower and more difficult if the chert is disseminated in small particles and veins in the limestone matrix. Both of these rocks would, of course, appear quite brittle if small pieces were struck with a hammer at the surface. Their reaction to the bit varies, however, in the protective surroundings at the bottom of a deep hole. Shales vary markedly in their fracture characteristics.

Plasticity. As a property of a rock, plasticity may be taken as one of the opposites of brittleness. In relation to the drilling action of toothed–rolling cutter bits, the plasticity of the drill cuttings and circulating fluid at the bottom of the hole must be considered. In this sense, plasticity refers to the adhesive and cohesive or sticky nature and quality of the drill cuttings in the surroundings at the bottom of the hole during drilling operations. These qualities depend upon the size and nature of the crushed rock particles. Sand grains, powdered chalk, powdered limestone, and powdered quartz exhibit but little plasticity and do not form strong, cohesive masses. On the other hand, powdered shales contain clays which readily hydrate and subdivide into even smaller particles which adhere strongly to each other and to most other substances, including the teeth of the bit under certain conditions.

The ultimate unfavorable result of these properties, which rarely occurs, is the packing of shale drill cuttings and/or caving into the spaces between the teeth of the bit so that, in effect, the rolling cutters become smooth rollers which rotate on the bottom of the hole and will not drill. This condition, commonly referred to as a *balled-up bit*, is most likely to result when a bit is first placed on bottom where shale drill cuttings and/or cavings have accumulated by settling during the round trip with the drill string, and weight is applied too rapidly to the bit before the circulation of drilling fluid has had a chance to remove the material from the bottom of the hole. The condition can occasionally be corrected, in the case of bits which jet fluid directly onto the cutters, by circulating drilling fluid over the bit for about half an hour with the bit held off bottom; but often it requires that the bit be pulled and the material removed, by means of a hammer and chisel from between the teeth of the bit.

Equivalent conditions can occur during the rapid drilling of soft shales. The rate of drilling-fluid circulation limits the weight that may be applied to the bit in such cases. The effectiveness with which the drilling fluid is jetted on the bottom of the hole, to keep the bottom of the hole and bit teeth clean, is important. It is apparent that the crushing strength of the rock fragments as well as their size is a factor in whether or not they may be crushed into a coherent mass that will fill the space between the teeth of a bit. The drilling fluid usually serves to increase the total clay or colloidal content at the bottom of the hole. This may serve as binder material between rock fragments. Other effects of the drilling fluid are discussed later.

Crushing strength, shear strength, tensile strength, and porosity. The strength of a rock undoubtedly is the most important factor in determining the rate at which it may be drilled. Tests have been devised to measure the shear and tensile strengths of materials. Drill bits measure the cohesive forces between the constituent rock particles in somewhat different ways from those by which they are measured in standard testing machines. That is to say that the drillability of a rock is a function of the bit which is being used to drill the rock and of all the drilling conditions. Broad correlations exist, since all standard tests indicate that quartzite, for example, is a stronger rock than a recently deposited shale. The pore space within a rock detracts from the strength of the rock and also decreases the amount of solid material which must be removed by the action of the bit. By far the best correlation between measurable rock properties and drilling rate is that with the fractional porosity of the rock. Particularly in such rocks as thick limestone beds, the porous zones may be located by the faster rates of penetration, which, of course, occur in these zones.

Theories of Rock Failure

When theories of rock failure are being considered in relation to the drilling of a well, it is necessary to keep uppermost in mind that systems or conditions of minimum total work required in producing rock fragments are not necessarily the most desirable or economical. As far as the power used in actual drilling operations is concerned, drilling rigs are equipped with an excess of horsepower, which is utilized in hoisting operations but which at present is not efficiently utilized in the drilling. Therefore, better balanced and more economical conditions can be achieved where theories or developments lead toward faster penetration rates or greater footage drilled between hoisting operations. The above statements apply particularly to the drilling of the harder shales, sandstones, limestones, dolomites, cherts, and quartzites. In drilling such rocks, rolling cutter bits of approximately eight inches in diameter utilize only about twenty-five to fifty horsepower. By way of contrast, the rig may be equipped with a total of one thousand horsepower, more or less.

Rittenger's Law. Where a well is drilled by mechanical means, small pieces of rock are produced by the action of the bit against the rock. In attempting to obtain a measurement of the useful work transmitted from the bit to the rock, it could be assumed that the energy required is proportional to the new rock surface produced during the drilling process. This assumption was used to obtain Rittenger's Law, which was developed primarily for rock-crushing machines. If D is the length of an edge of a large cube which is crushed into small cubes with an edge length of d, then there will be $(D/d) - 1$ surfaces of D^2 area sheared in each of three directions, so that

$$\text{Total surface produced} = 3\left(\frac{D}{d} - 1\right)D^2$$

If B equals the work required to produce a unit area of new surface, and since the volume of the large cube is D^3,

$$\text{Work per unit volume of rock} = \frac{3B\left(\frac{D}{d} - 1\right)D^2}{D^3}$$

$$= 3B\left(\frac{1}{d} - \frac{1}{D}\right)$$

Where D is large with respect to d and C is a constant to correct for actual rather than cubical fragments, the relation may be written as

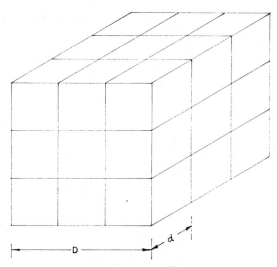

Fig. 15–1. Development of Rittenger's Law.

$$\text{Work per unit volume of rock} = \frac{3CB}{d}$$

$$= \frac{\text{const.}}{d}$$

where d is the effective size of the drill cuttings. This relation implies that if the energy absorbed in producing rock fragments is proportional to the new surface area produced, it will be inversely proportional to the effective diameter of the drill cuttings.

Kick's Law. Another rule proposed for rock-crushing operations is known as Kick's Law, which states that the energy required in producing smaller geometrically similar fragments varies as the volume or weight of the bodies varies. This law takes the form,

$$\text{Work per unit volume of rock} = \text{const.} \log \frac{D}{d}$$

Classical theories of failure. The classical theories of the failure or rupture of materials include the following:

1. *Maximum-stress theory*, which states that failure occurs at any point where the stress exceeds some critical value.
2. *Maximum-strain theory*, which states that failure occurs at any point where the stress (deformation) exceeds some critical value.

3. *Maximum-strain energy theory*, which states that failure occurs at any point where the strain energy stored in the body exceeds some critical value of strain energy.
4. *Maximum-stress difference* or *maximum-shear theory*, which has been developed graphically by Mohr, in which failure is determined by the largest differences between the principal stresses.

Mention should also be made of release fracturing, in which a portion of the rock spalls off after the release of a supporting or confining pressure. This phenomenon is observed at newly exposed rock faces in mines and quarries.

Percussion theories. The major efforts of Drilling Research, Incorporated, in the period 1949 to 1954 centered around the production of a percussion type of drilling tool, and these efforts are more fully described later in this book under the heading "Magnetostriction—Percussion Drilling." It was realized early in this investigation that adequate theories of rock failure were not available upon which to base the design of such a tool. Accordingly, a part of the efforts of the consultants and research group was directed toward developing theories of rock failure which could be applied in designing a percussion-type drilling apparatus. Such experiments as dropping a sharp-edged chisel on a piece of limestone were performed. Flat-edge dies and truncated chisels were also used. It found that a definite threshold energy level was required for rock failure to occur where a flat surface came into contract with the rock, but that the sharp chisel edge exhibited zero threshold energy. The wedge-shaped portion of the rock directly beneath the chisel was crushed to powder, and chips formed on both sides with fracture angles of about twenty degrees to the surface of the specimen. The greater the energy of the blow delivered to the chisel, the greater was the penetration and the amount of rock removed. The stresses on the chisels during penetration into the rock were studied by means of strain gauges attached to the chisels and the taking of high-speed motion pictures of the amplified output of the strain gauges as shown on an oscilloscope screen. Static tests of slowly applied loads showed identical stress patterns as the chisel was forced against the rock. It was therefore concluded that there is no inherent advantage in percussion as such, except in cases where it may be used in place of greater weights on the bit in order to drill a vertical hole.

Effects of the Drilling Fluid on Bit Performance

Several attempts, most of them at least partially successful, have been made to correlate such mud properties as viscosity and filtration loss with the drilling rates of rolling cutter bits. Both field data and laboratory experiments which used small bits of about one and one-fourth inches in diameter have shown generally similar correlations. In both cases a statistical analysis of the data is usually required to reveal any relationship. The rock strata in adjacent wells are never identical, and the teeth on the bits may not be exactly the same with respect to wear and the condition of their surface at the time of penetrating a particular stratum. Small blocks of rock used in the laboratory machine, even though cut side by side from a larger piece of apparently uniform rock, may drill at rates which differ up to 20 per cent or more. Nevertheless, statistical averages show definite relationships for any given set of conditions. Good correlations are generally obtained between mud viscosity and drilling rates, particularly for thinner muds. If muds having viscosities above about 45 seconds Marsh Funnel viscosity are included in the data, the water-loss quality may be best for the broader correlation. The drilling rate decreases as solids, particularly colloids, are added to the mud; and the viscosity increases and the filtration loss decreases with the addition of these solids. A good-quality mud may give a drilling rate less than half that obtained when circulating water.

There are several possible reaons why mud properties may affect the drilling rates of rolling cutter bits. However, these are not universally accepted or proved theories regarding the exact mechanisms whereby drilling rates are retarded and should be considered as speculation rather than proved facts. One theory, involving the removal of chips from their original position in the rock, holds that it is more difficult for a chip to be moved where a more viscous fluid must flow under the chip than if the chip moves into a less viscous fluid. Drilling fluids having a high filtration-loss quality would, of course, also aid in facilitating the initial movement of a chip, since the more abundant filtrate could more readily flow under the chip and at least partially relieve the hydrostatic pressure of the mud which tends to hold the chip in place. Less viscous fluids may be presumed to be better adapted to picking up the drill cuttings and carrying them away under the conditions of high turbulence caused by the discharge from the watercourses and jets of the bits. Fluids of greater density are generally presumed to have the advantage with respect to sweeping up the chips, but higher densities are unfortunately associated with high hydrostatic pressures, which tend to hold the chips in position in the formation and thereby retard drilling rate.

Another possible effect from the drilling mud is in relation to its lubricating quality between the bit and the bottom of the hole. Whether or not mud filter cake is formed on the bottom of the hole during drilling operations seems to be a controversial question at the present time. The greater hydrostatic pressure which exists in the mud, as compared with the pressure of the fluid inside the pores of the rock, would tend to cause filtration and therefore the formation of a filter cake. However, the formation of any filter cake would be limited by the action of the bit and by the flow of the drilling fluid, particularly where the flow of the fluid is directed to the bottom of the hole, as is the case with jet bits. A lubricating effect is apparently obtained from the colloidal fraction of the mud solids. Where larger-size bit cuttings are desired for geological examination, they are commonly obtained by adding more bentonite to the mud. The teeth of rock bits commonly make a track of small ridges and valleys as the bit rotates on bottom. Better lubrication on the bottom of the hole apparently permits the ridges to grow bigger before being broken off to form larger drill cuttings. Water used as a drilling fluid reduces the size of the larger drill cuttings and results in a faster rate of drilling, other conditions being equal. When air or gas is circulated as the drilling fluid, the drill cuttings obtained at the surface include a large amount of dust. In the latter case, as well as in the case of drilling with water, an opposing theory holds that the original cuttings are ground up or pulverized by the action of the drill string in their travel up the annulus to the surface.

The hydrostatic head of the drilling fluid has been shown to exert a strong influence on drilling rate.[1] In addition to laboratory data, field tests have been performed in which the drilling rig was so equipped that a back pressure could be held against the mud flow returning to the surface. The application of a back pressure caused a reduction in the rate of penetration of the drill. It is presumed that the hydrostatic head of the mud retards drilling rate because such pressure tends to hold the rock in place and therefore helps the rock to resist the fracturing which is necessary to produce a chip or drill cutting.

Two fluid-drilling systems. As a general rule, the drilling-fluid characteristcs desirable for controlling formation pressures and for preventing caving of exposed shales and damage to prospective producing horizons are opposed to those characteristics which permit a faster rate of penetration by the drill. Systems containing several strings of pipe for circulating or isolating columns of different fluids have been proposed, and early patents were issued on some of these systems. Some of the systems included pack-off elements between the bottom of a string of pipe and the drilled formation. The mechan-

[1] A. S. Murray and R. A. Cunningham, "Effect of Mud Column Pressure on Drilling Rates," *Trans. A.I.M.E. Petroleum Division*, Vol. 204 (1955), 196–204.

ical difficulties and additional expense of maintaining multiple strings of pipe have apparently precluded the development of such systems. The only two-fluid system which has been used in modern rotary operations is that utilizing aerated mud, in which compressed air is pumped into the top of the drill string along with liquid mud for the purpose of reducing the hydrostatic pressure of the mud and thereby increasing the rate of penetration of the bit.[2]

Metallurgical Influence

The performance of modern drilling tools is the result of developments both in mechanical design and in metallurgy. The bit manufacturers especially deserve commendation for maintaining continuous metallurgical research programs which have resulted in stronger, tougher, and more wear-resistant parts for the drag types and rolling cutter types of drill bits. The manufacturers of diamond bits have also contributed by providing a harder matrix metal for holding the diamonds, which has made possible the development of diamond bits that will drill out the entire hole as opposed to merely cutting an annulus. The need for even better materials is obvious in the light of desirable horsepower concentrations. It would appear to be desirable to concentrate five to ten times present amounts of horsepower in the area represented by the bottom of the hole provided drilling tools were available which would give sustained faster penetration rates.

Special Drilling Systems

The rotary system of drilling was introduced into the petroleum industry in 1901 at Spindletop by Captain Lucas. It was initially used for drilling wells which had to penetrate soft shale sections that could not be drilled or prevented from caving by cable-tool methods. For drilling such softer rocks, the rotary system has always been eminently satisfactory, especially when large volumes of mud—from ten to twenty barrels per minute—were supplied to the bit. The cable-tool system of drilling performs well in many areas of harder rock formations down to depths of about two to four thousand feet. At greater depths the total elasticity of the cable, which requires a slower drilling motion, the relatively longer time which must be spent in hoisting the drilling tools and the bailer, possible trouble from caving formations, the difficulties of controlling high or even normal pressure formations, and the necessity of running multiple strings of casing to seal off successive water sands all combine to make cable-tool drilling

[2] R. A. Bobo, G. S. Ormsby, and R. S. Hoch, "Phillips Tests of Air-Mud Drilling," *Oil and Gas Journal* (Feb. 7, 1954), 104.

unattractive. Rotary methods were soon applied to almost all of the deep-drilling operations. The fish-tail and other drag bits which were successful in drilling soft shales were replaced by rolling cutter bits, such as the cone bit invented by Howard R. Hughes in 1909. Penetration rates in the harder rocks such as hard limestones and dolomites and quartzites proved to be quite low, and thirty feet of hole might be drilled per day in some formations as compared with several hundred feet per day in softer formations. Such conditions have furnished an incentive for developing modifications of the rotary system which might be better adapted to penetrating the harder rock formations, particularly since excess horsepower is available at the rig during drilling operations.

Rotary percussion tools. Several percussion tools have been developed in which a section of round steel shafting three or four inches in diameter and ten feet more or less in length has been reciprocated vertically inside of the drill string just above the bit and has acted as a hammer delivering blows downward on top of the bit. The hammer is actuated by the flow of drilling mud through the tool, and the mud is subsequently discharged through the drill bit.

One of these tools was developed in the 1930's by Harry Pennington, of San Antonio, Texas. This tool used two tappet-type valves with a mud-flow passage down through the center of the hammer, all arranged so that the valves would seat and unseat at positions that would cause the mud pressure to raise and drop the hammer. The hammer was stopped in its upward course by compressing a coil spring. The bit was attached to the main body of the tool through a spline which permitted rotation of the bit, but the longitudinal motion in the spline caused the force of the hammer blows to be transmitted directly to the bit, and the mud-flow valves could not close nor could the tool operate until set on bottom so that the spline joint was retracted and the anvil forced up into the operating position. Many of these features are found in later tools. The Pennington tool was equipped with a star-type drill, which was slowly rotated on bottom while the mud-actuated hammer furnished essentially all of the drilling energy.

The Bassinger rotary percussion tool was developed in the 1940's and early 1950's. This tool uses cylindrical or sleeve-type valves, and the sleeve moves over the hammer during part of the stroke. Operating speeds of approximately two hundred blows per minute have been attained. Rolling cutter bits are used with this tool, and the action of the teeth of these bits combined with mud circulation readily removes from under the bit the pieces of broken-off bit teeth which accumulate in rotary-drilled wells. This tool combines the usual action of rolling cutter bits with a percussion hammer.

There appears to be some doubt, however, whether or not an entirely new drilling action is added to the bit, for the string of drill collars above the bit pick up a vibration from the rotation of the bit on bottom during ordinary drilling, and the string of drill collars usually has many times the size and weight of the hammer. Such tools appear to offer advantages in those cases where only a light weight may be carried on the bit in order to straighten or maintain a straight vertical hole. The disadvantage of such tools lies in the flow-pressure pulsations caused in the mud column by the reciprocating hydraulic tool at the bottom of the hole. Such pressure pulsations are undesirable where mud fluid may be lost into formations and may also cause some caving of exposed formations. A similar tool, which appears to have improved operating characteristics, has been developed by the Gulf Production Research Laboratories.[3]

Jet-pump pellet-implact drill. An experimental and theoretical investigation of the possibilities of an entirely different concept of drilling was conducted by the Carter Oil Company Research Laboratory in the early 1950's.[4] The system consisted of drilling a hole by means of steel pellets repeatedly impinging against the rock and thereby breaking and crushing it. The steel pellets used in drilling a nine-inch-diameter hole were one and one-fourth inches in diameter, and experiments were also conducted with pellets having smaller diameters. Drilling mud was pumped out through a primary nozzle under sufficient velocity to form a jet pump to draw fluid from the annulus down through a secondary nozzle. The pellets recirculated with the annulus fluid and were accelerated by the fluid flow through the secondary nozzle. It was determined that about 80 per cent of the kinetic energy of the pellets was utilized in disintegrating the rock. However, fluid friction in the jet pump permitted transfer of only about 35 per cent of the energy of the primary nozzle to the secondary nozzle, and since 32 per cent of the energy of the primary nozzle was utilized in accelerating and suspending the pellets, only 3 per cent remained to be applied to the formation. Both theoretical and experi-

[3] E. Topanelian, Jr., "The Application of Low Frequency Percussion to Hard Rock Drilling," presented at the Annual Fall Meeting of the Society of Petroleum Engineers of A.I.M.E., Dallas, October 6–9, 1957.

[4] "Development and Testing of Jet Pump Pellet Impact Drill Bits," by J. E. Eckel, F. H. Deily, and L. W. Ledgerwood, Jr. (The Carter Oil Company), presented at the Annual Fall Meeting of the Society of Petroleum Engineers of A.I.M.E., New Orleans, October 2–5, 1955.

mental studies indicated that the transfer of energy is very poor between the pellets and the flowing stream.

Magnetostriction-vibration drilling. During the years 1949 to 1955 an organized research program was sponsored by approximately forty-four oil well drilling and oil-producing companies organized under the name of Drilling Research, Incorporated. The general purpose of the research program was to survey all known methods of drilling, other than conventional, and investigate experimentally the more promising methods. Such methods as disintegrating or melting the rock by means of the down-hole combustion of acetylene were considered. This method has proved successful in drilling shallow shot holes for hard-rock quarrying. Similar use of the electric arc was investigated experimentally. However, the major research effort was directed toward producing a vibratory drilling apparatus and establishing theories of rock failure (disintegration) which might be applicable for such a drilling apparatus.

The principal mechanism investigated for producing vibrations was an electromechanical transducer. The operation of this machine is based on the magnetostrictive property exhibited by such materials as iron, nickel, cobalt, and particularly many alloys of these materials. Magnetostriction is the change of dimensions of a material caused by a change in the magnetic field surrounding that material. An alternating electric current flowing through solenoids surrounding a laminated core of resonant dimensions was used to induce vibrations in a star-type bit attached to the bottom of the drill string. The bit was rotated and mud was pumped down through the drill string and bit in the conventional manner. A power consumption of about 40 kilowatts was developed, and frequencies of about 230 cycles per second were employed. About 50 feet of $9\frac{7}{8}$-inch-diameter hole were drilled in a test well drilled to 278 feet by conventional rotary methods. The rate of penetration in the vibratory drilling was about twice that by conventional rotary drilling. However, laboratory experiments showed that the drilling rate of the star-type bits tends to decrease with wear, under conditions of low force per impact, and as the sharp chisel edges become dull, the power developed in the apparatus was disappointingly low.

Sonic drill. Experiments have been performed with other types of appratus for inducing longitudinal vibrations (sound-type vibrations) in the lower end of the drill string. One such apparatus consists of two eccentrically rotating masses mounted at the same height at some advantageous position in the string of drill collars. These masses are driven by a simple mud-flow turbine and rotate in opposite directions.

Vibration forces couple the two masses together so that they rotate at the same speed under such conditions that the horizontal forces cancel each other but produce maximum vertical forces. One objective of a device of this type is to produce standing vibration waves in the drill string. In standing waves, in which the nodes remain at the same position, the wave forces are at a maximum within the nodes where the velocity is zero, but the motion of the solid particles increases away from the nodes and passes through a maximum midway between the nodes. It is difficult for such devices to add any new mechanism to the drilling process, since the drill string naturally vibrates while rotating rolling cutter–type bits on bottom, and it is impossible to rotate such bits on bottom without causing drill-string vibrations.

Turbo drills. The concept of rotating the bit by means of a turbine located in the bottom of the drill string and driven by the circulating drilling fluid has long held great fascination for many persons associated with the drilling industry. Several arguments may be advanced in favor of such an arrangement. The drilling fluid must be circulated in any event to remove drill cuttings and maintain the hole. The rotary power to turn the bit is developed at the bottom of the hole adjacent to the bit. Probably the most valid argument is that much faster rates of rotation of the bit can be achieved and therefore greater horsepower can be supplied to the bit. The validity and applicability of the last statement depends entirely upon the capacity of the bit to utilize the increased rotary speeds and horsepower in an economical manner—which, in turn, probably depends largely upon the nature of the formation being drilled, along with other factors. Also, in deep-drilling operations, the rate of penetration tends to become less important in comparison with the total feet drilled per bit as a factor in over-all drilling-rate progress, since a round trip with the drill pipe must be made in replacing each worn-out bit. Again, for those cases in which the volume rate of drilling-fluid circulation must be increased for the sole purpose of developing power at the bottom of the hole, the additional flow-friction pressure losses occurring in the circulating system make the hydraulic transmission of power to the bottom of a deep hole inefficient as compared to mechanically rotating a string of pipe in the hole.

A single-stage turbine was patented by C. G. Cross, of Chicago, in 1873, possibly for the purpose of rotating a diamond bit. Some of the early turbines rotated on a horizontal shaft, and the bit was rotated through gears. Such gear arrangements are quite vulnerable to wear because of abrasive substances in the drilling mud, as proved by experiences in both the United States and Russia. The modern multistage axial-flow mud turbine was patented in 1924 by C. C. Scharpenberg of the Standard Oil Company of California. One of the early models had thirty stages, and the nine-inch-diameter unit developed 92 horsepower at 700 rpm from a flow of 550 gallons of mud per minute with a pressure drop of 580 psi across the turbine. In the 1930's, periodicals carried accounts of turbine developments in Russia. By 1949, Russian turbines were equipped with rubber bearings, a significant development, for mud with less than about 1 per cent sand is a fairly good lubricant for rubber bearings but will quickly wear out steel bearings. By 1954, 83 per cent of Russian oil wells were being drilled with mud turbines. In 1956, Dresser Industries arranged to have several of these turbines brought to the United States, and some experimental drilling has been conducted in relatively soft limestone.

In the period from about 1947 to 1950, a mud turbine was constructed in the United States under the direction of Critchel Parsons, of Dallas. Several hundred feet of hole were drilled in relatively soft rocks found at shallow depths in Louisiana and southern Arkansas. At that time sufficient mud volume could not be supplied to the turbine for sustained operation. More recent experiments with the turbodrill in rocks of medium hardness indicate some change in opinion regarding the best place where the turbodrill might be used to advantage. Whether or not changes and developments in bit design will permit fuller utilization of the high turbine rotary speeds of 500 to 800 rpm for faster drilling rates remains to be seen. Bits currently in use apparently will not stand up under these speeds. Advances in metallurgy and increasing knowledge of the mechanisms of rock failure will undoubtedly play their parts in all future developments.

High-speed–low-torque turbine. The development of a high-speed–low-torque drilling device was reported by the Humble Oil and Refining Company in 1957.[5] In this device, the bit and turbine rotated on a horizontal shaft. The cutting wheel was smaller in diameter than the hole, so that the drill pipe was rotated while drilling. The assembly provided for forcing the cutter wheel (and turbine) outward from the axis of the drill string, by means of mud pressure operating on a piston which was set at a slight angle to the axis of the drill string, when the tool was in operation. The turbine used was a single-stage impulse type. Limited success in field trials of the device was reported. Tungsten-carbide teeth were set in the rim of the turbine wheel to serve as cutting elements. Such use of higher speeds and lower torques is op-

[5] G. E. Cannon, "Development of a High-Speed Low-Torque Drilling Device," presented at the Annual Fall Meeting of the Society of Petroleum Engineers of A.I.M.E., Dallas, October 6–9, 1957.

posed to general trends of development in both drilling and machine-tool practice. The former has developed toward lower rotary speeds with greater weights applied to the bit. The latter has developed toward lower cutting speeds with a greater depth of cut in shaping materials like steel. Rocks vary considerably in their properties, and future developments of such high-speed cutting devices will be followed with considerable interest.

Trends in Development

The major developments in the drilling industry in the recent past have been concerned with increasing pentration rates and with drilling operations in the open water, as in the Gulf of Mexico. The mobility of rigs has been increased, and several types of mobile derricks have been introduced. Larger mud pumps and greater hydraulic and hoisting horsepower have been furnished to the drilling rigs. All associated industries and technologies have made their many contributions to the over-all efficiency of drilling operations. As far as future developments are concerned, however, the heart of the problem is centered around the drill bit or drill mechanism. All other developments are for the purpose of operating the drill itself.

The developments since about 1940 which have contributed most to faster penetration rates include (1) the use of heavier strings of drill collars and the application of greater weights on the bit; (2) recognition that thick muds greatly reduce drilling rates and the resulting use of thinner muds (including water) where possible as drilling fluids; (3) the development of jet bits which discharge mud directly on the bottom of the hole and the use of greater circulation rates; (4) the development of air-gas drilling, which has resulted in greatly increased penetration rates where this technique is feasible; and (5) the use of oil-emulsion muds, which usually result in more freely operating drill strings and apparently give somewhat better penetration rates in certain formations. Harder formations, such as hard limestones, have responded to greater bit weights and the use of water or gas as circulating fluids. Such formations have responded much less to jet bits than have the softer rocks such as most of the shales.

A significant trend in oil-production equipment is the development of tools with smaller diameters. Permanent completion techniques, which permit the abandonment of one zone and recompletion in another zone without moving in a drilling rig, are based on the use of tools which can be run into the well through the tubing. For example, perforating tools have been developed which can be run through 2-

inch-diameter tubing. Seven-inch casing was more or less standard size for the oil string, but more and more $5\frac{1}{2}$-inch casing is being used at present. If the trend continues, the use of even considerably smaller casing may be anticipated. The cost of such materials as tubular goods, including drill pipe, casing, tubing and drill bits, is an incentive to use holes of smaller diameter.

The oil-producing companies desire to have smaller-diameter holes drilled in order to realize savings in the amounts of drilling mud and cement required for the well. However, the $7\frac{7}{8}$- and $8\frac{3}{4}$-inch-diameter rolling cutter bits are preferred for drilling the major portion of the hole because rolling cutter bits of smaller diameters do not have comparable bearing strengths, so that less weight per inch of bit diameter must be carried and penetration rates are slower. Six-inch-diameter rolling cutter bits perform well and have been used to deepen a hole in which 7-inch casing has been set and cemented. Smaller bits are manufactured and used, such as the $4\frac{3}{4}$-inch diameter, but they are progressively less satisfactory. On the other hand, diamond bits of smaller diameter perform as well and possibly better than larger-diameter bits, particularly where they can be used to cut a thin annulus. However, for deep-drilling operations, development work is needed in the design of smaller-diameter bits and the technique of using them before full advantage can be taken of the smaller-diameter production tools. The total cost of the well is influenced more by the average rate of penetration than by any other single factor.

There are two interrelated problems to be considered in connection with smaller-diameter completions. One concerns the size of the drilled hole. Any reduction of hole size involves the reduction of drill-string-hole and casing-hole clearances which are necessary to prevent losses of drilling mud and cement into exposed geological formations, as discussed in Chapter 18. It also involves the efficiency of the drill bit. The second problem concerns the size of the oil-string casing which can be used in the well. A very considerable amount of work has been done in the development of completion tools which can be used inside of 2-inch tubing, including perforating and squeeze cementing. However, present-day fracture treatments of wells many times require the pumping of large volumes of fluid down through the $5\frac{1}{2}$-inch or 7-inch casing. Fracture fluids of the future, compounded so that they have a lower fluid-loss quality, may be adequate for treatment through 2-inch pipe.

CHAPTER 16

Coring

The objective in drilling an oil or gas well is to locate a hydrocarbon-bearing structure which will produce oil and/or gas in quantities sufficient to repay the cost of drilling and completing the well and also provide a nominal profit to the driller and owner. During the course of drilling, therefore, more precise information may be necessary concerning the lithologic and fluid-bearing characteristics of the formation before a decision can be made to complete the well and spend many additional thousands of dollars for completion equipment and services. One of the most reliable sources of information on the lithologic and fluid-bearing characteristics of a reservoir is an actual sample of the reservoir rock, with the fluid contained in it. *Coring* is the term applied to the technique whereby relatively large samples (by comparison with the normal size of the drill cuttings) of reservoir material are removed from their native state and brought to the surface for physical examination. Normally the most important information desired from the coring technique concerns (1) porosity, which is a measure of the fluid-carrying capacity of the formation; (2) permeability, which indicates whether the formation fluids will be able to flow at rates fast enough to permit economical production of the hydrocarbon fluids; (3) water saturation; and (4) hydrocarbon saturations, including the relative percentages of oil and gas. Cores are also obtained for a variety of other reasons, including geological studies, studies of fracture patterns in fractured formations, and studies of formations in order to obtain better well completions.

ROTARY CORING

Rotary coring was probably first introduced by the French engineer Leschat in 1863; however, it did not come into general use in the oil industry until the early 1920's. In order to obtain a core with rotary drilling tools, provision must be made for cutting the formation in the desired shape and retaining the core. The *rotary coring bit* is used to cut the core, and a *core barrel* is used to retain the core after it has been cut.

The *punch* type of coring equipment was one of the first rotary coring devices used to obtain cores in the oil industry. This equipment, shown in Fig. 16–1, is, as the name implies, essentially a punch. To obtain a core, the drill pipe was not rotated but spudded, i.e., alternately lowered and raised. The cored material was retained in the upper portion of the device. One of the earliest coring bits which utilized a rotary action to cut the core was the *Poor Boy*, or *Texas-type*, shown in Fig. 16–2. This core bit was made in the field by cutting, on the lower end of a section of pipe, teeth which would increase the penetration rate of the core bit. After a sufficient length of core had been cut, the

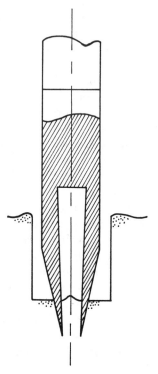

FIG. 16–1. Punch-type coring equipment.

teeth on the end of the bit could be turned in by rapidly rotating the drill pipe with increased weight on the bit and a reduced circulation rate. This prevented the core from being lost as it was brought to the surface. Various refinements of this Poor Boy type of bit were made in an effort to increase core recovery. One of the most important of these refinements was the isolation of the lower segment of the core bit from the drilling-mud circulating path, as shown in Fig. 16–3. Ports placed above the isolated section permitted normal drilling-mud circulation. Core recovery in soft formations was materially increased by the addition of the innovation.

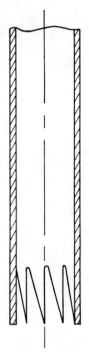 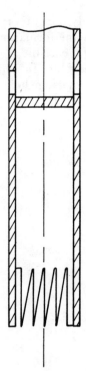

Fig. 16-2. "Poor Boy" or Texas-type coring device.

Fig. 16-3. Variation of "Poor Boy" coring device.

Rotary coring equipment has continued to keep abreast of technological developments in the drilling industry. At the present time there are basically three types of rotary coring equipment: (1) conventional coring, (2) wire-line retrievable coring, and (3) diamond coring.

Conventional Coring

Conventional coring is a continued development of the original rotary coring methods. Special equipment required in conventional coring includes a core bit, which is located on the extreme lower end of the drill stem, and a core barrel, for retaining the core after it has been cut, which is located immediately above the core bit. Figure 16–4 illustrates a typical conventional coring assembly.

The basic requirements for a good rotary core-cutting head are essentially the same as for a good drilling bit, because both are performing essentially the same function. Therefore the principles outlined in Chapter 10 on drilling bits also apply for coring bits. The problem of design of a good coring bit is somewhat more complicated, however, because only the outer rim of the formation is cut, leaving a maximum amount of the formation intact. Therefore, the cutting and bearing surfaces of a core bit are considerably smaller than the same surfaces on a drilling bit. In addition to the normal requirement that a bit should drill a gauge hole as fast as possible with minimum wear on the

FIG. 16–4. Conventional coring assembly.

cutting surfaces, the coring bit must also satisfy the additional requirements of (1) cutting an optimum-size core in such a manner that (2) the maximum amount of the core can be retained and brought to the surface for examination.

A coring bit with a drag-type cutter has been found to be a good soft-formation coring bit, while rolling cutters are better adapted to the harder formations. Typical core bits are shown in Fig. 16–5. The

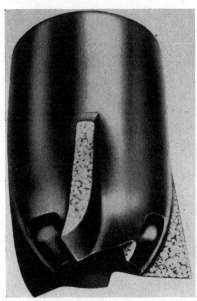

Fig. 16–5. Core bits.

arrangement and number of cutters on rolling cutter core bits can vary quite widely. Since in coring operations it is highly important that the cutting surfaces of the bit be kept clean and free of cuttings, particular attention must be directed toward the placing of the nozzles through which the drilling fluid circulates.

As one of the major objectives in coring is to recover and bring to the surface 100 per cent of the cored formation, the development of suitable core-retaining equipment is essential. The major item of such equipment is the *core barrel*. The core barrel has evolved from a very simple retaining space above the coring head, or bit, as shown in the punch-type core bit of Fig. 16-1, to the specialized circulating and retaining equipment shown in Fig. 16-4. As shown in Fig. 16-4, a conventional core barrel consists essentially of (1) an inner barrel, (2) an outer barrel, (3) a core retainer, or catcher, and (4) a vent or pressure-relief valve for venting pressure on the inside of the core barrel to the outside of the drill stem. Various innovations have been introduced in an attempt to increase core recovery, such as an inner core barrel which is not attached to the outer core barrel and therefore is free to either rotate or remain still, and various designs of core catchers for both hard and soft formations. Drilling fluid circulates between the inner and outer barrels but cannot pass through the inner barrel, with the result that there is increased core recovery and less flushing of the formation fluids by the drilling fluid.

The advantages of conventional coring are that a large-diameter core, as much as five inches or more, can be obtained in one operation. The principal disadvantages of the conventional coring technique are that the drill pipe must be removed from the hole and the special equipment attached before coring operations can begin; normal drilling operations cannot be resumed without removing the drill pipe from the hole and replacing the coring equipment with the normal drilling equipment; and the core cannot be recovered without removing the entire drill string from the hole.

Wire-Line Coring

In order to overcome the previously cited disadvantages of conventional coring, methods have been developed for obtaining a core, bringing the core to the surface, and proceeding with normal drilling operations, all without removing the drilling tools from the hole. This is accomplished by inserting the proper equipment in the lower part of the drill stem by means of a wire line which can be run inside the drill pipe. This *wire-line coring*, as it is called, has materially decreased the cost of obtaining cores, and thus many more cores may be obtained than would otherwise be possible. As the average depth of wells continues to increase, the time and money saved by not having to remove the drill pipe in order to obtain a core is substantial. A wire-line coring assembly is shown in Fig. 10–14. The only special equipment required on the lower end of the drill stem is a core bit. Before coring operations are initiated, the drill pipe must be removed from the hole and the conventional drill bit replaced with the core bit. After this has been accomplished, intermittent coring and drilling operations can be conducted without removing the drill pipe.

To obtain a core after the core bit is in place, the core-barrel assembly is forced down the inside of the drill pipe using drilling-mud pressure. When the core-barrel assembly reaches the lower end of the drill stem, a locking device holds the barrel in place. The core-barrel assembly, shown in Fig. 10–14, consists of a *cutter head, core catcher, core barrel, vent* or *inside pressure relief, locking device,* and a *retrieving head*. During coring operations, the circulating fluid passes between the core-barrel assembly and the drill collar. After the core has been cut, the core-barrel assembly with its core is retrieved by lowering through the drill pipe on a wire line a retrieving tool, or *overshot*, which is designed to engage the upper end of the core barrel. As the overshot is lowered over the upper end of the assembly, the locking devices are released, permitting removal of the entire assembly. As much as fifteen feet of core can be obtained in one operation.

When a core is not required, the core-barrel assembly is replaced with a center bit assembly, which allows conventional drilling operations to proceed until another core is desired. This center bit assembly, shown in Fig. 10–14 is essentially the same equipment as the coring equipment, except that the core cutter is replaced with a full cutting head. The center bit assembly is run and retrieved in exactly the same manner as the coring assembly.

There are so many variations in coring equipment that a complete description of all that is available cannot be included, here. However, this discussion should serve to point out the principles involved in coring operations.

The use of the wire-line coring technique is especially suitable for wildcat drilling operations where coring depths are not known in advance, because a program of alternate drilling and coring can be used without removing the drill pipe to change bits. The principal advantage of the wire-line coring method, in addition to this coring and drilling feature, is the great saving in rig time during coring operations. The principal disadvantage is the smaller-diameter core obtained.

Diamond Core Drilling

In order to increase both core recovery and penetration rate, use has been made of a diamond-faced coring bit. Diamond bits may be used to advantage in coring hard, dense formations where the cost of coring with rolling cutter bits is high. Although the cost of a diamond bit may be as much as fifteen to twenty times the cost of a conventional core bit, the reduction in the number of round trips and the increased penetration rate in many cases make the diamond core bit more economical. Because of the longer life of the diamond core bit, conventional cores as long as ninety feet can be obtained without difficulty in some areas. This reduces the number of round trips necessary by as much as one-third in formations where long coring intervals are desired. Wire-line coring equipment has also been developed for use with diamond coring bits. This equipment has essentially the same features as other wire-line coring equipment.

The diamond coring bit consists of a steel body to which a matrix is attached, which, in turn, holds a large number of diamonds. The diamonds, which form the cutting surface, usually extend across the face of the bit and cover a portion of both the inside and outside surfaces. The diamond bit must be designed to cause as little wear on the matrix as possible, for otherwise the matrix could be worn away and diamonds lost. Diamond core bits are specially designed for the formations to be cored. Relatively large stones are normally used to core formations where large cuttings are possible, such as large-grained sandstones, sandy shales, conglomerates, or fractured formations. For the fine-grained dense formations such as limestones, small stones are used in the core bit. The matrix must be as abrasion resistant as possible. At present the most commonly used method of attaching the diamonds to the core bit is by the use of a powdered-metal matrix, such as tungsten, cobalt, molybdenum, nickel, or other metals or alloys. The diamonds are placed in a mold with the powdered metal, and heat and pressure are applied until the mass becomes sintered. By this process a matrix is produced which will bond extremely well to the diamonds.

DESIGN OF EQUIPMENT

Many factors must be considered when designing satisfactory coring equipment. As has been pointed out, the goal in coring is to obtain 100 per cent recovery of formation and bring the cored formation to the surface without disturbing the fluid content of the core. In actual practice, it may not always be possible to achieve completely all the desired objectives. The one objective which has been sacrificed most often is keeping the original fluid content of the cores undisturbed, and this phase will be discussed in more detail later in this chapter.

A core of maximum diameter is desired; however, this objective must be tempered somewhat by other requirements: (1) the maximum possible bearing surface for the bit cutters, (2) sufficient clearance between the hole being drilled and the outer barrel assembly to prevent sticking the core barrel, (3) ample clearance between the outer barrel and inner barrel to permit free drilling-mud circulation, and (4) clearance between the inner barrel and the core to permit free entry of the core into the barrel.

Smooth drilling during coring operations is particularly desirable, because rough or uneven drilling will cause excessive core breakage with attendant poor recovery. Increasing the number of cutting surfaces will increase the smoothness of drilling operations, but in certain types of sticky formations, such as shales, close spacing may prevent proper cleaning of the cutting surfaces and thus materially reduce penetration rate.

Reverse-Circulation Coring

In an effort to increase core recovery and at the same time eliminate the necessity of retrieving the core, either by removing the drill pipe in conventional coring or retrieving the core barrel in wire-line coring, the technique of reverse-circulation coring has been developed. The principal feature in this method of coring is, as the name implies, the change in the flow path of the drilling fluid. In the reverse-circulation coring method the drilling fluid is circulated down the outside of the drill pipe and returned back to the surface through the interior. In this manner, at least theoretically, all of the formation which is being penetrated, including the small drill cuttings, will enter the inside of the drill pipe and can be collected. Originally, attempts were made to circulate the cores to the surface without the use of a core barrel within the drill pipe, but difficulties encountered in passing the core through surface connections resulted in placing an inner core barrel at the surface in some installations. After five to ten feet of core had been cut, the barrel would be removed and the core extracted. Having the core barrel at the surface is, of course, a distinct advantage. Reverse-circulation coring has not achieved widespread use because of the following disadvantages: (1) a streamlined path for the core must be provided and (2) because drilling-mud velocity at the cutting surface is less than in the conventional circulation method, a reduced penetration rate results.

Side-Wall Coring

Side-wall coring is a supplementary coring tool. It can be used in zones where core recovery by conventional or wire-line methods is small, or in zones where the latter cores were not obtained as drilling progressed. Side-wall cores can be obtained at any time after the formation from which a core is desired has been penetrated.

The introduction of the electrical log as a reliable formation-evaluation tool increased substantially the use of the side-wall coring device. The electrical log may indicate that a particular zone has possibilites for economical petroleum production; however, in order to evaluate this formation's potentialities more fully, an actual specimen is necessary. The side-wall coring device is lowered into the hole, usually on a logging cable, and a sample of the formation at the desired depth is obtained. A combination of these two tools has saved the oil industry untold millions of dollars by locating petroleum-producing zones that otherwise would have been bypassed and by revealing as nonproductive zones which might otherwise have been completed as possible producers.

Three basic types of side-wall coring devices have been used. The punch-type side-wall sampler which was used in conjunction with the drill pipe is no longer in use to any extent, nor is the rotary type of side-wall coring device, which used a rotating coring assembly.

The percussion type of side-wall coring device, illustrated in Fig. 16–6 operates on a principle similar to that of the bullet gun perforator. However, instead of sending a bullet into the formation, a short tube is forced into it. This instrument is run on an electrical cable, of the type used in logging or perforating operations. The tube is attached to the gun body with a steel cable. Core sizes range up to $1\frac{1}{4}$ inches in diameter and $2\frac{1}{2}$ inches in length. As many as thirty cores can be obtained in one run of the tool, although, to avoid sticking the tool, only one bullet is fired at a time. Shots can be accurately placed with this tool, as an SP curve can be run in conjunction with the sampler. If a previous electrical log is obtainable, the sampler can be positioned by direct log correlation.

Cable-Tool Coring

Cable-Tool Core Barrel

Cores recovered by the cable-tool method probably more nearly approximate the natural state of cores than those obtained by any other method, since the invading action of a high-pressure column of drill-

FIG. 16–6. Percussion-type side-wall coring device.

ing fluid is not usually present. A cable-tool core must be obtained by the churning action of the tools, and the design of a good cable-tool coring device must take these factors into consideration.

A typical cable-tool core device consists essentially of two parts, an outer drilling barrel and an inner core-retaining tube. The inner core-retaining tube rests on bottom at all times during the coring operation. The outer barrel consists essentially of three parts, the drill-barrel head, the drill barrel, and the drill-barrel shoe, which actually performs the cutting operation. The inner core-retaining tube has a relief valve that permits water trapped in the core tube to escape as the core enters the tube. This prevents fluid in the tube from building up pressure which would increase water invasion into the core and reduce the core recovery. In operation the cable-tool core barrel is attached to the lower end of the drill stem. The entire assembly is lowered to the bottom of the hole, which has previously been cleaned as much possible. Churning action is begun, care being taken to insure that the tools are not raised more than three feet in order to prevent the core-retaining tube from being lifted off bottom. Cores obtained with this barrel may be as large as 2 11/16 inches in diameter and over 7 feet in length. On account of the churning action of the bit, unbroken lengths of cores are very often only a few inches in length. Core recovery is normally very good with this equipment.

Chip Coring

Chip coring is the technique in which ordinary cable tools are used, and the large cuttings are saved for core analysis. When chip cores are being obtained, the cuttings must be removed quite frequently, normally once each foot. The theory of chip coring is that these large cuttings have been cut during the last stroke or two of the drilling bit and have not been reduced in size by churning. Thus the large chips represent approximately the last inch drilled, which is the principal reason the cuttings must be removed quite frequently. Frequently chips as large as 1.0 centimeters in diameter and 1.5 centimeters in length are recovered at the surface. The cuttings are recovered with a sand pump. Satisfactory techniques have been developed for analyzing these small cuttings for permeability, porosity, and fluid saturation.

Preservation of Cores

After the cores have been removed from the core barrel, proper preservation is essential for further evaluation. Immediately after removal from the core barrel, the core is normally laid out on a flat surface for visual examination. A geologist or engineer usually makes a detailed description of the core, noting any section of formation not recovered. Usually field tests will be made on a part of the core. As these cores may be needed later for reservoir studies of some type, the cores should be sealed as soon as possible to preserve their original fluid saturations and prevent weathering. Cores may be preserved for further analysis by quick freezing, coating in paraffin, sealing in jars or other suitable containers, or sealing in plastic bags. The use of quick-freezing is recommended only when the cores will be analyzed within a very short time.

The interpretation of the analysis of cores will be discussed in Chapter 17.

CHAPTER 17

Formation Evaluation

An accurate knowledge of the lithologic and fluid-bearing characteristics of a formation is necessary before an intelligent decision can be made regarding whether the well should be completed. A wrong decision may result in the bypassing of an entire field or in the spending of many thousands of unnecessary dollars in an unsuccessful completion attempt. When a potential hydrocarbon-bearing formation is encountered, in most instances additional information will be necessary to determine accurately the character of the formations and the nature of the fluids within the formation. The information sought in the process of formation evaluation includes (1) porosity, (2) absolute permeability, (3) effective permeabilities to the various fluids, (4) interstitial water saturation, (5) original oil and/or gas saturation, (6) formation pressure, and (7) composition of the reservoir rock. These data aid in the determination of the two actual factors of immediate interest: the volume of recoverable oil and/or gas and the rates at which these hydrocarbons can be produced. With this information, an intelligent decision can be made concerning the desirability of completing the well in a particular horizon. Many tools and services are available which will aid in getting this much-needed information. Three basic types of services are used in formation evaluation: drill-stem testing, core analysis, and logging.

Drill-Stem Testing

A drill-stem test is made to obtain more accurate information about the producing potentialities of an oil or gas reservoir. This test is actually a temporary completion of a well. It derives its name from the fact that the drill stem is used as a conduit to bring the formation fluids to the surface. The principal objectives of the test are to determine the types of fluids present in a particular formation and the rate at which these fluids can be produced. Drill-stem test equipment consists essentially of a packer and valve arrangement placed on the bottom of the drill pipe in such a position that all other formations can be isolated from the formation to be tested. By means of the packer and valve arrangement, the fluids from the formation being tested are directed to the inside of the drill pipe for the desired length of time. Additional features of drill-stem test equipment include recording pressure gauges, maximum-indicating thermometers, and additonal packers and valves to increase testing efficiency.

Drill-stem tests can be made either in open hole or in casing. A typical drill-stem testing assembly is shown in Fig. 17–1. The valves which control the entry of fluids into the testing tool and drill pipe are actuated at the surface by manipulating the drill pipe or by dropping a bar which ruptures a disk. Packers, which are used to isolate the formation to be tested from the remainder of the hole, are usually the wall type. Where an up-the-hole test is desired, it may be necessary to use two packers which "straddle" the formation, thus effectively isolating the formation to be tested from other formations both above and below. A typical straddle-packer test is shown in Fig. 17–2.

Drill-Stem Test Procedure

The equipment is assembled on the lower end of the drill pipe and lowered into the hole. Figure 17–3 shows the fluid-passage diagram for all phases of the test. In order to permit flow of the formation fluids into the drill pipe, the hydrostatic head of fluid within the drill pipe must be less than the pressure in the formation being tested. For this reason the drill pipe may be run completely dry, or a predetermined amount of fluid cushion, either water, mud, or gas pressure, may be used to reduce the extreme pressure differentials which might otherwise be encountered when the tool is first opened. In unconsolidated formations a large pressure differential may cause the formation to slough with consequent possible sticking of the fools. Fluid cushions may also be necessary to protect the drill pipe from collapse.

When the zone to be tested is reached, the packer is expanded against the sides of the wall. Special consideration should be given to the selection of a packer seat. The packer should provide a satisfactory seal in a hard formation which has not been eroded; in soft, eroded formations, a packer cannot be expected to provide a satisfactory seal. A caliper log is very helpful when selecting a packer seat. Electrical logs and cores are also of material benefit. Shales usually provide a very poor packer-seating formation, and since many producing formations are overlaid by shale, it may be necessary to seat the packer in the upper few feet of the formation being tested. After the packer has been set, the lower valve on the tool is opened so that the formation fluids can enter the test tool. If an original shut-in formation pressure is desired, a disk can be placed in the drill pipe one or two joints above the tester valve. Since there is no fluid cushion between the disk and the tester valve, when the tester valve is opened, enough fluid can move up below the disk to relieve the mud-column

FORMATION EVALUATION

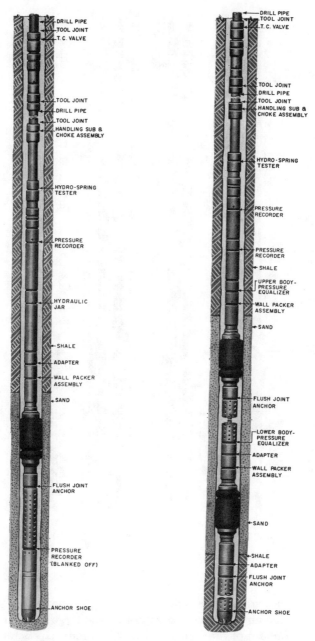

FIG. 17–1. Drill-stem test assembly.   FIG. 17–2. Straddle-packer drill-stem test.

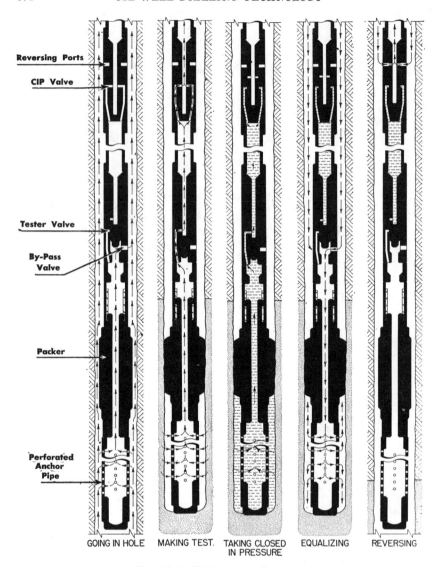

FIG. 17-3. Fluid-passage diagram.

pressure below the packer and show a true formation pressure. The volume of the air chamber should not be too large in order that only a limited amount of the formation fluids is produced. Before entering the test tool, these fluids usually have to pass through a perforated section of pipe, reducing the amount of debris that can enter the tool proper. After sufficient time has elapsed to obtain a representative initial shut-in pressure, the disk in the drill pipe is ruptured, usually by dropping a bar, and formation fluids are free to move to the surface inside the drill pipe. This period of the test is usually known as the flow period. It will vary in length from only a few minutes to several hours.

No drill-stem test should be run without utilizing at least two pressure recorders, for the application of two recorders is the only method by which an accurate analysis of the tool's behavior can be made. One of the recorders should be below the packer in the blanked-off position. No fluid will flow past this recorder during the test; therefore, it records the pressure directly from the annulus. The second recorder is in the flow-stream above the packer but below the bottom-hole choke, if it is used.

In order to analyze a drill-stem test chart correctly and safely, it is essential that two recorders be employed in the manner just described, so that the one may detect plugged chokes, plugged perforations, etc. In order to control the rate of entry of fluids into the tester, two chokes are used, one at the surface and one on the lower end of the test tool. At the end of the flow period the testing valve is closed and a final shut-in pressure is obtained. The pressure between the inside and outside of the tool is then equalized by aligning the proper ports in the testing valve. The packer is then unseated and the contents of the drill pipe are reverse-circulated to the surface.

An accurate measurement of all the produced fluids should be made, because this is one of the principal factors in determining the producing characteristics of a formation—the subsurface-pressure record is the other major factor in interpretation of drill-stem test results. A complete interpretation of the pressure records is beyond the scope of this book; however, a typical record, shown in Fig. 17–4 will be interpreted to illustrate the method:

At point S, the pressure is zero and the test tool is at the surface ready for lowering in the hole. At point A, the tool has been lowered into the hole and is now at the desired test depth. Increasing pressures from S to A show the increasing hydrostatic head of the mud column. At point B, the tester valve is opened and the pressure is reduced to the initial formation pressure. The pressure continues to increase to point C, which is the maximum initial closed-in pressure. At point C, the disk sub is ruptured, allowing the formation to produce into the drill pipe, which is at a still lower pressure. The pressure drops to a minimum at D almost immediately, and then the distance from point D to point E constitutes the flow period. At point E, the tester valve is closed for obtaining the final closed-in pressure, which is shown as point F. The line E–F is known as the pressure-build-up curve. At point F, the main tester valve is closed and the bypass valve opened, allowing hydrostatic mud pressure inside the tool. After the packer is released, the tool is removed from the hole, as is indicated by line F–G showing steadily decreasing hydrostatic pressure. At point G, the tool has reached the surface.

Absence of produced oil or gas during a drill-stem test does not necessarily prove that the zone will be non-productive. The perforated

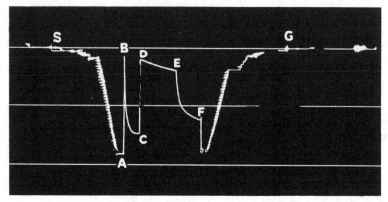

FIG. 17–4. Drill-stem test pressure record.

pipe may have become plugged—a circumstance which would be indicated by the difference in the two pressure recorders—or the permeability may have been so low that the formation would not yield its contents without artificial stimulation.

The annulus mud level should be watched carefully at all times. If the mud level in the annulus continues to drop, the packer is not performing properly, or a zone of lost circulation exists above the packer seat, or there is a hole in the drill pipe.

From the pressure history and fluid-production rates during a drill-stem test, methods have been developed for estimating permeability, permeability restrictions, and other useful reservoir-engineering information.

Logging-Cable Formation Testing

Methods have been developed for obtaining small samples of formation fluid by means of a tool lowered in the hole on a standard logging cable. One of these tools is shown in Fig. 17–5. In addition to the small sample of formation fluid, pressure build-up, static formation pressure, and mud-column pressure are also obtained. This tool can be accurately placed, as an SP curve of a conventional electrical log can be run in conjunction with the tool.

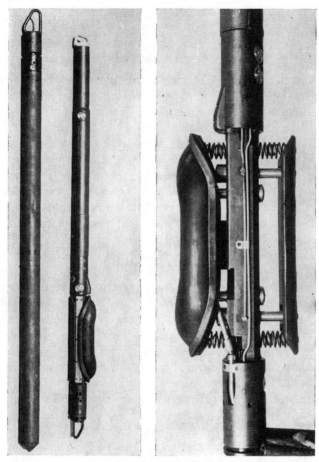

Fig. 17–5. Formation tester.

Essential elements of this tool include a pliable sealing pad which can be forced out against the formation hydraulically; a container having a capacity of approximately one gallon; and two bullets, which are fired through the pad into the formation, thus providing a channel for the formation fluids to enter the container. The pliable sealing pad is actuated by electrical controls at the surface and is held in position by a back-up shoe. This sealing pad provides a tight seal between the mud column and the formation, and when the bullets are fired through the pad, the perforations provide the flow path into the one-gallon container. The principal advantages of this tool over the conventional drill-stem test tool are its lower cost of operation and the rapidity of obtaining samples. The small sample, and particularly the conditions under which the sample is obtained, limit the application of the tool.

Core Analysis

As discussed in Chapter 16, the objective in coring is to recover 100 per cent of the cored interval and to bring this core to the surface without disturbing its fluid content. Unfortunately, this objective has never been completely realized, especially in rotary coring. Therefore, in analyzing cores, consideration must be given to changes which have been wrought on both the core and its fluid content in bringing it from its native state in the reservoir to the testing laboratory.

Two important phenomena occur in the coring process which materially affect interpretation of the laboratory core data: (1) As the core is being cut, filtrate from the drilling mud, which of necessity is being circulated at pressures which are greater than the formation pressure, displaces some of the reservoir fluids from the core; and (2) as the core is being brought to the surface, the gas, in the free state and/or in solution in the oil, expands, thereby driving out some of the oil and water which were native to the core, and the native oil shrinks as gas is evolved from solution. Thus, when the core reaches the surface, its original fluid content may have been altered drastically.

The flushing of cores is actually a displacing action, and the efficiency of flushing can be interpreted directly from the knowledge of fluid displacement. The factors which can potentially influence the degree of flushing of a core are (1) pressure differential (mud pressure over formation pressure), (2) permeability, (3) rate of penetration, (4) mud-circulation rate, (5) reservoir-fluid viscosities, (6) core size, (7) water loss of the mud, (8) properties of the reservoir rock, and (9) original fluids present in the rock. Each of these factors will be considered briefly.

Pressure differential. Given sufficient time, pressure differential will have no effect on the degree of flushing. However, since many coring operations may be for only a short period, as the pressure differential increases, the rate of invasion of filtrate will increase. As the pressure differential increases, complete flushing of the core will occur sooner.

Permeability. As the permeability increases, the rate of filtrate invasion will increase.

Rate of penetration. As the penetration rate increases, the time of flushing is reduced and the degree of flushing may be reduced. Here again, however, the degree of ultimate flushing is independent of the penetration rate.

Mud-circulation rate. Normally, as the circulation rate is decreased, the degree of flushing is also decreased. This may be attributed to less erosion of filter cake which may be built up, thereby increasing the resistance to invasion of filtrate.

Reservoir-fluid viscosities. As the reservoir fluid viscosities increase, the degree of flushing decreases. This is in accord with the established principles of fluid displacement. Displacement efficiency decreases as the viscosity of the displaced fluid increases.

Core size. As the size of the core is increased, the time required for displacement of the reservoir fluids is increased. Therefore, if the core can be obtained rapidly, the degree of flushing can be reduced as the core size is increased. It should be mentioned again, however, that the size of the core will have no effect on the ultimate flushing if sufficient time for complete displacement has elapsed.

Water loss of the mud. As the water loss of the mud increases, the volume of water available for flushing the core increases, and as a result complete flushing occurs much sooner. Therefore, if a core is obtained relatively rapidly, flushing will decrease as the water loss is decreased.

Properties of the reservoir rock. The relative permeability characteristics are an expression of several composite properties of the rock and fluids, and the degree of flushing will be governed by these relative permeability characteristics.

Original fluid content of the core. The type and composition of the native reservoir fluids will also greatly influence the degree of flushing. Oil which contains relatively large amounts of gas in solution will exhibit a correspondingly large degree of flushing on account of the liberation and expansion of the gas and the corresponding shrinkage of the oil.

As the reservoir oil saturation in a core may have been reduced to a minimum, indirect methods must be employed to approximate the original fluid content of the core. The method most generally used is to determine the connate-water saturation in the laboratory by capillary-pressure methods, and then if there are only two fluids present, as there would be if the cores were taken either in the oil zone or in the gas zone, the oil or gas saturation can be found by subtracting the connate-water saturation from unity. Expressed as an equation this is:

$$So = 1 - Cw \tag{17-1}$$

where

So = oil saturation, fraction

Cw = connate-water saturation, fraction.

The obvious disadvantage of this method is that the capillary-pressure method of determining connate water will yield a minimum water saturation, and if the water saturation is not at this minimum, then erroneous values of oil and water saturations will be obtained.

Oil, or a drilling fluid in which oil is the continuous phase, is frequently used. The core will be invaded, in this case, by oil instead of water, and the only condition which has actually changed is the type of invading fluid. Interstitial or connate-water saturation can therefore be determined with more accuracy if the coring fluid is oil. However, if the water saturation in the core is greater than the irreducible minimum, it is quite possible that the drilling-mud filtrate, in this case oil, will reduce the water saturation to the minimum value.

Statistical methods have also been developed to approximate the original fluid content of the cores. These methods are based on a knowledge of the forces which have been operating on the core to alter its original fluid content, together with statistical data showing the range of saturation for the various fluids for oil-, gas-, and water-producing zones. These statistical data are usually compiled separately for each reservoir in oil-producing areas where the lithologic character of the reservoirs is essentially the same. Gas- and condensate-producing formations exhibit very low oil saturations, varying from 0 to approximately 5 per cent saturation. Oil-productive formations also normally have low oil saturations, but these saturations will be higher than those found in gas- or condensate-producing formations.

Several methods have been proposed to preserve better the original core fluid contents. The use of balanced pressures has been suggested, with the hydrostatic head of the drilling fluid exactly equal to the formation pressure. If the formation pressure is not less than the drilling-mud pressure, then mud-filtrate invasion would be reduced substantially. This method is not practical in most areas because of the danger of blow-outs. Further, it would still not eliminate the flushing of the core because of expansion of the gas as the core is brought to the surface. Several pressure core barrels have been designed and used, in which the core, after it has been cut and while it is still at the bottom of the well, is sealed in a container at the pressure existing at the bottom of the well bore at the time the cut was made. This core can be brought to the surface without reducing the pressure on the core. The principal disadvantage of this method is that the core has been flushed by the invading drilling-mud filtrate as it was cut and the native reservoir fluids are therefore not recovered, even in the pressure core barrel. A combination of the pressure core barrel and balanced-pressure techniques, if practical, might better preserve the native core fluids.

Many additional coring techniques, designed to improve formation-fluid recovery, have been proposed. One involves the use of fluid at the bottom of the well which will not enter the pore structure of the core. However, a fluid medium that would meet this requirement satisfactorily has not yet been found.

The correct interpretation of core-analysis results requires a knowledge of all factors influencing the core and its native fluids from the time it was disturbed by the core bit until it is analyzed in the laboratory.

Cable-Tool Core Analysis

Coring by cable-tool methods usually eliminates one of the objectionable features of rotary coring—invasion of the core by a foreign fluid. Since at the present time most cable-tool cores are obtained in relatively low-pressure areas, very little hydrostatic head has to be maintained in order to control the formations encountered. The degree of flushing by gas expansion will also be reduced because of the lower pressures involved. For these reasons, oil saturations are normally much higher in cores obtained by cable-tool methods.

Logging

Another valuable method of analyzing the subsurface formation is by examination of the various logging records which are available: (1) driller's log, (2) cuttings log, (3) mud log, (4) electrical log, (5) radioactivity log, (6) microlog, (7) laterlog, (8) induction log, (9) microlaterlog, and (10) drilling-time log.

Driller's log. This is probably the oldest logging method and consists of the driller's interpretation of the formations he has encountered during his tour of duty. The value of driller's logs varies from almost worthless to very valuable, depending upon the experience and methods used by the driller in preparing the logs. A good driller's log is prepared after carefully observing drilling rate, cuttings, and action of the drilling tools. With the experience most drillers have accumulated, a valuable log can be compiled if it is based on over-all consideration of these factors.

Cuttings log. The cuttings which have been circulated to the surface or recovered by bailing in cable-tool operations provide a continuous record of the formations encountered. The analysis of cuttings from a rotary-drilled well requires considerable experience, as a great percentage of the cuttings recovered at a given depth may have been retained in the circulating system from other depths or may have fallen off the wall from some point up the hole. It is not unusual for as much as 98 per cent of the cuttings recovered at a given depth not to be representative of the formation actually being cut. There is a lag between the time the formation is cut and the time the cuttings reach the surface. This time lag depends on the depth of the well, velocity of mud flow, and efficiency of cuttings removal.

Mud log. Mud logging is a technique whereby the drilling fluid is continuously analyzed for small traces of hydrocarbons.

Electrical log. The electrical log is one of the most widely used logging tools. It provides a continuous record of the formations encountered in a bore hole. The log is obtained by lowering into the well bore on an insulated cable a tool which measures the spontaneous potential and resistivity of a subsurface formation. The electrical log may, under favorable circumstances, indicate the lithologic and fluid content of the formations. An electrical log cannot be obtained in a cased well bore.

Radioactivity log. Radioactivity logs can be run either in cased or open holes. These logs are obtained by lowering a tool, or sonde, into the well bore on an insulated cable. They measure radioaction intensity and the presence of hydrogen in the formations. This information will, in most instances, provide some indication of the nature of the formations and formation fluids.

Microlog. The microlog is a variation of the conventional electrical-logging system, where the area of investigation is quite small, normally one or two inches. It is used primarily to indicate permeable formations and is normally obtained in conjunction with the conventional electrical log.

Laterlog. The laterlog is also an electrical-resistivity measuring instrument. It is run on an insulated cable and provides a continuous record of the formations. It can be employed to good advantage where low-resistivity drilling muds, such as salt muds, are used. Thin beds can be readily identified with the laterlog.

Induction log. The induction log is another refinement of basic electrical-logging principles. It measures formation conductivity instead of resistivity. The scope of investigation is narrow, and for this reason adjacent beds have very little influence on the log. It is particularly adaptable to logging in holes drilled with oil or oil-base drilling fluids, or in empty holes, where there is no drilling fluid.

Microlaterlog. This is another refinement of the basic electrical log. The electrodes are spaced in such a manner that only a very small volume of the formation is investigated. It differs from the microlog in that the presence of a mud cake does not normally affect the measurements.

Drilling-time log. This log, which may be recorded mechanically or by the driller, is a record of the time required to drill each foot of formation. As different formations drill at different rates, this drilling-time log provides valuable information concerning the lithologic character of the formations being penetrated. Dull bits, changes in drilling speeds, and weight on the bit also affect the drilling-time log.

One of the most difficult problems encountered in drilling operations is the correct evaluation of the formations being penetrated by the drill bit. The use of all of these formation-evaluation tools would not be practical or economical. A careful study of the advantages of each technique and the applicability of these techniques in specific cases, along with the amount and type of information required, will determine which method can be most effectively used in a specific operation.

CHAPTER 18

Pressure Surges and Anomalies

CAUSES AND EFFECTS OF PRESSURE ANOMALIES

Pressure anomalies are considered to be all of the pressure variations by which the total pressure in the drilling fluid deviates either from the hydrostatic mud pressure when fluid is not being circulated or from the average flowing pressure at any point during fluid circulation. Pressure anomalies occur during a pulsating type of flow and are associated with rapid starting of drilling-fluid circulation and particularly with pipe movements in the hole. The rapid raising or lowering of a string of pipe in a hole filled with fluid is perhaps the most common cause of large pressure surges. The decreases caused in total mud pressure have been responsible for blow-outs and the loss of wells. The increases in total mud pressure have been responsible for expensive losses of mud into exposed geological formations. Where such mud losses have occurred into producing formations, they have undoubtedly in many cases injured the producing ability of the well.

BLOW-OUTS CAUSED BY PRESSURE REDUCTIONS DURING HOISTING

The first important paper on the subject of pressure changes associated with pipe movement, "Changes in Hydrostatic Pressure Due to Withdrawing Drill Pipe from the Hole," was presented before the American Petroleum Institute in 1934 by George E. Cannon.[1] This paper cited a study of twenty-seven blow-outs which occurred in the drilling of 891 wells. Of these, twenty-two were definitely associated with the withdrawal or subsequent lowering of the drill string into the hole. A total of eleven wells had to be abandoned.

Since it was known that the hydrostatic head of the mud was sufficient to overbalance the formation pressures, a review of the statistics had prompted an experimental investigation to measure pressure reductions which might occur during the raising of the drill pipe. The measurements were made by means of a subsurface pressure gauge placed in the bottom of the hole or in the lower part of the drill string. Some of the results are illustrated in Figs. 18–1, 18–2, and 18–3. This investigation showed that serious pressure reductions can occur when the drill pipe is being raised and that such reductions are related to

[1] Published in *API Drilling and Production Practice* (1934), 42–47.

the annular area between the drill pipe and the hole. The only other correlation obtained was that the pressure reductions were directly related to the gel strengths of the muds. Muds of zero gel strength have subsequently been preferred for most drilling operations, particularly for drilling in such areas as the Gulf Coast.

The fact that no correlations between pressure reductions and such quantities as the mud viscosity and rate of pipe movement were obtained has been discussed by Cardwell.[2] He pointed out that, for the high gel strength muds in use at the time, once fluid (or pipe) movement was started, no further pressure changes would be expected. The systems were controlled by gel strength so far as pressure reductions were concerned. Where muds have lower gel strengths, as recommended from the results of the investigation, other factors become of governing importance in determining the magnitude of the pressure surges.

Mud Losses Caused by Pipe Movement

The second paper on the general subject of pressure changes associated with pipe movement was presented by Goins and others in 1951.[3] Their investigation was concerned primarily with pressure increases, although pressure reductions were also noted. A total of forty-five

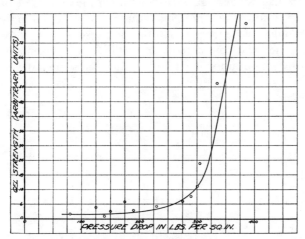

Fig. 18–1. Relation of gel strength of mud to pressure drop due to swabbing in 7-in. casing.

[2] W. T. Cardwell, Jr., "Pressure Changes in Drilling Wells Caused by Pipe Movement," *API Drilling and Production Practice* (1953), 97–112.

[3] W. C. Goins, Jr., J. P. Weichert, J. R. Burba, Jr., D. D. Dawson, Jr., and A. J. Teplitz, "Down Hole Pressure Surges and Their Effect on Loss of Circulation," *API Drilling and Production Practice* (1951), 125–32.

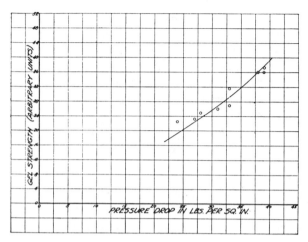

Fig. 18-2. Relation of gel strength of mud to pressure drop due to swabbing in 10¾-in. casing.

separate losses of mud had occurred during the drilling of a well. In the drilling of an adjacent well, in which twenty-two losses occurred, the possible causes were noted. Of these losses, three were associated with balled drill collars, six with downward movement of balled collars, five with downward pipe movement where no balling occurred, and eight could not be traced to a specific cause. Again an experimental investigation was conducted wherein direct measurements of pressure surges were obtained from bottom-hole pressure gauges. Some of

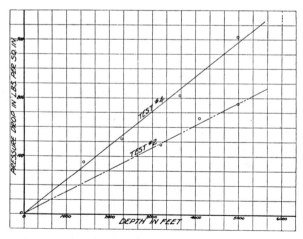

Fig. 18-3. Relation of pressure drop due to swabbing in 7-in. casing to depth.

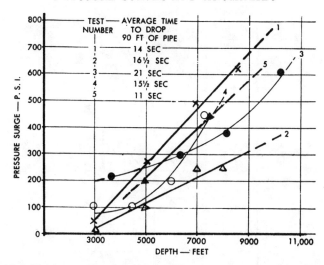

Fig. 18-4. Pressure surges caused by rapidly running one stand of drill pipe.

the results are shown in Figs. 18-4 through 18-9. While the effects of mud viscosity, gel strength, and density could not be determined, it was established that the velocity of pipe movement was a controlling factor. The muds used were low gel strength muds whose 10-minute gel strength varied from 0 to 8 grams Stormer. Whether or not the bit was plugged (a plugged bit would prohibit upward flow through the

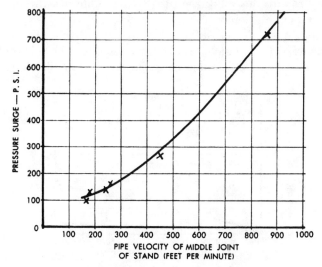

Fig. 18-5. Variation of surge pressure with rate of pipe movement when running one stand in Test 1 at 8,000 ft.

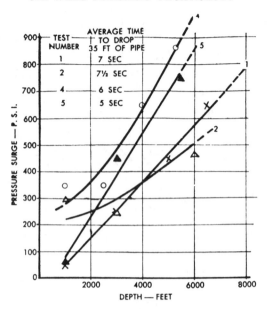

Fig. 18-6. Pressure surges caused by rapidly spudding 35 ft. of pipe while circulating.

drill pipe) was shown to affect the magnitude of the pressure surge. As might be expected, the greatest increases in pressure were obtained by rapidly lowering the drill string while maintaining a fast rate of mud circulation.

In drilling a fourth well, adjacent to the first three wells, the corrective measures adopted included the use of Hydril slim-hole tool joints in order to minimize annulus constrictions. Oil-emulsion mud was

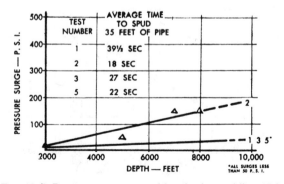

Fig. 18-7. Pressure surges caused by slowly spudding 35 ft. of pipe while circulating.

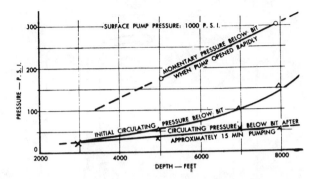

Fig. 18-8. Circulating pressures as measured below the bit in Test 2 (surface pump pressure 1,100 psi).

used to minimize balling, such as the sticking of mud cake or shale fragments on the bit, drill collars, or tool joints. The drill pipe was lowered at rates of 35 to 40 seconds per stand as compared to the normal 14 to 15 seconds per stand. This practice added less than one hour to round-trip time even as deep as 10,000 feet. The well was drilled to total depth without a loss of circulation, although there had been a total of seventy-nine losses while drilling the first three wells.

Pressure–Gel Strength Relationships

A definite relationship between pressures and gel strengths can be derived from the geometric conditions illustrated in Fig. 18–10. It is assumed that the annulus and the inside of the smaller pipe are filled with a material such as a gelled mud which has a definite measurable shear strength. Shear strength is the force which is required to shear a definite area. The force applied is determined by the pressure and the cross-sectional area over which the pressure is applied. On the

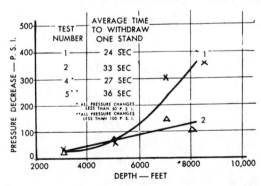

Fig. 18-9. Pressure decrease caused by rapidly withdrawing one stand.

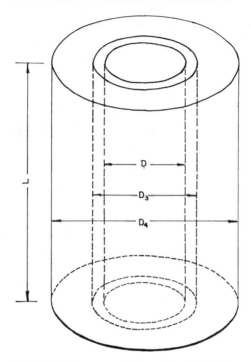

Fig. 18-10. Pressure–gel strength relationships.

other hand, the sheared area depends on the length of the pipe section and is best designated as the "lateral sheared area." It depends upon the total circumferences sheared and the length of the pipe section.

The pressure required to cause movement of the gelled fluid inside of a circular pipe may be derived as follows:

Force Causing Flow = Force Resisting Flow

(Pressure)(Cross-sectional Area)

= (Shear Strength)(Lateral Sheared Area)

$$P \frac{\pi}{4} D^2 = s \frac{\pi}{12} DL$$

$$P = \frac{sL}{3D} \qquad (18\text{-}1)$$

where

P = pressure, psi
D = ID of pipe, internal diameter, in.
s = unit shear strength of the gelled mud, lb./ft.^{2}

PRESSURE SURGES AND ANOMALIES 391

= 0.00326 × Stormer gel strength in grams
L = length of pipe section, ft.

The pressure required to cause movement of the gelled fluid in the annulus may be derived in a similar manner.

(Pressure)(Cross-sectional Area) = (Shear Strength)
(Lateral Sheared Area)

$$P \frac{\pi}{4}(D_4^2 - D_3^2) = sL \frac{\pi}{12}(D_4 + D_3)$$

$$P = \frac{sL}{3(D_4 - D_3)} \qquad (18\text{--}2)$$

where

D_4 = large diameter of annulus, in.
D_3 = pipe OD, outside diameter, in.

Example: Mud circulation has been shut down on a drilling well for sufficient time for the mud to develop a gel strength of 40 grams, Stormer. If the drill pipe is not moved longitudinally or rotated, what pump pressure will be required to break (start) circulation?

Additional data:

	Depth,	8,000 ft.
	Drill pipe,	3.8-in. ID
		4.5-in. OD
	Drill collar,	2.25-in. ID
		6.25-in. OD
		450-ft. length
	Hole diameter, 8.75-in.	

Solution: The shear strength of the mud is

$$s = (0.00326)(40 \text{ grams})$$
$$= 0.1304 \text{ lb./ft.}^2$$

The pressure required inside the drill pipe is

$$P = \frac{s}{3} \sum \frac{L}{D}$$

$$= \left(\frac{0.1304}{3}\right)\left(\frac{8{,}000 \text{ ft.} - 450 \text{ ft.}}{3.8 \text{ in.}} + \frac{450 \text{ ft.}}{2.25 \text{ in.}}\right)$$

$$= 95 \text{ psi}$$

The pressure required inside the annulus is

$$P = \frac{s}{3} \sum \frac{L}{D_4 - D_3}$$

$$= \frac{0.1304}{3}\left(\frac{450 \text{ ft.}}{8.75 \text{ in.} - 6.25 \text{ in.}} + \frac{8,000 \text{ ft.} - 450 \text{ ft.}}{8.75 \text{ in.} - 4.5 \text{ in.}}\right)$$

$$= 85 \text{ psi}$$

Total pump pressure required = 95 psi + 85 psi

= 180 psi

Example: Referring to the preceding example, if the drill string were raised slowly, so that acceleration effects would not rupture the gelled mud, calculate the height to which the mud would rise inside of the drill pipe. Drilling-mud density is 10.4 ppg.

Solution: Since the smaller diameter is at the bottom of the inside of the drill string, the problem is not complicated by losses of continuity (cavitation inside of drill string). The same considerations apply in the annulus. It has been determined in the preceding example that the annulus will break down (or flow) prior to any fluid movement inside of the drill string. Therefore the pressure equilibrium, between gel rupture and hydrostatic head, will be determined from conditions inside the drill string. It was determined in the preceding example that the pressure required for gel rupture inside the drill string is 95 psi.

Rupture Pressure = Hydrostatic-Head Pressure

95 psi = (0.052)(10.4 ppg)(h ft.)

h = 176 ft.

Under these conditions, a 93-foot stand of the drill pipe, as raised up into the derrick, would remain full of mud. A mud gel strength of about (93/176) (40 gr.) or about 20 grams would be sufficient to cause pulling a "wet string" of pipe at a depth of 8,000 feet, with the pipe remaining full of mud as it is raised up into the derrick. In such cases, the mud squirts out of the pipe when it is unscrewed at the derrick floor.

Example: Referring to the preceding examples, calculate the reduction in bottom-hole pressure caused by slowly raising the drill pipe.

Solution: As the drill string is raised, a displacement occurs at the position of the lower end of the string which must be filled by flow either down through the annulus or down through the interior of the drill string. Flow of the gelled mud will occur first in the channel which breaks down (starts to flow) at the lower differential pressure. It has been determined that 85 psi will cause flow in the annulus, whereas 95 psi are required to cause flow inside the drill pipe. Therefore the pressure reduction is 85 psi. This is equivalent, at 8,000 feet, to a reduction of 0.2 ppg in drilling-mud density.

Viscous-Flow Effects

The pressure reductions which may be expected in withdrawing pipe, or the pressure increases from running pipe into hole, which may be caused from purely viscous effects of Newtonian-type liquids, was studied by Cardwell.[4] His derivation follows the classical derivation of Pouiseuille's Law, with the model including a small thin-walled tube centered inside of the larger tube. He obtained the Pouiseuille-type equation:

$$P = \frac{4L\eta U}{R^2} \left[\frac{1}{\frac{Z^2}{Z^2-1}\ln Z - \frac{Z^2-1}{Z^2}} \right] \qquad (18\text{-}3)$$

where

$$Z = \frac{R}{r}$$

R = radius of large tube (hole radius)
r = radius of small tube (such as thin wall drill pipe)

For calculating the magnitude of pressure changes which might be caused by purely viscous-flow conditions, Cardwell used a viscosity of 300 centipoises. This might well represent a drilling mud in a semi-gelled condition. Calculated pressure drops are shown in Fig. 18–11.

Acceleration Effects

The pressure necessary to set into motion a column of fluid may be derived directly from Newton's Second Law of Motion. Generally the column of fluid in the annulus is the one of interest, since the pressure exerted against the geological formations is of concern. The acceleration pressure is derived in the following manner:

$$F = Ma \qquad (18\text{-}4)$$

$$pA = \frac{Ah\rho}{g} a$$

[4] "Pressure Changes in Drilling Wells Caused by Pipe Movement," *API Drilling and Production Practice* (1953), 97–112.

$$p = \frac{h\rho a}{g} \qquad (18\text{-}5)$$

$$p = \frac{h\rho}{gA}\frac{dq}{dt} \qquad (18\text{-}6)$$

$$P = 0.0002155\frac{h\rho}{A}\frac{dq}{dt} \qquad (18\text{-}7)$$

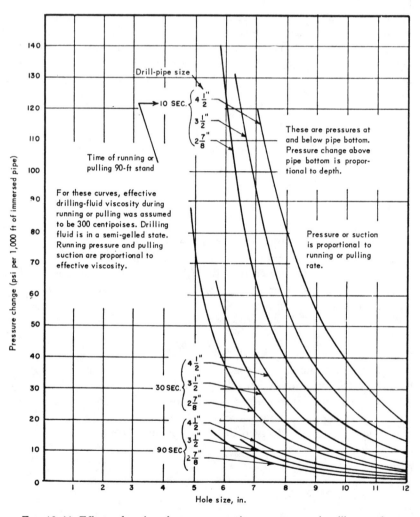

FIG. 18-11. Effects of various factors on running pressures and pulling suction.

where

F = force, lb.
M = mass, slugs
a = acceleration, ft./sec.^{2}
g = accleration of gravity, 32.174 ft./sec.^{2}
ρ = fluid density, lb./ft.^{3}
  = (7.48) (density in lb./gal.)
h = length of section, ft.
A = cross-sectional area of section, ft.^{2}
$\dfrac{dq}{dt}$ = instantaneous change in rate of flow, ft.^{3}/sec.^{2}
p = pressure, lb./ft.^{2}
P = pressure, psi

Equation (18–7) was used by Clark[5] in calculating pressure surges caused by running pipe.

Plastic- and Turbulent-Flow Effects

The calculation of pressure changes caused by plastic and turbulent flow effects was discussed by Ormsby[6] in 1954. The need for determining whether the flow is of plastic nature or turbulent nature in each particular flow section was pointed out. For example, as drill pipe is being lowered into the hole, the flow section inside the drill collars might be in turbulent flow while the inside of the drill pipe and the several annuli might be in plastic flow. The annulus was divided into sections for calculation purposes, such as the drill-collar-hole annulus, and the total lengths of the tool joints were combined and treated as a separate flow section. It was pointed out that the average flow velocity at any point in the annulus, with respect to the stationary casing or wall of the bore hole, is determined by the volume rate of flow divided by the cross-sectional area of the annulus at the particular point being considered. The average flow velocity with respect to the pipe being raised or lowered is the velocity with respect to the stationary wall plus the velocity of the pipe movement. The arithmetical average of these is the velocity with respect to the stationary wall plus one-half of the velocity of pipe movement.

---

[5] E. H. Clark, Jr., "Bottom Hole Pressure Surges while Running Pipe," Paper No. 54-PET-22, ASME Petroleum Mechanical Engineering Conference, Los Angeles, September 26–29, 1954; also published in *The Petroleum Engineer*, January, 1955.

[6] G. S. Ormsby, "Calculation and Control of Mud Pressure in Drilling and Completion Operations," *API Drilling and Production Practice* (1954), 44–55.

The equations of Dunn, Nuss, and Beck[7] for plastic flow and the equations of Pigott[8] for turbulent flow were used to calculate flow-friction pressures both inside the circular drill pipe (or casing) and in the annulus. The equivalent flow velocity used in calculating pressure losses in the annulus was the arithmetical average of the flow velocities with respect to the stationary wall and the moving pipe. This equivalent velocity appears to be correct in the limiting case where the moving pipe completely fills the stationary bore hole. Accordingly, the results calculated for running casing in the hole would tend to be accurate, and this was the special case of interest in Ormsby's investigation.

Similar calculations of pressure changes caused by pipe movement were made by Clark.[9] The modified equations for equivalent annular velocity, for use in conventional flow-friction-loss equations, which show a decreased influence of the velocity of the moving drill pipe or casing as its diameter becomes smaller, were given as follows:

For laminar flow, casing-hole annulus,

$$U = -\frac{q}{A} + 0.46\, U_p$$

For laminar flow, drill-pipe-hole annulus,

$$U = -\frac{q}{A} + 0.39\, U_p$$

For turbulent flow, casing-hole annulus,

$$U = -\frac{q}{A} + 0.181\, U_p$$

For turbulent flow, drill-pipe-hole annulus,

$$U = -\frac{q}{A} + 0.166\, U_p$$

For inside of circular pipe, laminar or turbulent,

$$U = -\frac{q}{A} + U_p$$

---

[7] R. W. Beck, W. F. Nuss, and T. H. Dunn, "Flow Properties of Drilling Mud," *API Drilling and Production Practice* (1947), 9–22.

[8] R. J. S. Pigott, "Mud Flow in Drilling," *API Drilling and Production Practice* (1941), 91–103.

[9] "Bottom Hole Pressure Surges while Running Pipe."

PRESSURE SURGES AND ANOMALIES

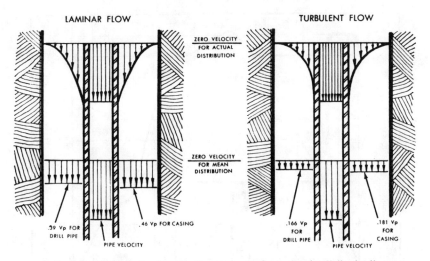

FIG. 18–12. Velocity distribution and mean velocity for "clinging" (velocity due to moving boundaries in bottomless hole).

where

U = equivalent velocity for flow-pressure calculations, ft./sec.
q = rate of flow, with respect to stationary bore hole, ft.^{3}/sec.
A = cross-sectional area of flow for each uniform flow section considered, ft.^{2}
U_p = velocity of pipe movement, ft./sec.

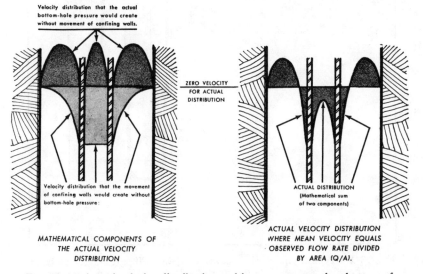

FIG. 18–13. Actual velocity distribution and its components when bottom of hole and displacement are added to conditions of Fig. 18–12.

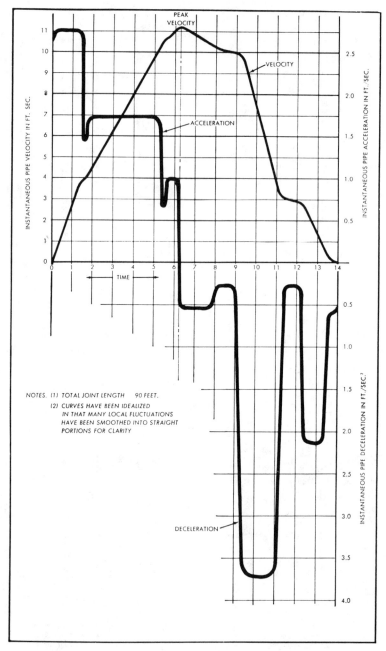

Fig. 18–14. Typical pipe motion curves for $4\frac{1}{2}$-in. drill pipe lowered 90 ft. in 14 sec. at 10,000 ft.

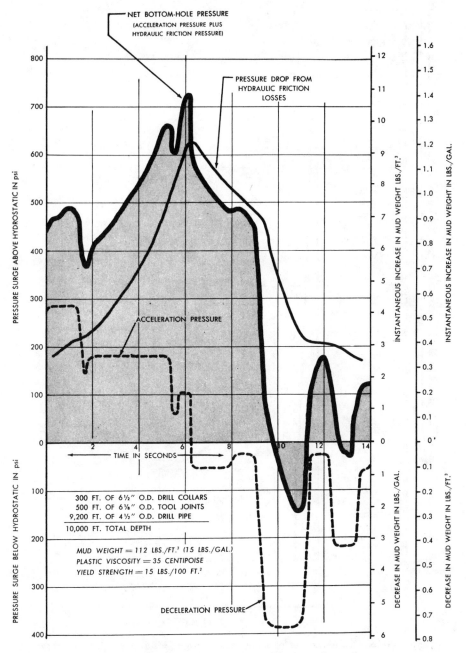

Fig. 18-15. Bottom-hole pressure surges for 4½-in. OD drill pipe with plugged bit lowered 90 ft. in 14 sec in 8¾-in. hole.

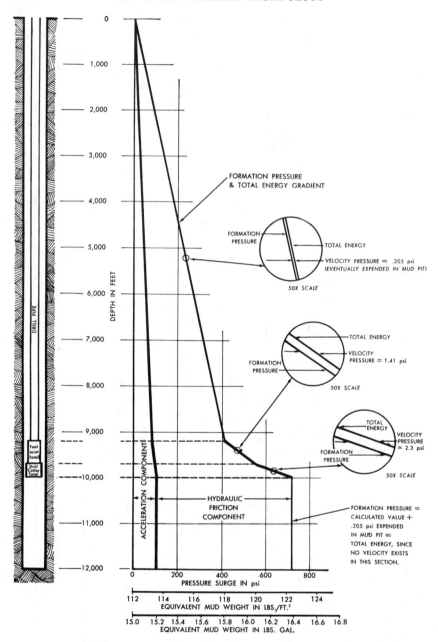

FIG. 18–16. Distribution of pressure with depth at instant of peak pressure on Fig. 18–15—4½-in. drill pipe with plugged bit lowered 90 ft. in 14 sec. in 8¾-in. hole.

All quantities are taken as positive in the downward direction. Accordingly, upward flow in the annulus or in the pipe would be a negative quantity.

The volume rate of flow $(-q)$ at any point is equal to the net downward displacement divided by the cross-sectional flow area. Consider the drill-string-hole annulus where the string is being lowered with a plugged bit, so that no mud will flow upward inside of the drill pipe. If the drill string is being lowered with a velocity U_p, in the drill-

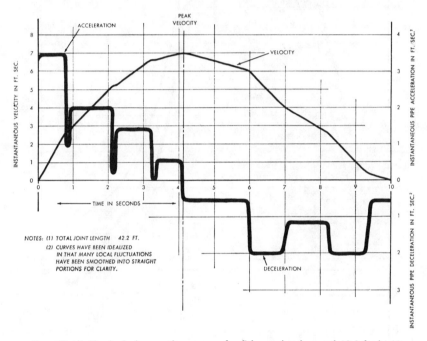

FIG. 18–17. Typical pipe motion curves for 7-in. casing lowered 42.2 ft. in 10 sec. at 10,000 ft.

collar-hole annulus the upward flow of mud would be the product of U_p and the total (outside) cross-sectional area of the drill collars. At the same instant, the upward flow of mud in the drill-pipe-hole annulus would be the product of U_p and the outside cross-sectional area of the drill pipe, as computed from the OD of the pipe. At any instant, q varies along the drill string according to the displacement at the particular depth point.

For the more general case where mud can flow up inside the drill string as it is lowered into the well, or where the casing shoe permits mud to flow into the bottom of the casing as the casing is lowered, the

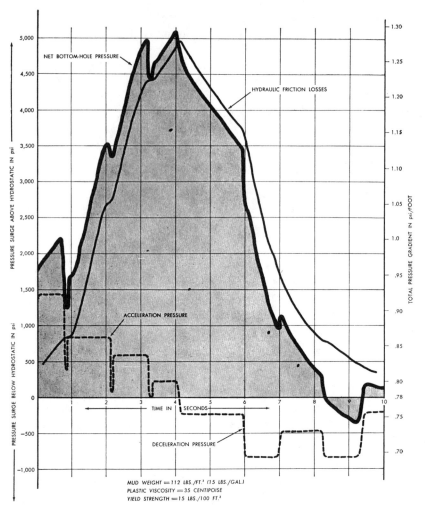

FIG. 18–18. Bottom-hole pressure surges when lowering 10,000 ft. of 7-in. flush joint casing equipped with conventional float shoe in 8¾-in. hole (joint length = 42.2 ft.).

upward flow of mud is divided. If the level of the mud inside the pipe remains at the same height as the mud level in the annulus as the pipe is lowered into the well, the condition is spoken of as "100 per cent fill-up." If the upward velocity of mud flow inside the top of the pipe is 80 per cent of the downward velocity of the pipe, the condition would be 80 per cent fill-up. In the usual case, the percentage of fill will not be known (unless measured at the mud-flow line). The mud which does not flow upward inside the drill pipe increases the upward flow in the annulus. Clark suggested that various values of fill be assumed for each pipe velocity of interest in order to permit the calculation of upward flow velocities in all uniform flow sections and of flow pressure losses in all sections, whether the individual flow sections be in plastic or in turbulent flow. With such calculated values at hand, the percentage of fill which will occur at any particular pipe velocity can be determined by adding the flow pressure losses in the

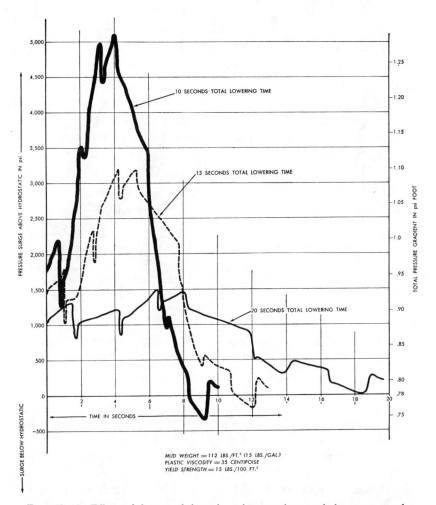

FIG. 18–19. Effect of increased lowering time on bottom-hole pressure when lowering 10,000 ft. of 7-in. flush-joint casing equipped with conventional float shoe in 8¾-in. hole.

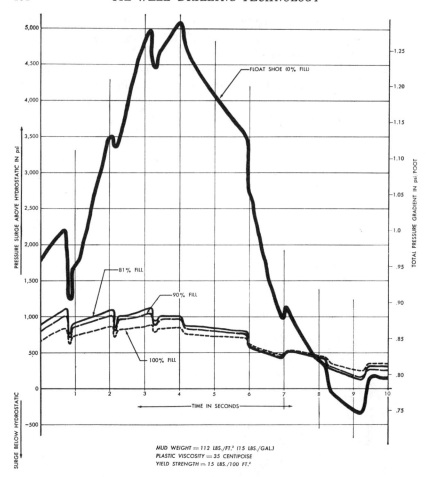

Fig. 18–20. Effect of fill-up equipment on bottom-hole pressure when lowering 10,000 ft. of 7-in. flush-joint casing in 8¾-in. hole (joint length = 42.2 ft.).

annulus to those inside the pipe. The difference in hydrostatic heads in the pipe and annulus must, of course, be included in the calculation. If there is no restriction in the bottom of the hole or drill string, the pressure inside the bottom of the drill pipe equals the pressure in the bottom of the annulus. The percentage of fill which will occur at any pipe velocity is found by this consideration of a point of common equal pressure. Figures 18–12 to 18–20 were constructed by Clark.[10]

[10] *Ibid.*

Summary

Both pressure increases and pressure decreases of considerable magnitude have been measured at the bottom of wells. Calculations and actual measurements indicate that the pressures caused by pipe movement are linear between the bottom of the drill string and the top of the hole. It seems apparent that pressure increases as high as those which can be calculated cannot in fact occur, because the exposed formations will break down and mud losses will take place at lower total pressures. Similarly, pressure reductions tend to be relieved by gas, oil, or water flowing into the bore hole from exposed formations. It has been suggested that the elasticity of the geological formations through which the well is drilled may also tend to dampen or lessen such pressure surges. Nevertheless, the fact that mud losses and blowouts have originated from such causes is indisputably established.

CHAPTER 19

Casing and Casing-String Design

FUNCTIONS AND REQUIREMENTS OF CASING

Casing is an essential part of the drilling and completion of an oil well. It consists of lengths of steel pipe either screwed or welded together to form a continuous tube to the desired depth. One or more of the following strings of casing is required in every well:

1. conductor casing
2. surface casing
3. intermediate or protective casing (one or more strings)
4. oil-string casing

The total length of a continuous tube used in a well is called a *string* of casing, and may or may not have the same dimensions throughout its length. *Conductor casing* is the largest-diameter casing used in a well, and is required only where the surface soils are so incompetent that the washing and eroding action of the drilling mud would create a large cavity at the surface. The conductor casing controls this eroding action. The next smaller-diameter casing which is required is the *surface casing*. Its principal function it to protect fresh-water sands and provide an anchor for blow-out-preventer equipment. The amount of surface casing that may be required will depend primarily on the depth of the fresh-water sands in the area, which may vary from only a few feet to several thousand feet. In most oil-producing states, some state regulatory body determines the amount of surface casing which must be run in any particular area.

Following the setting of adequate surface casing, one or more strings of casing will still be required, the number depending on the depth of the well and the problems found in drilling. If the well is unusually deep or severe drilling problems are encountered, such as abnormal pressure formations, heaving formations, or lost circulation zones it may be necessary to set an *intermediate string* of casing to seal off the long open hole or the zones causing trouble. After all necessary intermediate strings of casing have been set, the final string of casing required is the *long string* or *oil string*. This is the string of casing which is set immediately above, or through, the producing formation. If the oil string is set on top of the producing formation, an *open-hole* completion results. If the bottom of the casing is set below the producing horizon, it becomes necessary to perforate the casing to permit communication between the inside of the casing and the producing formation. This results in a *perforated casing* completion. A schematic diagram of a typical cased well is shown in Fig. 8–1. The oil string of casing is required to prevent cave-ins and maintain a clean hole, seal off or separate oil-, gas-, or water-bearing zones, and seal off thief zones into which the formation fluids might migrate.

The term *casing* is generally applied to strings of pipe which extend from the surface downward to the bottom of the particular string, which is referred to as its *setting depth*. In contrast, the term *liner* is generally applied to strings of pipe which do not extend to the surface. Liners may serve as the oil string through several hundred feet in the bottom part of a well. They have been used where drilling problems required that the intended oil string of casing be set before the total depth of the well was reached and it was deemed unnecessary to extend to the surface the smaller pipe, which was later set and cemented in the bottom part of the well. There are also liners for special purposes, such as the slotted liners, perforated liners, and gravel-packed liners which are set opposite producing zones for the purpose of preventing sand from entering the well. Liners are sometimes cemented in place or set with packers at their top or bottom or both, or sometimes merely set on the bottom of the well.

Properties of Casing

Many sizes, types, and grades of casing are required to meet the varied needs of the petroleum industry. In order to standardize as much as possible the specifications for commonly used types of casing, the American Petroleum Institute (API) has developed minimum standards for much of the tubular goods used by the oil industry. Thus, when a prospective purchaser of tubular goods orders a particular "API specification" pipe, he knows the pipe has been manufactured and tested according to the particular API specification established for this pipe.

Casing manufactured according to API specifications is available in three ranges of lengths. These ranges of length are:

Range 1: 16 to 25 feet inclusive: 95 per cent or more of any carload shall not vary more than 6 feet in length, with a minimum length of not less than 18 feet.

Range 2: 25 to 34 feet inclusive: 95 per cent or more of any carload shall not vary more than 5 feet in length, with a minimum length of not less than 28 feet.

Range 3: Minimum 34 feet: 95 per cent or more of any carload shall not vary more than 6 feet in length, with a minimum length of not less than 36 feet.

There are at present four different approved grades of API casing and one other grade which has been given tentative approval. The mechanical properties of these grades of casing are shown in Table 19-1. The numerical part of casing grade designation is actually the minimum yield strength in thousands of psi, which is convenient for remembering the strength of the various grades.

TABLE 19-1

MECHANICAL PROPERTIES OF API CASING

Grade	F-25	H-40	J-55	N-80	P-110
Yield strength, min., psi	25,000	40,000	55,000	80,000	110,000
Yield strength, avg., psi	*	50,000	65,000	85,000	123,000
Tensile strength, min., psi	40,000	60,000	75,000	100,000	125,000
% Elongation in 2 in., min.					
Strip specimens	40	27	20	16	15
Full section specimens	45	32	25	18	17

All API casing is normally furnished with threads and couplings. Couplings must be specified as either "short threads and couplings" (ST&C) or "long threads and couplings" (LT&C), the latter having about 30 per cent greater strength in tension. Casing is defined on the basis of its nominal outside diameter (OD) and nominal weight per foot, which is an average weight that includes the couplings. The wall thickness, therefore, cannot be calculated from the nominal weight per foot and the outside diameter, but must be obtained from data furnished by the manufacturer. The weight per foot controls the strength, wall thickness, and inside diameter (ID) of the casing

Drift diameter is a concept which should be clearly understood. The drift diameter is the guaranteed minimum inside diameter of any part of the casing. The numerical values of drift diameters are published by the manufacturer in accordance with API specifications for every nominal size and nominal weight per foot classification. Casing is a commercial product made by rolling and piercing operations, and the actual internal diameter may vary a few thousandths of an inch on account of slight changes in wall thickness or slight "out-of-roundness." The drift diameter is the diameter which should be used to determine the largest-diameter equipment (bit, packer, etc.) which can safely be run inside the casing. On the other hand, the internal diameter (ID) should be used for such things as volume capacity calculations. The drift diameter, as well as the ID, should be obtained from a casing table.

Testing of Properties

The API has established specifications for casing, and when a manufacturer has a product which meets all the specifications, he is entitled to place the official API stamp on the product. This applies not only to casing but also to many other products used by the oil industry. Complete API specifications for casing, tubing, and drill pipe are listed in the latest editon of *API Standard 5A*. The manufacturing process is stipulated, along with chemical requirements, tests and testing procedures, dimensions, weights, lengths, tolerances, and many other specifications which are important if the consumer is to obtain a reproducible, consistent product. For example, casing can be made by the seamless, electric-weld, or lap-weld process. However, only Grade F–25 casing can be made by the lap-weld process, and Grade P–110 casing can be made only by the seamless process.

Casing is one of the most expensive items of equipment used in completing a well. Therefore, the casing used for any well should be properly designed to eliminate unnecessary casing costs. Important factors in casing design are the type and size of equipment which will be used in the well. Sufficient clearance should be allowed between the casing and the equipment not only to allow free passage of the equipment, but also to permit the use of fishing tools to recover parted or stuck tools and equipment. Smaller and smaller casing is being used, principally because of the development of small producing equipment and fishing tools. Not too many years ago 7-inch OD casing was more or less standard for the oil string in many areas. Within recent years, $5\frac{1}{2}$-inch casing has become more or less standard. In addition to the price advantage on the casing, in some areas the smaller hole required for the $5\frac{1}{2}$-inch casing can be drilled faster and more economically than the larger hole required for the 7-inch casing. In all events, the casing and drilling-bit programs must be developed simultaneously, as the drilling bit must pass through the upper strings of casing, and the hole drilled for a particular size of casing must include sufficient clearance for a satisfactory cement job. Typical minimum hole sizes for various casing sizes are shown in Table 8–1. Clearances may vary from those shown in Table 8–1 in certain areas where experience has shown that more or less clearance can or must be used.

A knowledge of the forces involved in running and setting casing in a well bore is necessary in order to select the proper size, grade, and weight of casing.

In accordance with sound engineering principles and economy, casing must be designed so that it will not (1) fail under tension, (2) fail by collapse from external pressure, or (3) fail by bursting from internal pressure.

In addition to the physical requirements which must be satisfied at the time of well completion, there are other requirements, particularly the chemical requirement that the casing will not be corroded to the extent of leaking or undergoing other failure before the zone has been depleted of oil or gas. This requirement cannot always be met. For example, many wells drilled in a field whose expected life is longer than about fifteen years (depending upon the particular locality) may be cased with 7-inch oil strings, with the expectation of later running and cementing smaller casing inside the 7-inch casing when it fails on account of corrosion. Wells of comparatively short life or in noncorrosive areas may be cased with the less expensive $5\frac{1}{2}$-inch casing.

In the past there has been a tendency to use large casing sizes in dually completed wells. However, adequate artificial lift equipment and tubular products are now available to permit the use of $5\frac{1}{2}$-inch and smaller casing in dual-completion wells.

Design Safety Factors

Designing a casing string consists of choosing the proper weight and grade of steel of a particular diameter casing, and the type of threads and couplings, for the various sections which will most economically perform their intended functions. Casing tables furnished by the manufacturers in accordance with API specifications list the properties of the various weights and grades of casing, such as their collapse resistance, bursting resistance, and tensile load capacity. All of the values so listed contain adequate safety factors based on the minimum yield strength of the steel, which itself is less than the average yield strength. In designing casing strings, the published values of properties such as the tensile load capacity are used in conjunction with design safety factors. In contrast to the safety factors contained within the listed values of the properties, the design safety factors make allowance for well conditions which cannot be accurately predicted. For example, the design safety factor in tension is usually applied to the weight of the pipe below any point considered in the string; but if it becomes necessary to pull a string of casing out of a well, it would be difficult to foretell the fuctional forces which would be operating between the casing and the wall of the drilled hole. Design safety factors, therefore, are based largely upon experience.

Tensile strength. Any given joint of casing in the casing string must support the weight of all the casing suspended below it. This is strictly true at the time the casing is run in the hole, although the top of the casing may be lowered slightly after the casing is cemented, so that the set cement will support a portion of the weight of the string. The safety factor used may be based on the weight of the casing in air, or it may be based on the weight of the casing as buoyed up by the mud in the hole. In the former case, the safety factor will tend to be lower, for the casing will generally lose roughly one-eighth of its weight when immersed in mud. Safety factors varying from 1.5 to 2 are commonly used. It may be noted that if tensile requirements alone are considered, the wall of the pipe would be thick at the top and taper down to the thinness of paper at the bottom of the hole. In API seamless casing, the joint strength at the couplings is the weakest place, and the joint strength is used when designing for tensile strength.

The formula recommended by the API for determining the tensile stresses in casing is shown below:

Short couplings:

$$P = 0.80\left[C(33.71 - D)\left(\frac{1}{t - 0.07125} + 24.45\right)Aj\right] \quad (19\text{--}1)$$

Long couplings:

$$P = 0.80\left[C(25.58 - D)\left(\frac{1}{t - 0.07125} + 24.45\right)Aj\right] \quad (19\text{--}2)$$

where:

P = minimum joint strength, lb.
D = OD of casing, in.
d = ID of casing, in.
t = wall thickness, in.
Aj = area under root of last perfect thread, in.^{2}
   = $0.7854\,[(D-0.1425)^2 - d^2]$
C = constant for steel grade as shown in Table 19–2.

TABLE 19–2

VALUES OF C FOR EQUATION (19–1) AND EQUATION (19–2)

Grade	Short Couplings	Long Couplings
F–25	53.5	—
H–40	72.5	—
J–55	96.5	159
N–80	112.3	185
P–110	146.9	242

Collapse resistance. Design for collapse resistance is commonly based on the hydrostatic head of the drilling mud in the hole at the time the casing is run into the well. To find the reason for this, suppose that the casing is cemented and the cement rises for a short distance in the annulus behind the casing and there becomes set. The casing may then be perforated opposite an oil zone and tubing with a packer run in, and the packer set at a point slightly above the depth to which the cement made a perfect seal in the bottom of the annulus. Then, in attempting to start producing oil in the zone, the inside of the tubing might be swabbed empty so that there will be no internal pressure in the casing below the setting depth of the packer. If the cement does not make a perfect seal on the outside of the casing, the full hydrostatic head of the mud standing in the annulus behind the casing will be exerted on the outside of the casing and tend to collapse the casing. A safety factor of 1.125 has long been used to guard against collapse from this cause. In contrast to the requirements for tensile strength, if the requirements for resistance to collapse alone are considered, the wall of the pipe would be thick at the bottom and taper to paper thinness at the top.

In analyzing the factors which affect the collapse resistance of casing, the yield strength of the steel has been found to be one of the basic elements. As the yield strength of the steel increases, the collapse resistance of the casing increases. However, the collapse resistance of a particular grade of steel casing is materially altered when stresses are applied in more than one direction. When casing is placed in a well, the forces tending to collapse the casing are due not only to the external pressure applied but also to the weight of the casing below the design point. This biaxial load, as it is called because the loads are at right angles to each other, in effect reduces the yield strength of the steel.

The formulae recommended by the API for computing collapse resistance are shown below. These formulae have been empirically developed for elastic or plastic failure and do not include the effect of additional weight of casing below the design point.

Collapse, for Grades H–40, J–55, N–80, and P–110

Elastic failure:

$$P = 0.75 \left[\frac{62.6 \times 10^6}{(D/t)(D/t - 1)^2} \right] \qquad (19\text{-}3)$$

Plastic failure I:

$$P = 0.75 \left[(YP) \left(\frac{2.503}{D/t} - 0.046 \right) \right] \qquad (19\text{-}4)$$

(For D/t values of 14 to intersection with elastic curve)

Plastic failure II:

$$P = 0.75\left[\frac{(2YP)(D/t - 1)}{(D/t)^2}\right] \qquad (19\text{--}5)$$

(For D/t values less than 14)

Collapse, for Grade F–25

Stewart's formula:

1. For values of P less than 581 or t/D less than 0.023:

$$P = 0.75[50{,}210{,}000\ (t/D)^3] \qquad (19\text{--}6)$$

2. For values of t/D greater than 0.023:

$$P = 0.75[86{,}670\ (t/D) - 1{,}386] \qquad (19\text{--}7)$$

where:

P = minimum collapse pressure, psi
D = OD of casing, in.
t = wall thickness of casing, in.
YP = average yield strength, psi

Bursting strength. Normally, at the bottom of the well the pressure on the outside of the casing in equal to or greater than the pressure inside the casing. This external pressure is derived either from the hydrostatic head of the drilling mud or, perhaps, the pressure of the water in the pores of the rock adjacent to the cemented area of the annulus behind the casing. However, at the top of the hole, there is no hydrostatic fluid head to exert external pressure. Any internal pressure there must be held by the casing itself. For any given formation pressure which enters at the bottom of the well, the internal pressure exerted at the top of the casing will be greatest if the casing is filled with gas. This is true because gas exerts much less hydrostatic head than oil or water, which might help to balance the formation pressure at the bottom of the well. Often the requirements for tensile strength are such that the string of casing is heavy enough at the top to withstand any internal pressures which will be encountered. In other cases, particularly in regions of abnormally high formation pressures, the top of the string must be designed for bursting strength. Pressures of 8,000 psi or more occasionally must be allowed for; that is, such pressures would normally be confiend in the tubing which is run inside the casing, but the possibility of tubing leaks must be allowed for.

The length of time that a well is expected to maintain high pressure and the corrosive qualities of the well fluids are considerations. A safety factor of 2 has been used in some cases.

In most cases, however, where corrosion is not excessive and where formation pressures will normally decline, design safety factors of about 1.10 have been considered adequate.

The bursting or maximum safe internal pressure is determined by Barlow's formula:

$$P = 0.875(2st/D) \tag{19-8}$$

where:

P = minimum internal yield pressure, psi
t = nominal wall thickness, in.
D = OD of casing, in.
s = minimum yield strength, psi

In general, tension failures will occur at the top of the string or at the top of a uniform section of casing; collapse failures will occur at the bottom of the string or at the bottom of a uniform section in the casing string; and bursting failures will occur near the top of the string or near the top of a uniform section of the string which is above the top of cement. It should be noted that all these formulae have a safety factor included in them.

Biaxial Stresses

Tests have conclusively proved that biaxial loads (i.e., loads on casing caused by both external pressure and the suspended weight) acting at a point reduce the effective collapse strength of casing, the magnitude of this reduction being rather large in many instances. These stresses may become very important when designing combination strings of casing. A combination string of casing is a string of casing having the same OD throughout its length, but composed of sections of various weights and grades, according to the requirements at the different depths in the hole, so that the most economical combination, consistent with good engineering design for the various stresses which will be encountered, results.

As has been said in connection with designing a casing string for tension, the pipe should be thick at the top and thin at the bottom. When designing for collapse, the pipe would be thick at the bottom and thin at the top; and when designing for bursting strength, the pipe must be strong in areas where it is unsupported by cement or fluids—normally near the top of the hole. Obviously the design demands of these three forces are in conflict with each other, and it is necessary to design a casing string that will satisfy all the requirements of the various forces and combinations of forces. An apparently simple solution would be to use uniformly thick-walled pipe from top to bottom. However, not only would the cost be excessive, but in deep wells the heavy casing throughout the entire length would cause ex-

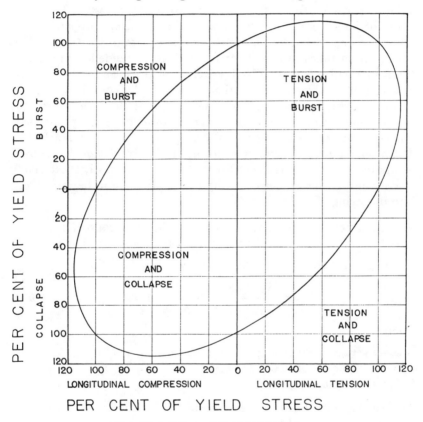

FIG. 19–1. Ellipse of biaxial yield stress.

cessive tensile stresses that would require increased thickness of the casing, which could lead to an endless cycle of continuously increasing the wall-thickness requirements of the casing.

It is necessary, therefore, that a string of casing be designed to provide a satisfactory margin of safety for every operating condition expected to be encountered, and still use the lightest-weight casing possible. The combination string of casing is at present the most satisfactory method of attaining this objective.

The starting point for designing a combination casing string is the bottom of the hole. From tables of casing properties, the lightest weight and least expensive grade of casing is selected which will withstand collapse from the particular weight of mud used in the well, using a safety factor which is customarily $1\frac{1}{8}$. This weight and grade of casing is then carried up the hole to the point where the next lighter or less expensive grade will withstand the collapse pressure. This point may not be selected directly from a table of casing properties, but must be computed according to the *ellipse of biaxial yield stress*. This computation takes into account the additional tendency to collapse the casing which is caused by the weight of the lower sections of the casing suspended from any particular point under consideration.

As the design progresses from the bottom of the hole toward the surface, and perhaps one or more sections of lighter casing have been introduced into the casing string, a point will finally be reached at which the requirements for strength in tension outweigh the requirements for collapse resistance. From this point upward, successively heavier and better grades of casing will again have to be introduced into the string of casing.

MATHEMATICAL RELATIONS USED IN CASING-STRING DESIGN FOR COLLAPSE RESISTANCE

The general equation of the biaxial yield stress ellipse is

$$\frac{S_c^2}{S_o^2} - \frac{S_c S_t}{S_o^2} + \frac{S_t^2}{S_o^2} = 1 \qquad (19\text{-}9)$$

where:

S_o = average yield strength of the material, psi
S_c = peripherial or hoop stress, psi
S_t = tensile stress, psi

This equation has been verified experimentally and has been found to describe accurately the capacity of the casing to resist deformation, particularly in the region where collapse may be caused by a combination of tensile loading, from the weight of the lower portion of casing, by the application of external pressure. The external pressure results from the hydrostatic head of mud standing in the annulus outside of the casing when the internal pressure in the casing might be reduced to zero by effectively swabbing the mud out of the inside of the casing.

The equation of the biaxial yield stress ellipse may be written in a much more convenient form as

$$r^2 + rt + t^2 = 1 \qquad (19\text{-}10)$$

where:

$$r = -\frac{S_c}{S_o}$$

= fractional collapse resistance

$$t = \frac{S_t}{S_o}$$

= fractional tensile yield stress

This form of the biaxial yield stress ellipse equation is more convenient, because both the fractional collapse resistance and the fractional tensile-yield stress can be expressed in terms of commonly known quantities. Casing manufacturers furnish collapse-resistance charts for the various sizes, weights, and grades of casing which are essentially a portion of the biaxial yield stress ellipse made specifically for one particular size, weight, and grade of casing. Such charts are plots of the weight suspended by the casing versus the external collapsing pressure. They may be arrived at by the following substitutions:

$$r = \text{fractional collapse resistance}$$
$$= \frac{C}{K}$$

$$t = \text{fractional tensile yield stress}$$
$$= \frac{W}{AS_o}$$

where:

C = collapse resistance under tensile stress, psi
K = minimum collapse resistance with no tensile load, psi
W = weight supported by the casing, lb.
A = cross-sectional area of the steel in the casing, sq. in.
S_o = average yield stress of the steel, psi

The biaxial yield stress ellipse equation then becomes

$$\frac{C^2}{K^2} + \frac{CW}{KAS_o} + \frac{W^2}{A^2 S_o^2} = 1 \qquad (19\text{-}11)$$

It may be noted that when $W=0$, $C=K$. Solved for C, the collapse resistance under tensile load W, the relation is

$$C = \frac{WK}{2AS_o}\left[\sqrt{4\left(\frac{AS_o}{W}\right)^2 - 3} - 1\right] \qquad (19\text{-}12)$$

The curve for any particular size, weight, and grade of casing may be constructed by substitution of the proper values in this equation. Such values are listed in handbooks published by the various casing manufacturers. For example, the curve for 7-inch, 29 lb./ft. Grade N–80 casing may be constructed by using the following values:

$S_o = 85{,}000$ psi; $A = 8.449$ sq. in.; $K = 6{,}370$ psi

In additional to such numerical data, the curves are also published in chart form.

Graphical Solution of Collapse Change Points

The graphical solution for collapse change points is a direct solution which makes use of the *effective collapse resistance versus tensile load* charts which are furnished by the various casing manufacturers. Such charts are usually made for a particular size of casing, such as 7-inch casing, and the several curves on the chart represent the various weights and grades of that size of casing which are available.

It is customary to use a small design safety factor, usually $1\frac{1}{8}$, for resistance to collapse. This small design safety factor is permissible because the pressure inside the casing ordinarily helps to balance the external pressure on the casing, and the design condition of no internal pressure is not encountered extensively. This collapse safety factor may be handled as follows:

Let

$$C = Fhp \qquad (19\text{-}13)$$
$$= 0.052\, FGh \qquad (19\text{-}14)$$

where:

$C =$ the collapse resistance of the casing, psi
$F =$ collapse design safety factor, usually taken as 1.125
$h =$ depth from the surface to any particular point in the casing string which is being considered for collapse resistance, ft.
$p =$ mud pressure gradient, psi/ft.
$G =$ mud density, lb./gal.

It may be noted that the external mud pressure in psi is given either by the product hp or by the expression $0.052\, Gh$, and collapse resistance actually used exceeds the external pressure by the design safety factor. This relationship is very convenient for calculating the actual depth in the well which is equivalent to any collapse resistance of the casing, for any particular density of drilling mud which will stand behind the casing. That is, the effective collapse resistance of the casing, as plotted versus tensile load, may be converted to a depth point in the well by the relation

$$h = \frac{C}{Fp} \qquad (19\text{-}15)$$

$$= \frac{C}{0.052\ FG} \qquad (19\text{-}16)$$

If the collapse change points are determined on a chart in terms of C, the effective collapse resistance, the corresponding depth points may be found in terms of feet by using the above relationship.

The starting point for designing a section of a combination string of casing is at the bottom of the section. The weight hanging below that section must be known. The starting point for designing the string is at the bottom of the hole, for at that point the suspended weight below the section is known to be zero. If a point in any section of uniform weight per foot is considered, the weight suspended below that point is the tensile load at the point and is derived from the weight of the casing string below that point. The suspended weight may be expressed as

$$W = W_i + Bw(h_i - h) \qquad (19\text{-}17)$$

where:

W = tensile load on the casing at depth h, lb.
W_i = weight of the casing suspended below the depth h_i, lb.
B = buoyancy factor, less than one, dependent on mud density
w = weight (in air) per foot of casing in the section of casing being considered, lb./ft.
h = depth in the well to the point being considered, ft.
h_i = depth in the well to the lower end of the uniform section of casing being considered, ft.

The quantity h may be eliminated by using the former equation to yield the following:

$$W = W_i + BWh_i - \frac{BwC}{0.052\ FG} \qquad (19\text{-}18)$$

The point where the lower end of a section of the casing string falls on the effective collapse resistance versus tensile load chart will always be known. Therefore, to draw a straight line representing a uniform weight per foot section of the casing on the chart, it will be necessary only to know the slope of the line. This may be obtained by differentiating the above equation to obtain

$$\text{Slope} = \frac{dW}{dc}$$

$$= -\frac{Bw}{0.052\,FG} \qquad (19\text{--}19)$$

Since the buoyancy factor B is a function of mud density, and, for steel, may be written as

$$B = 1 - \frac{7.48G}{488} \qquad (19\text{--}20)$$

where 488 is the density of steel in pounds per cubic foot, the relation reduces to

$$\text{Slope} = -\frac{w}{F}\left[\frac{19.25}{G} - 0.295\right] \qquad (19\text{--}21)$$

Where the customary collapse factor of safety of 1.125 is used, the relation reduces to

$$\text{Slope} = -w\left[\frac{17.11}{G} - 0.2623\right] \qquad (19\text{--}22)$$

The negative sign before the slope indicates that the line will slope upwards towards the left.

Introducing the same safety factor into the other working equation gives

$$C = (0.052)(1.125)\,Gh$$

$$= 0.05844\,Gh \qquad (19\text{--}23)$$

or

$$h = 17.11\,\frac{C}{G} \qquad (19\text{--}24)$$

Examples of Casing-String Design

The following examples illustrate the method of designing casing strings. In these examples, a stright line drawn on the tensile load versus collapse resistance chart represents a section of a string of casing. In general, the area to the left and underneath the curve for a particular weight and grade represents the area where that weight and grade of casing may be used. Although some special types of joints furnished by manufacturers have joint strengths equal to or greater than the cross-sectional strength throughout the length of the casing, the listed values of the joint strength of regular API casing is less than the strength throughout the length. Therefore, a horizontal line at a tensile load equal to the published joint strength reduced by a suitable safety allowance cuts off the top portion under the curve in defining the usable area on the chart. In general, the lines representing the solution proceed from the bottom of the chart upward toward the left-hand side of the chart. As a general rule, when the line drawn on the chart intersects from the left the line representing a lighter weight or lesser grade of casing, it is recommended to change to that weight and grade of casing, under the assumption that it is available at the time it is needed. The reason for this is not necessarily apparent from API code legend on the chart. The reason for changing to the lighter or lesser grade, when it will satisfactorily perform the job, becomes most apparent when the API symbols on each curve are replaced by the cost in dollars per foot of pipe. Factors governing the cost include

TABLE 19-3

Design Data for 7-inch API Casing

Nominal Weight lb./ft.	Wall Area sq. in.	Grade	Collapse psi	Tension		Burst, Internal yield psi
				Short threads 1,000 lb.	Long threads 1,000 lb.	
17	4.912	F	1,100	118	—	1,440
		H	1,370	160	—	2,310
20	5.749	H	1,920	191	—	2,720
		J	2,500	254	—	3,740
23	6.656	J	3,290	300	344	4,360
		N	4,300	350	400	6,340
26	7.549	J	4,060	345	395	4,980
		N	5,320	402	460	7,240
29	8.449	N	6,370	454	520	8,160
32	9.317	N	7,400	505	578	9,060
35	10.172	N	8,420	554	635	9,960
38	10.959	N	9,080	600	688	10,800

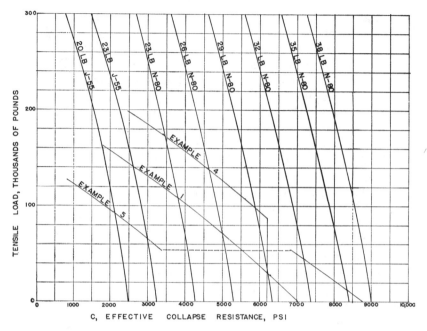

Fig. 19-2. Solution of collapse change points in casing-string design.

the size, weight per foot, grade of steel, type of coupling, price at the steel mill, transportation, and transportation-tax charges.

Example 1: Design a combination 7-inch OD casing string for a setting depth of 10,000 feet with 12 lb./gal. mud in the hole. A collapse design safety factor of 1.125 is to be used. API seamless casing will be used, and a design factor of safety of 2 will be used in tension, with the buoyant force of the mud considered as reducing the weight of the string of casing.

The starting point for the design is at the bottom of the casing, or 10,000 feet. The required collapse resistance is

$$C = 0.05844\ Gh$$
$$= (0.05844)(12.0)(10,000)$$
$$= 7,013 \text{ psi}$$

This point may be plotted on the tensile load versus effective collapse resistance chart for 7-inch casing at W (load) $= 0$, C (effective collapse resistance) $= 7,013$ psi. Figure 19-2 illustrates the solution of this example. Table 19-3 shows casing data used in the solution, and the results of the calculations for the example are shown in Table

19-4. It may be seen that 32 lb./ft. Grade N–80 casing is the lightest and least expensive casing that will withstand this collapse pressure at the bottom of the hole. The first section of casing will, therefore, be composed of 32-pound N–80 casing. This first section can be represented on the W vs. C chart by a line whose slope is

$$\text{Slope} = - w \left[\frac{17.11}{G} - 0.2623 \right]$$

$$= - 32 \left[\frac{17.11}{12.00} - 0.2623 \right]$$

$$= - 37.23 \text{ lb./psi}$$

Beginning at the point $W=0$ pounds, $C=7,013$ psi, a line with a slope of -37.23 lb./psi is drawn until it intersects the curved line representing 29 lb./ft. N–80 casing.[1] It is not necessary to start using the lighter grade of casing at this point; the heavier grade will not fail if it is carried farther up the hole. But it is certainly more economical to start using the lighter casing at the greatest depth where it will withstand the collapse pressure. Therefore, the top of the first section will be taken at the intersection of the straight line and the 29-lb. N–80 curve, where $C=6,240$ psi. The depth at this point is found by the relation

$$h = 17.11 \frac{C}{G}$$

$$= \frac{(17.11)(6,240)}{12.0}$$

$$= 8,898 \text{ ft.}$$

The first section, therefore, consists of 32 lb./ft. N–80 casing, starts at 10,000 feet, and is carried up the hole to 8,898 feet for a total length of 1,102 feet. In practice, the length would be approximately 1,102

---

[1] It is convenient to lay off this slope on the tensile load versus effective collapse resistance chart by using two drawing triangles. Since the C scale on the chart goes to a value of 10,000, multiply the slope of -37.23 by $-10,000$ to obtain a value of $W=372,300$ pounds. Set one triangle on the chart so that one edge of it passes through the point $C=10,000$ on the *effective collapse resistance* axis and through the point $W=372,300$ pounds on the *tensile load* axis. Then place the second triangle against another edge of the first triangle so that the slope of the line can be maintained as you slide the first triangle along the edge of the second triangle. Then slide the first triangle downward until its edge passes through the starting point of the section, which in this case was $C=7,093$ psi and $W=0$. Draw the line from the starting point upwards towards the left until it interesects the curved line representing the next higher section of casing.

feet, either slightly greater or slightly less, depending upon how the lengths of the joints of pipe available came out as the string would be made up.

The second section of the string of casing will start where the straight line representing the first section intersected the curve for 29-lb. N–80 casing. The second section will consist of 29-lb. N–80 casing. It may be represented by a straight line with a slope of

$$\text{Slope} = -w\left[\frac{17.11}{G} - 0.2623\right]$$

$$= -29.0\left[\frac{17.11}{12.00} - 0.2623\right]$$

$$= -33.74 \text{ lb./psi}$$

A line of this slope is drawn until it intersects the curve for 26-lb. N–80 casing. This same procedure is used for designing the first, or lower, four sections of the casing string. Notice that no 26-lb. J–55 casing was used in this string and for purposes of this example it may be assumed that it was not available. The fifth section consists of 23-lb. J–55 casing. Notice that the line drawn representing this section intersects the 20-lb. J–55 curve at a point above its ultimate joint strength, using a safety factor of 2 in designing for tension. Therefore, no 20-lb. J–55 casing can be used. The fifth section, composed of 23-lb. J–55 casing with long threads and couplings can be carried up to the point where the weight, suspended below the top of the section, equals the ultimate joint strength reduced by a suitable safety factor. Then it will be necessary to use a stronger weight and grade of casing. These calculations are best carried out as follows:

First section	(1,102 ft.)(32 lb./ft.)(0.816) =	28,765 lb.
Second section	(1,768 ft.)(29 lb./ft.)(0.816) =	41,837 lb.
Third section	(1,683 ft.)(26 lb./ft.)(0.816) =	35,706 lb.
Fourth section	(1,682 ft.)(23 lb./ft.)(0.816) =	31,568 lb.
Total weight of first four sections		=137,876 lb.

(*Note:* The safety factor of 2 in tension is used in this example, with the buoyant effect of the mud considered in reducing the weight of the pipe. The buoyancy factor is (weight of steel in mud)/(weight of steel in air), which reduces to $(1-0.01533G)$, in this case, $[1-(0.01533)(12.0 \text{ lb./gal.})]$ or 0.816.)

The ultimate joint strength of 23-lb. J–55 API seamless casing with long couplings is **344,000 pounds**. Using a safety factor of 2, the usable joint strength is **172,000 pounds**. The length of the fifth section, which is determined by tension requirements, is

$$\frac{172{,}000 \text{ usable joint strentgh} - 137{,}876 \text{ lb.}}{(0.816)(23 \text{ lb./ft.})} = 1{,}817 \text{ ft.}$$

TABLE 19-4

7-INCH API SEAMLESS CASING STRING DESIGNED TO RESIST COLLAPSE FROM 12.0 PPG MUD SETTING DEPTH OF 10,000 FEET
(bursting requirements not considered)

Section	Weight and Grade	C, Required Effective Collapse Resistance at Bottom of Section psi	Depth of Bottom of Section ft.	Length of Section ft.	Effective Weight of Section lb.	Cumulative Effective Weight to Top of Section lb.	Usable Joint Strength in Tension lb.	Depth to top of Section ft.	Slope of Section of Collapse Chart lb./psi
Seventh	26-N LT&C	—	456	456	966	200,966	230,000	0	—
Sixth	23-N LT&C	—	1,948	1,492	28,000	200,000	200,000	456	—
Fifth	23-J LT&C	2,640	3,765	1,817	34,103	172,000	172,000	1,948	−26.76
Fourth	23-N ST&C	3,820	5,447	1,682	31,568	137,876	175,000	3,765	−26.76
Third	26-N ST&C	5,000	7,130	1,683	35,706	106,308	201,000	5,447	−30.25
Second	29-N ST&C	6,240	8,898	1,768	41,837	70,602	227,000	7,130	−33.74
First	32-N ST&C	7,013	10,000	1,102	28,765	28,765	252,500	8,898	−37.23

The fifth section therefore consists of 1,817 feet of 23-lb. J-55 casing furnished with long couplings.

The sixth section will be composed of the next stronger grade, 23-lb. N-80 furnished with long couplings. The ultimate joint strength is 400,000 pounds, and the usable joint strength for a tension factor of 2 is 200,000 pounds. The length of the sixth section is

$$\frac{200{,}000 \text{ lb.} - 172{,}000 \text{ lb.}}{(0.816)(23 \text{ lb./ft.})} = 1{,}492 \text{ ft.}$$

The combined effective weight of the first six sections then is 200,0000 pounds. The usable joint strength of 26-lb. N-80 casing with long couplings is 460,000/2 or 230,000 pounds. The length of the seventh section then may be

$$\frac{230{,}000 \text{ lb.} - 200{,}000 \text{ lb.}}{(0.816)(26 \text{ lb.})} = 1{,}414 \text{ ft.}$$

However, only 456 feet of 26-lb. N–80 need be used in the seventh section to reach the top of the hole. The design of the string of casing is now completed.

The results of the foregoing calculations are summarized in Table 19–4. The requirements for bursting strength have not been analyzed in this first example. However, the bursting-strength capacities of the several sections are listed in Table 19–3, and for the purposes of this example they will be assumed to be adequate for known producing conditions.

Example 2: This example is an alternate solution of the tensile load requirements of Example 1. This alternate solution would be applicable in cases where an intermediate string of casing had been set at a depth greater than 1,948 feet, which is the top of the 23 lb./ft. J–55 casing, or in other cases where such hole conditions prevailed that the highest possible stuck (or frozen) point in the casing string would be below the top of the 23 lb./ft. J–55 casing. In such cases there would be no justification for providing a greater excess load-carrying capacity higher in the string of casing than exists at the top of the 23 lb./ft. J–55 casing.[2] The top of that section was picked according to a design safety factor of 2 in tension. The excess load-carrying capacity provided for at the top of the 23 lb./ft. J–55 casing with long threads and couplings is the ultimate joint strength minus the effective weight of the casing below that point, 344,000 pounds minus 172,000 pounds, or 172,000 pounds.

The next higher section is to be composed of 23-lb N–80 casing with long threads and couplings with an ultimate joint strength of 400,000 pounds. The length of this section may be computed as follows:

400,000 lb. ultimate joint strength − 172,000 lb. excess capacity

= 228,000 lb. usable joint strength

$$\frac{228{,}000 \text{ lb. usable joint strength} - 172{,}000 \text{ lb. load at bottom}}{(0.816)(23.0 \text{ lb./ft.})} = 2{,}980 \text{ ft.}$$

This is greater than the 1,948 feet needed to reach the top of the hole. The top section would then consist of 1,948 feet of 23-lb. N–80 casing with long threads and couplings.

Again in this example, the requirements for bursting strength have not been analyzed but are assumed adequate for known producing conditions.

Example 3: This example introduces the design calculations for

[2] C. N. Bowers, "Design of Casing Strings," a paper presented at the 30th Annual Fall Meeting, Petroleum Branch, A.I.M.E., New Orleans, October 2–5, 1955.

bursting strength into the data used in Example 1. The bursting-strength requirements are negligible at the bottom of the hole. The first four sections are designed for collapse resistance as given in Example 1.

In designing for bursting resistance, it will be assumed that the pressure inside the casing is equal to the hydrostatic mud head at the total depth of 10,000 feet, which is 6,240 psi for 12 ppg mud. This is based on the assumption that the formation pressure is not greater than the hydrostatic mud head and that the casing becomes filled with gas which transmits the pressure to the top inside of the casing, neglecting the hydrostatic head of the gas. The pressure on the outside of the casing will be assumed to be equal to the hydrostatic head of a column of fresh water extending downward from the surface. This is based on the lowering of the mud level outside the casing immediately after cementing, when the excess water mixed in the cement slurry filters into porous formations, and it allows for filtration bridges and gelling of the mud behind the casing. According to these assumptions, the bursting-strength requirements may be expressed as follows:

$$\text{Required bursting strength} = 6{,}240 \text{ psi} - 0.433\, h,$$

where h is the depth in feet.

It is assumed that the formation pressure will decline in a short time, so that a design safety factor of 1 may be used for bursting requirements for the assumed conditions. By equating the required bursting strength and the listed bursting capacity, the following is obtained:

$$\text{Bursting capacity, psi} = 6{,}240 \text{ psi} - 0.433h$$

or

$$h = \frac{6{,}240 \text{ psi} - \text{Bursting capacity, psi}}{0.433}$$

The 23 lb./ft. J–55 casing has a bursting capacity of 4,360 psi. The depth to the top of the section composed of this pipe should then be

$$h = \frac{6{,}240 \text{ psi} - 4{,}360 \text{ psi}}{0.433}$$

$$= 4{,}350 \text{ ft.}$$

This depth is below the point where it was intended to change over to the 23 lb./ft. J–55 pipe. Therefore, no pipe of this weight and grade can be included in the string and meet the specified bursting requirements.

The next step is to go back and pick the top of the fourth section according to bursting or tensile-strength requirements, whichever may be applicable. The fourth section is composed of 23-lb. N–80 casing, which has a bursting strength of 6,340 psi. Since this exceeds the bursting requirement at the top of the hole, this pipe could be run from the depth of 5,447 feet to the surface, so far as bursting requirements are concerned. The top of the fourth section is therefore picked according to tensile-strength requirements. It is composed of 23-lb. N–80 casing with short threads and couplings with a joint strength of 350,000 pounds. Using a safety factor of 2 results in a usable joint strength of 175,000 pounds. The effective weight of the lower three sections is 106,308 pounds, and the buoyancy factor is 0.816. The length of the fourth section is computed as follows:

$$\frac{175{,}000 \text{ lb. usable joint strength} - 106{,}308 \text{ lb. load on bottom}}{(0.816)(23.0 \text{ lb./ft.})} = 3{,}660 \text{ ft.}$$

The fifth section must be composed of the next stronger pipe. An increase in tensile strength can be achieved most economically by switching from short threads and couplings to long threads and couplings. Following the reasoning of Example 2, the excess load-bearing capacity at the top of the fourth section is 350,000 pounds $-175{,}000$ pounds, or 175,000 pounds. If it is deemed sufficient to carry this excess load-bearing capacity to the top of the hole, the length of the fifth section could be

$$\frac{400{,}000 \text{ lb. ultimate} - 175{,}000 \text{ lb. allowance} - 175{,}000 \text{ lb. load}}{(0.816)(23.0 \text{ lb. ft.})}$$

$$= 2{,}660 \text{ ft.}$$

However, only 1,787 feet are required to extend to the surface.

The results of these calculations are shown in Table 19–5. In the table, the usable joint strength listed for each weight and grade is the recommended ultimate minus 175,000 pounds.

Example 4: Design a combination 7-inch OD casing string for a setting depth of 12,000 feet with 12.5 lb./gal. mud in the hole, with the condition that the mud will not be swabbed out to a depth greater than 8,500 feet in bringing the well into production nor internal pressures subsequently reduced below the corresponding pressure in the bottom of the hole. A collapse design safety factor of 1.125 is to be used, and a design safety factor of 2 is to be used in tension, with the buoyant force of the mud considered as reducing the weight of the string of casing. Design for bursting is to be based on the assumption

TABLE 19-5

7-INCH API SEAMLESS CASING STRING DESIGNED TO RESIST COLLAPSE FROM 12.0 PPG MUD—SETTING DEPTH 10,000 FEET
(Bursting requirements based on formation pressure equal to hydrostatic mud head)

Section	Weight and Grade	C. Required Effective Collapse Resistance at Bottom of Section psi	Depth to Bottom of Section ft.	Length of Section ft.	Effective Weight of Section lb.	Cumulative Effective Weight to Top of Section lb.	Ultimate Joint Strength in Tension, lb.	Usable Joint Strength in Tension lb.	Slope of Section on Collapse Chart lb./psi	Depth to Top of Section ft.	Expected Bursting Requirement at Top of Section psi	Bursting Strength psi
Fifth	23-N LT&C		1,787	1,787	33,538	208,538	400,000	225,000	−26.76	0	6,240	6,340
Fourth	23-N ST&C	3,820	5,447	3,660	68,692	175,000	350,000	175,000	−26.76	1,787	5,470	6,340
Third	26-N ST&C	5,000	7,130	1,683	35,706	106,308	402,000	227,000	−30.25	5,447	3,880	7,240
Second	29-N ST&C	6,240	8,898	1,768	41,837	70,602	454,000	279,000	−33.74	7,130	3,160	8,160
First	32-N ST&C	7,013	10,000	1,102	28,765	28,765	505,000	330,000	−37.23	8,898	2,390	9,060

that the pressure within the casing may reach a value equal to the hydrostatic mud head at total depth, and the external pressure on the casing may be reduced to the head of a column of fresh water extending downward from the surface.

(*Note:* The pressure equivalent of the mud within the casing below 8,500 feet would be (12,000 ft. −8,500 ft.) (0.052) (12.5 lb./gal.) or 2,275 psi bottom-hole pressure at 12,000 feet with zero internal bottom-hole pressure assumed in testing or producing any zone above 8,500 feet, with a linear pressure relationship between 8,500 and 12,000 feet.)

By way of explanation, it may be noted that the casing is assumed to be surrounded by 12.5 lb./gal. drilling mud and to contain 12.5 lb./gal. mud inside the casing from 8,500 feet to the total depth of 12,000 feet. Therefore, the external mud pressure, which would tend to collapse the casing, increases from the surface down to 8,500 feet, but the pressure difference remains constant below that depth because increases in external mud pressure are counterbalanced by equal increases of the hydrostatic mud head inside the casing. It may be noted also that the pressure reduction of 6,209 psi, calculated below, which is permitted in the bottom part of the hole, would in turn permit a pressure reduction in excess of 5,000 psi for testing or producing any formation below 8,500 feet, if the reasonable assumption is made that the hydrostatic mud head does not exceed the formation pressure by more than 1,000 psi. It may also be calculated that if the pressure inside the casing at the 12,000-foot depth were reduced to zero, the required collapse resistance would be 8,766 psi, which would necessitate the use of 38 lb./ft. N–30 casing, whereas 32 lb./ft. is the heaviest casing requred in the following solution. Therefore, a saving in cost is possible where the present design conditions are applicable.

The starting point for the design is at the bottom of the casing, or 12,000 feet. The required collapse resistance is

$$C = 0.052\ FGh$$
$$= (0.052)(1.125)(12.5\ \text{lb./gal.})(8,500\ \text{ft.})$$
$$= 6,209\ \text{psi}$$

This point is plotted on the tensile load versus effective collapse resistance chart for 7-inch casing at $W=0$ and $C=6,209$ psi on Fig. 19–2. It may be seen that 29 lb./ft. N–80 casing is the lightest casing that may be used in the first section. This section is represented on the chart by a vertical line, since in this section the tensile load increases while required collapse resistance remains constant. This vertical line passes through the 29 lb./ft. N–80 line at a tensile load of 36,000 pounds. The length of the first section, corresponding to a weight of 36,000 pounds, is calculated as follows:

$$\text{Mud buoyancy factor} = 1 - \frac{(7.48)(12.5 \text{ lb./gal.})}{488}$$

$$= 0.8084$$

$$\text{Length of first section} = \frac{36{,}000 \text{ lb.}}{(0.8084)(29 \text{ lb./ft.})}$$

$$= 1{,}536 \text{ ft.}$$

The second section of casing is to consist of 32 lb./ft. N–80 casing. Its length is determined in two steps, as follows: First, the vertical line representing the casing from 12,000 feet to 8,500 feet, which is a distance of 3,500 feet, is extended to the tensile load corresponding to the 8,500-foot depth point.

Tensile load at 8,500 feet

$$= 36{,}000 \text{ lb.} + (3{,}500 \text{ ft.} - 1{,}536 \text{ ft.})(0.8084)(32 \text{ lb./ft.})$$

$$= 86{,}806 \text{ lb.}$$

Above 8,500 feet, the required collapse resistance decreases as the tensile load increases. By Equation (19–22), the slope of this line may be calculated as

$$\text{Slope}_2 = -w\left(\frac{17.11}{G} - 0.2623\right)$$

$$= -w\left(\frac{17.11}{12.5 \text{ lb./gal.}} - 0.2626\right)$$

$$= -35.408$$

The line drawn on Fig. 19–2 with this slope intersects the 29 lb./ft. N–80 casing line at a required collapse-resistance value of 5,885 psi and a tensile-load value of 98,000 pounds. The corresponding depth is found from Equation (19–24).

$$h = 17.11\frac{C}{G}$$

$$= 17.11\frac{(5{,}885 \text{ psi})}{12.5 \text{ lb./gal.}}$$

$$= 8{,}055 \text{ ft.}$$

The second section of the string then consists of 32 lb./ft. N–80 and extends from 10,464 feet to 8,055 feet for a length of 2,409 feet.

The third section of the string consists of 29 lb./ft. N–80 (the same as the first section), and the slope of the line on the tensile load versus collapse chart is:

$$\text{Slope}_3 = -(29.0 \text{ lb./ft.}) \left(\frac{17.11}{12.5} - 0.2623 \right)$$

$$= -32.089$$

The third section extends to the intersection of this line with the 26 lb./ft. N–80 line, at a required collapse resistance of 4,645 psi and the corresponding depth in the well of 6,358 feet.

The results of the design calculations are given in Table 19–6. It may be observed that the tops of sections 1, 2, 3, and 4 are picked by collapse-resistance calculations. The top of section 5, which is 23 lb./ft. N–80 casing with long threads and couplings, would be found at the depth of 3,236 feet according to the permissible tensile load on the coupling using a design safety factor of 2. However, the top of this section was actually picked according to bursting-strength requirements.

In designing for bursting strength, it was assumed that the pressure in the hole outside of the casing would never become less than the hydrostatic head of fresh water extending downward from the surface. It was further assumed that the internal pressure to which the casing might be subjected would be equal to the hydrostatic mud head at total depth. The depth to the top of the fifth section, which is composed of 23 lb./ft. N–80 casing, was calculated as follows:

Assumed pressure inside casing

$$= (12{,}000 \text{ ft.})(0.052)(12.5 \text{ lb./gal.} = 7{,}800 \text{ psi}$$

Bursting-strength capacity of 23 lb./ft. N–80 = 6,340 psi

Assumed minimum pressure outside casing = $0.433 h$

The bursting strength is equated to the difference between the assumed inside and outside pressures and solved for the corresponding depth.

$$6{,}340 \text{ psi} = 7{,}800 \text{ psi} - 0.433 h$$

$$h = 3{,}372 \text{ ft.}$$

Therefore, the top of the fifth section was set at 3,372 feet.

The sixth section is composed of 26 lb./ft. N–80 casing with long

CASING AND CASING-STRING DESIGN 433

TABLE 19-6

7-INCH API SEAMLESS CASING STRING DESIGNED TO RESIST COLLAPSE FROM 12.5 PPG MUD FOR SETTING DEPTH OF 12,000 FEET BUT NOT TO BE SWABBED BELOW 8,500 FEET

Section	Weight and Grade	C, Required Effective Collapse Resistance at Bottom of Section psi	Depth to Bottom of Section ft.	Length of Section ft.	Effective Weight of Section lb.	Cumulative Effective Weight to Top of Section lb.	Usable Joint Strength in Tension lb.	Depth to Top of Section ft.	Bursting Strength psi	Slope of Section on Collapse Chart	Expected Bursting Requirements at Top of Section psi
Eighth	32–N LT&C	—	544	544	14,073	274,073	289,000	0	9,060	−35.408	7,800
Seventh	29–N LT&C	—	1,824	1,280	30,000	260,000	260,000	544	8,160	−32.089	7,564
Sixth	26–N LT&C	—	3,372	1,548	32,536	230,000	230,000	1,824	7,240	−28.769	7,010
Fifth	23–N LT&C	3,490	4,777	1,405	26,123	197,464	200,000	3,372	6,340	−25.450	6,340
Fourth	26–N ST&C	4,645	6,358	1,581	33,230	171,341	201,000	4,777	7,240	−28.769	5,732
Third	29–N ST&C	5,885	8,055	1,697	39,784	138,111	227,000	6,358	8,160	−32.089	5,047
Second	32–N ST&C	6,209	10,464	2,409	62,318	98,327	252,500	8,055	9,060	−35.408	4,312
First	29–N ST&C	6,209	12,000	1,536	36,009	36,009	227,000	10,464	8,160	8	3,269

threads and couplings, the next lightest casing which will satisfy tensile-load requirements. It may be observed in Table 19–6 that the tops of the sixth and seventh sections were picked according to their joint strengths, and neither collapse nor bursting requirements again entered into the calculations.

Study of the biaxial yield stress ellipse will show that a safety factor is automatically associated with using the listed bursting capacity, as was done in the preceding example, where the calculations concern a point in the casing string which will be under tension. Tension on the pipe slightly increases its bursting capacity.

Example 5: Design a combination 7-inch OD casing string for a setting depth of 7,000 feet with 10.5 lb./gal. mud in the hole, with the condition that the lower 1,500 feet between 5,500 feet and 7,000 feet will be designed to resist the theoretical overburden pressure of 1.0 psi/ft. with a design safety factor of 1.25 for collapse. This design is to resist the flow into the well of a bed of bentonite located near the bottom of the hole. Buoyant effects in the bottom 1,500 feet are to be ignored, but the buoyant force of the mud may be considered as reducing the effective weight on the casing above that depth. From 5,500 feet to the surface, the casing is to resist the collapsing force of the 10.5 lb./gal. mud with a design safety factor of 1.125 in collapse. A design safety factor of 2 in tension is to be used throughout.

The starting point for the design is at the bottom of the hole. The required collapse resistance may be calculated by Equation (19–13).

$$C = Fhp$$
$$= (1.25)(7,000 \text{ ft.})(1 \text{ psi/ft.})$$
$$= 8,750 \text{ psi}$$

This indicates that 38 lb./ft. N–80 casing must be used in the first section. This point is plotted on the collapse versus tensile load chart at $C = 8,750$ psi and W (tensile load) $= 0$. The slope of the line representing the first section may be calculated by the following relation, which is equivalent to Equation (19–19):

$$\text{Slope}_1 = -\frac{Bw}{Fp} \qquad (19\text{--}25)$$
$$= -\frac{(1)(38.0 \text{ lb./ft.})}{(1.25)(1 \text{ psi/ft.})}$$
$$= -30.40$$

A line drawn on the chart, Fig. 19–2, interests the 35 lb./ft. N–80 line at $C=8,360$ psi and $W=12,000$ pounds. The corresponding depth may be found from Equation (19–15).

$$h = \frac{C}{Fp}$$

$$= \frac{8,360 \text{ psi}}{(1.25)(1 \text{ psi/ft.})}$$

$$= 6,688 \text{ ft.}$$

The second section consists of 35 lb./ft. N–80 casing, and the slope of the line representing the second section is calculated by Equation (19–25):

$$\text{Slope}_2 = -\frac{(1)(35.0 \text{ lb./ft.})}{(1.25)(1 \text{ psi/ft.})}$$

$$= -28.00$$

The line drawn on Fig. 19–2 representing the second section intersects the 32 lb./ft. N–80 line at $C=7,180$ psi and $W=45,000$ pounds. It is therefore possible to change to 32 lb./ft. N–80 at that point. The corresponding depth is

$$h = \frac{7,180 \text{ psi}}{(1.25)(1 \text{ psi/ft.})}$$

$$= 5,744 \text{ ft.}$$

The third section consists of 32 lb./ft. N–80 casing, and the slope is calculated as before at a value of -25.60. A line drawn with this slope on Fig. 19–2 would intersect the 29 lb./ft. N–80 line at a value of $C=6,020$ psi and a corresponding depth of 4,816 feet. However, it is desired to carry the design-resisting overburden pressure only to the depth of 5,500 feet. Therefore, the top of the third section is taken as 5,500 feet. The corresponding value of C is calculated:

$$C = Fhp$$

$$= (1.25)(5,500 \text{ ft.})(1 \text{ psi/ft.})$$

$$= 6,875 \text{ psi}$$

On the line representing the third section on Fig. 19–2, this occurs at $W=52,800$ pounds. At this point the design changes to resist the collapsing force of the drilling mud. The corresponding value of C is calculated:

$$C = (0.052) FGh$$
$$= (0.052)(1.125)(10.5 \text{ lb./gal.})(5,500 \text{ ft.})$$
$$= 3,375 \text{ psi}$$

The horizontal line drawn on Fig. 19–2 from $C=6,875$ to $C=3,375$ at the tensile-load value of 52,800 pounds represents this change in design conditions.

The tensile-load and collapse-resistance requirements of $W=52,800$ pounds and $C=3,375$ psi are satisfied by 23 lb./ft. N–80 casing with short threads and couplings. Accordingly, this pipe is used for the fourth section commencing at 5,500 feet. The slope of the line representing the fourth section is calculated by Equation (19–22):

$$\text{Slope}_4 = -w\left(\frac{17.11}{G} - 0.2623\right)$$
$$= -23 \text{ lb./ft.}\left(\frac{17.11}{10.5 \text{ lb./gal.}} - 0.2623\right)$$
$$= -31.45$$

A line drawn with this slope intersects the 23 lb./ft. J–55 casing line at $C=3,020$ psi and $W=64,000$ pounds. The corresponding depth value is calculated by Equation (19–24):

$$h = 17.11 \frac{C}{G}$$
$$= 17.11 \frac{3,020 \text{ psi}}{10.5 \text{ lb./gal.}}$$
$$= 4,921 \text{ ft.}$$

The design then proceeds on up the hole to the surface by using the same methods illustrated in Example 1. The results are given in Table 19–7.

Analytical Solution of Collapse Change Points

The analytical solution for collapse change points has been developed by Curran.[3] The biaxial yield stress ellipse is written as

$$r^2 + rt + t^2 = 1 \qquad (19\text{–}10)$$

---

[3] B. E. Curran, "Direct Method for Calculating Lengths of Section, Simplified Design of Combination Casing Strings," *Oil and Gas Journal*, Vol. XLVII, No. 24 (October 14, 1958), 80.

TABLE 19-7

7-INCH API SEAMLESS CASING STRING DESIGNED TO RESIST COLLAPSE FROM OVERBURDEN PRESSURE FROM SETTING DEPTH OF 7,000 FEET TO 5,500 FEET AND COLLAPSE PRESSURE OF 10.5 PPG MUD FROM 5,500 FEET TO SURFACE

Section	Weight and Grade	C, Required Effective Collapse Resistance at Bottom of Section psi	Depth to Bottom of Section ft.	Length of Section ft.	Effective Weight of Section lb.	Cumulative Effective Weight to Top of Section lb.	Usable Joint Strength in Tension lb.	Depth to Top of Section ft.	Bursting Strength psi	Slope of Section on Collapse Chart	Expected Bursting Requirements at top of Section psi
Eighth	23–N ST&C	—	186	186	3,590	153,590	175,000	0	6,340	−31.45	3,822
Seventh	23–J ST&C	—	1,377	1,191	23,000	150,000	150,000	186	4,360	−31.45	3,741
Sixth	20–J ST&C	2,130	3,471	2,094	35,142	127,000	127,000	1,377	3,740	−27.34	3,226
Fifth	23–J ST&C	3,020	4,921	1,450	27,983	91,861	150,000	3,471	4,360	−31.45	2,319
Fourth	23–N ST&C	3,375	5,500	579	11,174	63,878	175,000	4,921	6,340	−31.45	1,691
Third	32–N ST&C	7,180	5,744	244	7,808	52,704	252,500	5,500	9,060	−25.60	1,440
Second	35–N ST&C	8,360	6,688	944	33,040	44,896	277,000	5,744	9,960	−28.00	1,335
First	38–N ST&C	8,750	7,000	312	11,856	11,856	300,000	6,688	10,800	−30.40	929

where

r = fractional collapse resistance
t = fractional tensile-yield stress

The following substitutions are made:

$$r = \frac{pF(h_i - x)}{K} \qquad (19\text{–}26)$$

$$t = \frac{W_i + wx}{AS_o} \qquad (19\text{–}27)$$

where

p = drilling-mud pressure gradient, psi/ft.
F = collapse safety factor
h_i = depth to the bottom of the section whose length is to be calculated, ft.
x = length of the section which is being calculated, ft.
K = minimum collapse resistance of the pipe in the next higher section, unadjusted for tension, psi
W_i = weight of casing hanging below the bottom of the section whose length is being calculated, lb.
w = weight per foot of casing in the section whose length is being calculated, lb./ft.
A = cross-sectional area of the steel in the pipe in the next higher section, sq. in.
S_o = average tensional-yield stress in the steel in the next higher section, psi

The solution for the length of the section is given by Curran as

$$x = \frac{h_i}{a} + \frac{n}{b} - c\sqrt{1 - \left(\frac{h_i + n}{d}\right)^2} \qquad (19\text{–}28)$$

$$= \frac{h_i}{a} + \frac{n}{b} - c\sin\left[\cos^{-1}\left(\frac{h_i + n}{d}\right)\right] \qquad (19\text{–}29)$$

where

$$a = 1 - \frac{m-2}{m(2m-1)}$$

$$m = \frac{pFAS_o}{wK}$$

$$n = \frac{W_i}{w}$$

$$b = \frac{m(2m-1)}{m-2}$$

$$c = \frac{AS_o}{w\sqrt{1+m(m-1)}}$$

$$d = \frac{2K}{pF\sqrt{3}}\sqrt{1+m(m-1)}$$

It may be noted that Curran used no buoyancy effect; therefore, a buoyancy factor of $(1-0.01532\,G)$, where G is the mud density in pounds per gallon, should be introduced into the weight terms, in order to eliminate an otherwise concealed safety factor. If this is done, the various terms become

$$a = 1 - \frac{m-2}{m(2m-1)}$$

$$m = \frac{GFAS_o}{(19.23 - 0.295\,G)wK}$$

$$n = \frac{W_i}{w}$$

$$b = \frac{m(2m-1)}{m-2}$$

$$c = \frac{AS_o}{w(1-0.01532\,G)\sqrt{1+m(m-1)}}$$

$$d = \frac{22.206K}{GF}\sqrt{1+m(m-1)}$$

G = mud density, lb./gal.

The analytical solution, while theoretically correct, has two serious disadvantages. First, the calculations are quite involved, and mistakes are easily made in performing such calculations. Second, the answer is determined as the difference between two large numbers, which inherently introduces inaccuracy, even if a calculating machine is used. It seems probable that the most accurate solution can be obtained by constructing a large-scale chart of effective collapse resistance versus tensile loading and obtaining a graphical solution from such a chart.

Casing Landing Practices

After casing has been run in the well bore and cemented, it is attached at the surface to the well-head equipment. Casing is secured by the use of slips, in a bowl which has been attached to the next larger casing string. The slips have serrated edges, or teeth, which engage the outside of the casing. The weight of the casing tends to pull the slips lower into the bowl, which decreases in internal diameter in the downward direction, thus causing the slips to engage the casing even more firmly. An important element of this equipment is a pressure-sealing device. Since the slips do not provide a pressure-tight seal, some means must be provided to control effectively the maximum pressure which may be encountered in the annular space between the casing string being landed and the next larger casing string. This is normally provided for by using O-ring seals or packing which can be tightened against the outside of the casing by application of outside mechanical force to the packing. After the casing has been attached, any excess length above the casing bowl is normally cut off with a welding torch.

The procedure followed in securing or fastening the casing at the surface is known as "landing" the casing. Several methods of landing the casing can be used, each of which will create different stresses in the casing. Inasmuch as the casing has been designed to resist specific stresses, the landing method must not create stresses greater than the casing is capable of withstanding, or casing failure will result.

Factors that will affect the stresses in casing after it has been cemented are (1) formation pressure, (2) formation temperature, (3) earth movements, (4) internal pressure, and (5) internal temperature.

The magnitude of formation pressures and temperatures are normally known before the casing is used, and there will usually be little or no change in these factors during the useful life of the casing; therefore, the effects of these factors can be considered in the design of the casing string.

The stresses induced by earth movements are rare and, when the problem of earth movement becomes critical, as it is in several areas of the world—e.g., Venezuela and California—special casing and techniques are required to control the problem. The two general types of earth movement are lateral earth movements along fault planes and vertical movement of earth formations caused by compaction of sediments. Even small earth movements will cause stresses greater than normal casing can withstand. Special types of casing and techniques which have been used to prevent casing collapse include (1) installing devices in the casing string which will permit shortening of the casing string as earth compaction occurs, and (2) elimination of the bond between the casing and the formation.

Internal temperatures and pressures are subject to variation during the producing life of the well; therefore, it is important to determine the changes in stresses caused by variations in these factors. Increasing the temperature of casing will cause it to expand both laterally and vertically. A vertical expansion will cause a reduction in the tensile stresses already imposed on the casing. Likewise, reducing the temperature below that which prevailed when the casing was set causes the casing to shrink, and this shrinkage will, in turn, increase the tensile load placed on the casing when it is landed. Increasing the internal pressure will cause increased tensile stresses on the casing, since, as the internal diameter is increased, the length must decrease. It should be pointed out that both external and internal temperature changes have the same effect on the casing; however, external and internal pressures create tensile stresses exactly opposite in direction. Increasing the external pressure acts to reduce the tensile load, or, in effect, causes a reduction in internal pressure.

The various methods of landing casing which have been used are listed below:

1. Landing the casing immediately after completion of the cementing operation.
2. After cement has set, the elevators are removed from the casing and the casing landed exactly as it stands, unsupported.
3. After the cement has set, the casing is landed in such a manner that the casing at the top of the cement is (a) in tension, (b) in compression, or (c) completely balanced so far as tensile and compressive stresses are concerned.
4. After cement has set, the casing is landed with a certain arbitrary weight (as measured by the weight indicator) which is (a) less than or (b) more than the weight of the pipe just prior to cementing.
5. After cement has set, the casing is landed with a certain number of inches of (a) stretch or (b) slack, per 1,000 feet of casing, measured from the top of cement.
6. After cement has set, the casing is landed in such a manner that the casing at the top "freeze" point (determined by stretch measurements) is (a) in tension, (b) in compression, or (c) completely balanced so far as tensile and compressive stresses are concerned.

The principal objective in landing casing is to land it in such a way that minimum stresses will be retained in the casing, and therefore a maximum amount of additional stress can be tolerated. A brief analysis of the advantages and disadvantages of the various methods of landing casings is presented below:

Method 1. There is a disadvantage in that, if the pipe is landed before the cement has set, all of the pipe will necessarily be in tension, and in long strings, this tensile stress may be large.

Method 2. The principal disadvantage is that the tensile stress at the casing hanger immediately after landing the casing is zero, and if additional forces, later in the life of the well, cause a net decrease in the tensile load, i.e., increasing temperature or decreased internal pressure (or increased external pressure), the casing may become unseated.

Method 3. The principal disadvantage is that the effective point in the casing string where the casing is held firmly, or frozen, is usually well above the top of the cement. This is caused by cave-in of the open hole surrounding the casing. Actually, in many instances, the casing is probably frozen just below the base of the next larger casing string. Thus any landing practice based on the assumption that the casing is free down to the top of cement would be erroneous and would result in excessive tensile or compressive stresses being applied to the casing.

Method 4. The principal disadvantage is that, unless the length of free casing is known, impressing of excessive tensile or compressive stresses in the casing may result from this method also.

Method 5. This method has the same disadvantage as methods 3 and 4, in that measurements are based on the assumption that the uppermost frozen spot in the pipe is at the top of the cement. This assumption is, in most cases, incorrect, and as a result severe stresses are liable to be imposed on the casing.

Method 6. This is probably the most satisfactory of the methods listed. Actually, whether method 6a, 6b, or 6c will produce better results is dependent upon the grades of casing above the freeze point and the depth to the freeze point. For most satisfactory results, the freeze point should be determined just prior to landing the casing, followed by an analysis of the anticipated additional stresses on the casing during the life of the well. With this data, the casing should then be landed in such a manner that the maximum safety factor for it will be available for both tensile and compressive stresses. In actual practice this may dictate the selection of method 6c when temperature increases are not expected to be large. In casing strings where long, unsupported columns may exist, with large increases in internal temperature expected, method 6a would in all probability be the best landing method. Method 6b would probably have little use in actual practice. It could conceivably have some value in situations where the internal pressure was expected to increase substantially and internal temperature was expected to decrease substantially, thus increasing the tensile loading and reducing the compressive stress.

CHAPTER 20

Cementing Operations

The cementing of casing in oil wells is almost a universal practice and is done for a number of reasons, depending on the particular casing being cemented. Where conductor casing is required, it must be cemented in order to prevent the drilling fluid from circulating outside the casing and thus causing the surface erosion which the casing was designed to prevent. Surface casing must be cemented in order to seal off and protect fresh-water formations, provide an anchor for blow-out-preventer equipment, and give support at the surface for the deeper strings of casing. Intermediate strings of casing are cemented in order to seal off abnormal pressure formations effectively, isolate incompetent formations which would cause excessive sloughing unless supported by casing and cement, and shut off zones of lost circulation in order to allow drilling to progress further. Oil strings are cemented in order to prevent migration of fluids to thief zones and sloughing of formations which would cause a reduction in well productivity. Cement also effectively protects the casing from corrosive environments, notably corrosive fluids, which may exist in the subsurface formations.

With a few notable exceptions, Portland cement is the principal constituent of most cementing materials. It is the ordinary cement which has been used by the construction industry for many years. However, with the advent of its use for cementing casing in oil wells, the additional requirement of pumpability at increased temperature and pressure necessitated some revisions in specifications. Additives have been developed which change the specifications of Portland cements to adapt them for use in oil well cementing.

In order for an oil well cement to perform satisfactorily the task allotted it, certain requirements must be met:

1. The cement slurry must be capable of being placed in the desired position by means of pumping equipment at the surface.
2. After being placed, it must develop sufficient strength within a reasonably short time, in order that the waiting-on-cement (WOC) time can be kept to a minimum.
3. The cement must provide a positive seal between the casing and the formation.
4. Sufficient strength must be developed in the cement to avoid mechanical failure.

5. The cement must be chemically inert to any formations or fluids with which it may come into contact.
6. The cement must be stable enough that it will not deteriorate, decompose, or otherwise lose its qualities of strength for the length of time it will be in use, which may be many years.
7. The cement must be sufficiently impermeable that fluids cannot flow through it when it has set.

Composition of Oil Well Cements

Portland cement, which has been the principal constituent of most oil well cements, obtained its name from its similarity to a building stone found on te Isle of Portland, off the coast of England. It is a combustion product, and its principal constituents are limestone, clay, shale, slag, bauxite, and miscellaneous iron-bearing materials. In the manufacture of Portland cement, the proper quantities of the materials are mixed together and heated to approximately 2,700 degrees Fahrenheit in an oven or "kiln." After being subjected to the kiln temperature, the material is converted to "clinker," and it is this clinker which is ground and becomes Portland cement. The chemical composition of the cement varies, but in general it is composed of varying percentages of the materials shown in Table 20–1. Tricalcium silicate and tricalcium aluminate, which react quite rapidly with water, are the constituents principally responsible for high early-strength characteristics of cement. Dicalcium silicate and tetracalcium aluminaferrite react more slowly and contribute to the slow increase in strength of cement. Gypsum is used to control the reaction rate of tricalcium aluminate. Magnesia is actually an undesirable element and the percentage of it is kept as low as possible. Magnesia will react with water, although very slowly, to form magnesium hydroxide $[Mg(OH)_2]$ and, especially in construction cements, will eventually cause cracks in the concrete if too much of it is present. Free lime is usually present to some extent in Portland cements. Because it too will react quite slowly with water to cause expansion of the cement, the quantity is also kept to a minimum.

The American Society for Testing Materials (ASTM) has adopted specifications for ten different types of Portland cements used in construction work, but these specifications do not adequately define the properties required in oil well cements. Therefore, specifications for seven classes of API cements have been established. Table 20–2 shows the different API classes of cements. The principal difference among them is the pumpability time (thickening time). API specifications for the thickening times of the various classes of cements are

TABLE 20-1
Hydration Products of Cement

Products in Dry Cement	Hydration Products Atmospheric Temp. & Press.	Hydration Products Temperatures of 200 to 375 deg. F. at Elevated Pressures
Tricalcium Silicate ($3CaO \cdot SiO_2$)	$1.5CaO \cdot SiO_2 \cdot 1.0$ to $2.5H_2O$ $+Ca(OH)_2$	$2CaO \cdot SiO_2 \cdot 1.0$ to $1.25H_2O$ (Hydrate) $+Ca(OH)_2$
Dicalcium Silicate ($2CaO \cdot SiO_2$)		$2CaO \cdot SiO_2 \cdot 1.0$ to $1.25H_2O$ (Hydrate)
Tricalcium Aluminate ($3CaO \cdot Al_2O_3$)	$3CaO \cdot Al_2O_3 \cdot 3CaSO_4 \cdot 31H_2O$ $+3CaO \cdot Al_2O_3 \cdot 6H_2O+KOH$ and NaOH	$3CaO \cdot Al_2O_3 \cdot 6H_2O$
Gypsum ($CaSO_4 \cdot 2H_2O$)		$CaSO_4$
Alkalis (ss $Na_2SO_4 - K_2SO_4$		
Tetracalcium Aluminoferrite ($4CaO \cdot Al_2O_3 \cdot Fe_2O_3$) or ss' ($6CaO \cdot Al_2O_3 \cdot Fe_2O_3 - 2CaO \cdot Fe_2O_3$)	$3CaO \cdot R_2{}^3O_3 \cdot 3CaSO_4 \cdot 31H_2O$ $+3CaO \cdot R_2O_3 \cdot 6H_2O$	$3CaO \cdot Al_2O_3 \cdot 6H_2O$ $+Ca(OH)_2 + Fe_2O_3$
Gypsum ($CaSO_4 \cdot 2H_2O$)		
Calcium Hydroxide [$Ca(OH)_2$]		
Alkalis: ($23CaO \cdot K_2O \cdot 12SiO_2$)	$1.5CaO \cdot 1.0$ to $2.5H_2O$ $+KOH+Ca(OH)_2$	
($8CaO \cdot Na_2O \cdot 3Al_2O_3$)	$3CaO \cdot Al_2O_3 \cdot 6H_2O + NaOH$ $+Ca(OH)_2$	
Magnesia (MgO)	$Mg(OH)_2$	$Mg(OH)_2$
Free Lime (CaO)	$Ca(OH)_2$	$Ca(OH)_2$

shown in Table 20-3. In addition to minimum thickening-time requirements, API specifications include requirements for (1) chemical compositions for both regular and high sulfate-resistant cements, (2) soundness, (3) fineness, and (4) 8-hour and 24-hour compressive strengths.

Ordinary Portland cements were originally used in oil-well cementing operations; however, as early as 1917, "oil well cements" were available. The principal differences between construction and oil well cements are that (1) no aggregate is added to the oil well cements (in other words, a "neat" slurry is used) and (2) large volumes of water

TABLE 20-2
API Classes of Cement

Class	Recommended Use
A	Intended for use to 6,000-ft. depth,* when special properties are not required. Available in regular type only (similar to ASTM C 150, type I).
B	Intended for use to 6,000-ft. depth.* Available in the regular type (similar to ASTM C 150, type II) for conditions requiring moderate sulfate resistance, and in the high sulfate-resistant type.
C	Intended for use to 6,000-ft. depth,* for conditions requiring high early strength. Available in the regular type (similar to ASTM C 150, type III) and in the high sulfate-resistant type.
N	Intended for use to 9,000-ft. depth,* for conditions of moderate temperature and pressure. Available in the regular type (having moderate sulfate resistance) and in the high sulfate-resistant type.
D	Intended for use to 12,000-ft. depth,* for conditions of moderately high temperature and moderately high pressure. Available in the regular type (having moderate sulfate resistance) and in the high sulfate-resistant type.
E	Intended for use to 14,000-ft. depth,* for conditions of high temperature and high pressure. Available in the regular type (having moderate sulfate resistance) and in the high sulfate-resistant type.
F	Intended for use to 16,000-ft. depth,* for conditions of extremely high temperature and extremely high pressure. Available in the regular type (having moderate sulfate resistance) and in the high sulfate-resistant type.

* These depth limits are based on the conditions imposed by the casing-cementing well-simulation tests (Schedules 1–9, incl., *RP 10B*), and should be considered as approximate values.

are used in oil well cements in order to permit the cement slurry to be pumped. Water-cement ratios may vary from 25 per cent to more than 65 per cent by weight for conventional Portland cements. In-

TABLE 20-3
Minimum Thickening Times for API Cements*

Simulated Depth, ft.	Minimum Thickening Time—Minutes						
	Class A	Class B	Class C	Class N	Class D	Class E	Class F
6,000	80	80	80	80	80	80	80
8,000	—	—	—	80	80	80	80
9,000	—	—	—	80	80	80	80
10,000	—	—	—	—	80	80	80
12,000	—	—	—	—	82	82	82
14,000	—	—	—	—	—	92	92
16,000	—	—	—	—	—	—	102

* Pan American pressure-thickening time.

creasing the water-cement ratio will increase the pumpability time, reduce the weight of the slurry, and increase the setting time of the Portland cement. Figure 20–1 shows neat cement slurry weights and volumes for various water-cement ratios using Portland cement.

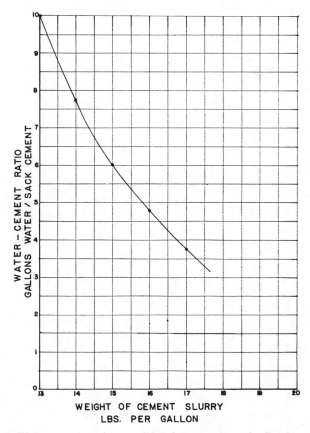

Fig. 20–1. Water-cement ratio *vs.* weight of cement slurry for Portland cement.

Major considerations in oil well cements are the thickening time and strength development. Thickening time is important because it determines the length of time the slurry can be pumped. Strength is important only to the extent that the cement must perform satisfactorily its required functions.

Farris,[1] in research work directed toward the determination of minimum strength requirements for oil well cements, found that a

[1] R. Floyd Farris, "A Practical Evaluation of Cements for Oil Wells," *API Drilling and Production Practice* (1941), 283.

tensile strength of only 8 psi (approximately 100 psi compressive strength) is satisfactory. Based on this finding and other data, there is general agreement in the industry that the development of compressive strengths of only a few hundred psi will be satisfactory.

Thickening time can be controlled by the fineness to which the clinker is ground. In more finely ground cement, the specific surface area is greater and the reaction with water will take place at a much faster rate; therefore, the pumpability time can be controlled somewhat by controlling the fineness of grinding. However, the reaction of the cement with water occurs principally at the surface of the cement particle, and as the fineness of the cement decreases, the individual particles become larger and the reaction is not as complete. For this reason other methods are normally used to control the reaction time or setting time of cements. Two other methods can be used to increase the setting time of cements: (1) the components of cement which hydrate rapidly can be reduced, or (2) additives which retard the setting time can be used. Cements in the latter category are called *retarded slow-set cements* and those in the first category are called *unretarded slow-set cements*.

Effects of High Pressures and Temperatures on Cement Properties

Increasing temperatures and pressure above atmospheric conditions will result in decreased thickening time (pumpability time) for most oil well cements. Increasing the pressure under isothermal conditions will increase the compressive strength. The effect of increasing temperature is somewhat more complicated. The compressive strength of most cements will increase slightly up to some critical temperature, usually between 200 and 240 degrees Fahrenheit, after which any increase in temperature will cause marked reduction in strength. In some cases this strength may be wholly or partially regained after a period of approximately thirty days. The effects of temperature and pressure on the strength and pumpability of oil well cements are shown in Figs. 20–2 through 20–4 and in Table 20–4.

Ludwig[2] has very concisely summarized the research work concerned with the setting and hardening of cements. The hydration products of cements at both atmospheric and elevated temperatures and pressures are shown in Table 20–1.

[2] N. C. Ludwig, "Portland Cements and Their Application in the Oil Industry," *API Drilling and Production Practice* (1953), 183.

TABLE 20-4

Effect of Pressure or Pumpability of Oil Well Cements

Well-Depth API Casing Cementing Conditions*	Temperature		Pumpability Time*	
	Static	Cementing	Portland Cement Water, 5.2 gal./sk.	Slow-Set Cement Water, 4.5 gal./sk.
2,000 ft.	110°F	91°F	6:00+	—
4,000 ft.	140°F	103°F	6:12	—
6,000 ft.	170°F	113°F	3:22	—
8,000 ft.	200°F	125°F	2:07	6:00+
10,000 ft.	230°F	144°F	1:34	4:09
12,000 ft.	260°F	172°F	1:07	2:55
14,000 ft.	290°F	206°F	1:00	2:15

* API Testing Code *RP-10B*.

Special Oil Well Cements

Several special types of oil well cements have been evolved over a period of years where specific conditions required modification of the conventional cements. Several of these will be described below.

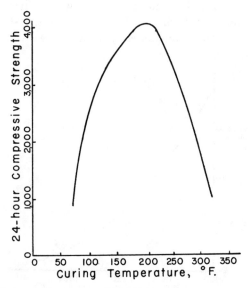

Fig. 20-2. Relationship between 24-hour compressive strength and curing temperature for Portland cement.

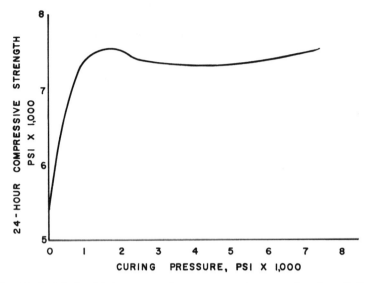

Fig. 20–3. Relationship between curing pressure and compressive strength for Portland cement.

Bentonitic Cements

Conventional oil well cement slurries weigh fifteen to sixteen pounds per gallon, have high water-loss properties, and develop much higher early strengths than necessary. In many cases incompetent

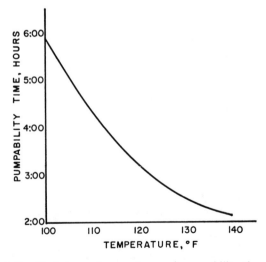

Fig. 20–4. Relationship between temperature and pumpability time for Portland cement.

formations will "break down" when a cement of relatively high density is placed behind the casing. This is especially true where long columns of cement are required. The high filtration rate of conventional cements results in relatively large volumes of water being lost to permeable formations, leaving a thick cement cake on the formation which may build up until circulation past this zone cannot be continued. The high filtration rate may also cause excessive water to enter potential oil-producing reservoirs. The water in the vicinity of the well bore will reduce the flow of oil and may react with clays in the reservoir rock to cause expansion of the clays with consequent reduction in permeability. Tests have shown conclusively that as the strength of cement increases, penetration by perforating guns, particularly bullet perforators, is markedly reduced.

Bentonite was first added to cements to increase the consistency of the slurry, and when it is added, a lower-density slurry of lower unit cost can be used. Cements with as much as 25 per cent bentonite have been successfully used in oil well cementing operations. Where large quantities of bentonite are used, a dispersant such as calcium lignosulfonate is usually added to the bentonite cement mixture. Slurries weighing twelve to fourteen pounds per gallon have been placed several thousand feet above the bottom of the casing, in many cases eliminating the need for stage cementing. Because bentonitic cements have lower strengths than conventional oil well cements, deeper penetration results where perforating is required. The cost of materials can be reduced as much as 25 per cent by using bentonitic or modified cement. The API water loss of a conventional oil-well cement slurry may exceed 1,000 c.c. in thirty minutes, while bentonitic, or "modified" cements may have water losses of only about 100 c.c. in thirty

TABLE 20-5

THICKENING TIME FOR BENTONITIC CEMENTS

(containing varying amounts of bentonite and calcium lignosulfonate)

Portland Cement

Well Depth API Casing Cementing Conditions*	8% Bentonite 0.4% HR-7†	10% Bentonite 0.5% HR-7†	12% Bentonite 0.6% HR-7†
6,000 ft.	3:00+ hrs.	3:00+ hrs.	3:00+ hrs.
8,000	2:20	3:00+	3:00
10,000	2:16	2:12	2:26
12,000	1:55	1:44	2:19
14,000	1:52	1:37	2:14

* API Testing Code *RP-10B*.
† Calcium lignosulfonate.

minutes. As has been said, this may be an important consideration in many instances. Table 20–5 shows thickening times versus depths for cements containing varying amounts of bentonite and calcium ligno-sulfonate.

The addition of bentonite to Portland cement will reduce the thickening time of the cement; however, the addition of bentonite to some brands of slow-set cements causes serious reduction in the thickening time, and its effect on the thickening time varies from brand to brand. Because of this variable effect on the thickening time of slow-set cements, most bentonitic cements are prepared with ordinary Portland cement, modified with retarders if necessary.

Pozzolanic Cements

Pozzolans, siliceous rocks of volcanic origin, first found near Pozzeroli, Italy, have long been used as cementing materials. Within recent years, other siliceous materials resembling pozzolans have been used in oil well cementing. Originally these so-called pozzolanic materials were mixed with Portland cement. A new cementing material has been developed which uses fly ash—a combustion product resulting from the burning of pulverized coal in power plants—hydrated lime, and chemical agents to control the setting time. An analysis of a

TABLE 20–6
Chemical Analysis of Typical Fly Ash

Constituent	Per Cent by Weight
SiO_2	43.20
Al_2O_3 and Fe_2O_3	42.96
CaO	5.92
MgO	1.03
SO_3	1.70
Loss on ignition	1.98
Undetermined	2.21

typical fly ash is shown in Table 20–6. These pozzolan-lime slurries weigh approximately fourteen pounds per gallon, about two pounds per gallon less than conventional cement slurries. Thickening times can be governed in such a way that the pozzolan cements will have longer pumpability times than most conventional cements. Moreover, larger columns of cement can be placed at lower pump pressures. Thickening times for various simulated well depths are shown in Fig. 20–5.

Perlite Cements

Perlite cements are prepared by adding to Portland cement an expanded perlite material. Perlite is made by heating a certain crushed

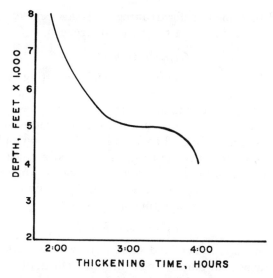

Fig. 20–5. Relationship between API simulated well depth and thickening time for Portland cement.

volcanic ore to its fusion point, at which point a cellular product of very low density is formed. The density of bulk perlite is approximately thirteen pounds per cubic foot. Because of its cellular structure, perlite is capable of absoring large quantities of water. When it alone is added to a cement slurry, it tends to separate from the cement and rise to the top of the slurry. To prevent this separation, small percentages of bentonite are usually added to perlite cement slurries, because bentonite disperses the perlite more uniformly through the slurry. The low densities of perlite cements give them the same basic advantages of bentonitic cements, but the unit cost of perlite cement is somewhat greater than either bentonitic or neat cement.

Diatomaceous Earth Cement

The development of diatomaceous earth cement is discussed here principally to point out some of the ideas and materials which are being considered for oil well cementing applications. A low-density cement is desirable for several reasons, and recent research has led to the development of diatomaceous earth cement, in which a special grade of diatomaceous earth has been added to Portland cement. This results in a reduced slurry density, partly because of the lower density of the diatomaceous earth and partly because of the much larger quantities of water which can be used without causing separation of the solids. The principal disadvantages of this cement are the increased thickening time and the reduced strength.

Gypsum Cements

Gypsum ($CaSO_4 \cdot 2H_2O$) is used in special cementing operations requiring a quick-setting, hard cement. Its principal advantages are that it expands upon setting and that movement does not affect its setting properties. The expansion quality of the set gypsum cement offers advantages in sealing off zones of lost circulation and in placing bridging plugs. Gypsum cements also have good application in controlling blow-outs, as the cement will set properly while it is in motion. The high cost of gypsum cements, however, limits their use to special types of remedial cementing operations.

Resin Cements

A resin cement is a Portland cement to which a resin has been added. It is a special-purpose cement that has limited application principally on account of its high cost. Its principal advantage is the ability of the resin to penetrate a mud filter cake and bond to the formation.

Diesel Oil Cement

Diesel oil cement is Portland cement to which has been added a surface active agent. This cement is designed for mixing with diesel oil rather than with water. It will not set and harden until it comes in contact with water. The principal use of diesel oil cement is in shutting off water production from the completion interval of a well.

CONTAMINATION OF CEMENT

After the cement has been mixed to form a slurry, it is possible for foreign materials to become mixed with it, and any material mixed with the cement slurry will have some effect on it. The most common contaminants are fresh water, drilling mud, and saline water.

Fresh water is normally used in preparing the cement slurry. The maximum amount of water which can be added and not interfere with a satisfactory set cement is normally used to insure thorough mixing and adequate pumpability time. Any additional water which becomes mixed with the cement slurry may result in excessive settling of the solid particles in the cement and open void spaces in the annular column of slurry.

As most well bores are full of drilling mud when they are ready for cementing, preventing contamination of the cement slurry with the mud may be very difficult. The development of plugs to isolate the slurry from the placement fluid was a significant advance in preventing cement contamination. These plugs, made of rubber, wood, or other easily drillable material, not only separate the slurry from the displacing fluid, but also remove mud from the wall of the casing. However, after the cement has left the casing and is moving up the annular space outside of the casing, the possibility of mud contamination increases. The mud cake on the sides of the well bore and the mud remaining on the outside of the casing are the principal sources of contamination by this agent. Drilling mud is composed of a liquid phase, which is normally fresh water; a colloid phase, bentonite; a weighting phase, which is normally barite and sand; and a chemical phase, which includes the large number of chemicals used to treat the mud. The effect of each of these mud phases on the cement slurry will be considered briefly, with the exception of fresh water, which has already been discussed.

Contamination of the cement with bentonite from the drilling mud will have essentially the same effect on the slurry that bentonite has on modified cements. Bentonite reduces the strength and decreases the setting time of cements. Indeed, sufficiently large concentrations of bentonite may reduce the strength of the cement to the extent that it never sets properly.

The weighting phase of the drilling mud is chemically inert and is not deleterious to the cement slurry. In many cases where abnormal pressures are encountered, it is necessary to increase the density of the cement slurry in order to control the formation fluids. Barites or other weighting materials can be added to the slurry for this purpose.

The chemical phase of the drilling mud may have variable effects on the cement. Compounds which are adsorbed on the surfaces of the individual cement grains will retard the setting time of the cement. Many of the organic compounds used in drilling muds will react in this way. Starches, calcium lignosulfonate, quebracho, and sodium carboxymethylcellulose (Driscose) are examples of mud additives which retard the setting time of cement. Certain inorganic chemicals, according to Ludwig,[3] increase the concentration of salts in solution in the liquid phase of the slurry, thereby accelerating the setting time of the cement. Sodium hydroxide (caustic soda) and sodium carbonate will provide this accelerating effect.

Other compounds, such as the alkali pyrophosphates, sodium silicates, and some soaps will have variable effects on the properties of the slurry.

[3] *Ibid.*, 204.

Saline waters are the most common of the formation fluids which contaminate cement slurries. Sodium chloride is the principal mineral constituent of most saline waters and will reduce the setting time of the cement, unless the sodium chloride content is several hundred thousand parts per million, in which case the thickening or setting time will be increased by this unusually large salt concentration. Seldom, if ever, however, would these large salt concentrations be encountered, as most formation waters have saline contents of considerably less than 100,000 parts per million. Sodium chloride has exactly the opposite effect on the strength development of cements. Normal salt concentrations encountered in actual practice will cause an increase in strength development of the set cement, while the trend is reversed at unusually high concentrations, resulting in a reduction of strength development.

Squeeze Cementing

Squeeze cementing is the name given to the cementing operation in which relatively large pressures are used to force cement into specific places. It is principally a remedial type of operation, and originally involved the overcoming of the overburden pressure and actually forcing cement slurry into the formation. In recent years, however, it has been demonstrated that satisfactory squeeze cement jobs can be accomplished without hydraulically fracturing the formation. Experiments which led to these conclusions were necessitated by the so-called permanent completion practices, in which the use of high pressures was not possible.

The squeeze-cementing technique also has some application in the sealing off of zones of lost circulation. Many times the use of lost-circulation materials in the drilling mud is not effective, in which case squeeze cementing may effectively isolate the trouble zone. Cement is squeezed into the lost-circulation zone, and then, after the cement has set, drilling continues. In many cases this action effectively isolates the lost-circulation zone. Additives, such as flexible fibers and ground walnut shells, may be used in the cement to plaster the formation and provide an additional seal. Perlite cements have also been used in lost-circulation problems, because the granular material tends to bridge across the face of the formation, thus aiding in establishing a permanent seal.

Cementing Methods

The basic objective in cementing casing is the placing of an uncontaminated cement slurry at the proper position in the annular space between the casing and the hole in such a manner that an effective, impermeable seal is obtained between the formation and casing. The series of illustrations in Fig. 20–6 illustrate the procedure followed in a routine cementing operation.

The cement is mixed with water to form a slurry of the desired density. Mixing is accomplished on the surface with the use of specially

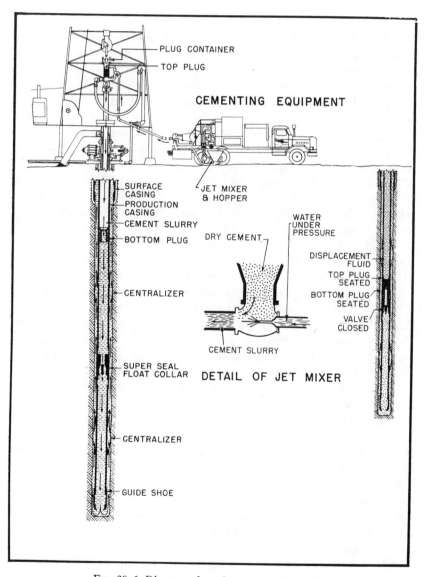

FIG. 20–6. Diagram of a primary cementing job.

designed mixing vessels, or "hoppers," and is then pumped down the inside of the casing. Either sack cement or bulk cement can be used. After the proper volume of cement has been mixed, some fluid must be used to place the cement in the desired position. Most bore holes are filled with drilling fluid when the cementing operation begins, and for this reason drilling mud is normally used as the displacing fluid. In Fig. 20-6 (left) the cement is being placed in the casing and drilling mud is being used to place the cement. It should be noted that as the cement is pumped into the casing, it displaces drilling mud from the annular space outside the casing. This return mud will flow back into the mud pits. The returning mud should be watched carefully, as it indicates that the cement is displacing the mud. If mud "returns" are not obtained at the surface while the cement is being pumped into the casing, then some fluid, either cement or mud, is being lost to the formations. When this occurs, there is always some doubt about the proper placement of the cement. Figure 20-6 (right) shows the cement as it is properly placed. After determining the capacity of the casing, the volume of displacing fluid should be measured carefully to insure proper placement of the cement. It is especially important that good-quality cement be placed at the bottom, or shoe, of the casing, and if too much displacing fluid is used, there will be no cement at the bottom of the casing.

In order to achieve the desired objective in cementing, special equipment has been designed to minimize contamination of the cement. The use of plugs to clean the casing and separate the cement slurry from the displacing fluid has already been discussed. Figure 20-7 shows special cementing equipment which may be used in cementing equipment which may be used in cementing casing. Each of these items of special equipment will be discussed briefly.

Casing centralizers. There must be a predetermined minimum thickness of cement throughout the cemented interval. However, the bore hole will not be perfectly straight, for there will be minor deviations where the drill bit was deflected as it penetrated the different types of formations. Therefore, when casing is run in the hole, it will be in contact with the bore hole in some places. If some provision is not made to overcome these points of contact between casing and formations, the probabilities of a poor cement job are great. Casing centralizers are flexible springs attached to the casing to insure the centering of casing in the hole and provide a more uniformly cemented casing string.

Wall cleaners or "scratchers." As discussed in more detail in Chapter 7, a filter cake is formed on the wall of the bore hole as drilling pro-

Scratchers

Guide Shoe

Float Collar

Centralizer

FIG. 20–7. Cementing equipment.

gresses. The removal of this filter cake is desirable before the cement is placed as the cake reduces the effectiveness of the cement. In order to isolate the formations, the cement must bond to them as well as to the casing. If the cement cannot bond properly to the formation, future remedial cementing operations may be required. Wall cleaners, or "scratchers," were developed to remove the filter cake from the wall of the bore hole. They are composed of many short, flexible wires arranged either in vertical, spiral, or circular patterns. As the casing is lowered into the well bore, these wires come into contact with the bore-hole wall, "scratching" the wall and removing the filter cake. The casing may be rotated to increase the effectiveness of the scratchers.

Casing guide shoes. Guiding the lower end of the casing past ledges and irregular places in the casing may be hazardous if special methods are not used. A guide shoe is basically a short section of steel pipe with the lower end rounded to facilitate passage of the casing through irregular places in the bore hole. The lower portion of the guide shoe usually contains cement to increase the shock-absorbing characteristics of the shoe. It also usually contains a back-pressure valve arranged to permit circulation from the inside of the casing to the outside. However, reverse circulation, from the outside of the casing to the inside, is not possible as the back pressure valve will seat, preventing fluids from entering the casing from the outside. The primary purposes of this valve assembly are to prevent the cement slurry from re-entering the inside of the casing after it has been placed, and to allow the casing to be "floated" down the hole—that is, the inside of the casing is left empty, or only partly filled, the additional buoyant force on the casing reducing the load on the derrick.

Float joint and float collar. It is very important that a good cement bond be effected at the lower end of the casing. In many cases where the casing is set through the producing formation, the lower end of the casing will be in a water-bearing formation. If a good cement bond is not attained, this water may enter the casing, necessitating remedial cementing operations. If the lowest joint of casing is left filled with cement, the hazards of a poor cement job at the bottom of the casing are reduced. If the cement has become slightly contaminated with the displacing fluid, the contaminated portion will probably be left in this last joint of casing. A float collar is placed immediately above the float joint of casing. This collar, in addition to performing its regular duty of attaching two joints of casing, has a back pressure valve, similar to the one described in the guide shoe, which prevents circulation back into the casing. The internal diameter of the float collar is re-

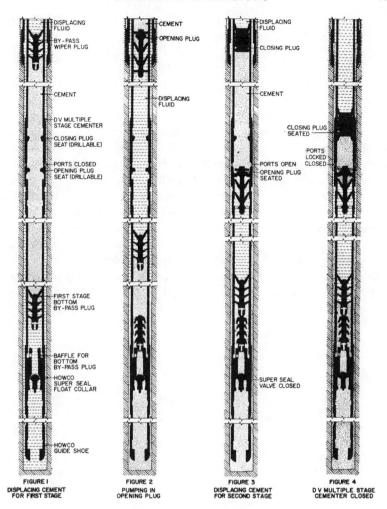

FIG. 20-8. Stage cementing.

duced by cement, plastic, or other drillable material to provide a positive seat for the cementing plugs.

In many cases the bottom of casing is set so near the producing interval that some of the cement in the float joint will have to be drilled out. An experienced driller can determine whether the cement has set properly by the manner in which it drills. Also, in many instances the well may be deepened, in which case any special equipment used in cementing must be easily drilled. For this reason the internal construction of all float collars and guide shoes should contain material which can be easily drilled with conventional drilling bits. A float joint is a joint of the normal casing used.

Stage Cementing

In some wells long columns of cement must be placed in the annular space behind the casing. Excessive pressures caused by placing these long columns of slurry may cause fluid losses to incompetent formations, or the cement slurry, in its unusually long travel length, may become contaminated. To eliminate these possibilities, stage-cementing equipment has been developed. In this technique the desired amount of slurry is placed around the lower part of the casing in the conventional fashion. Then, by the use of a shut-off plug and a stage-cementing collar, as shown in Fig. 20–8, the cement slurry is diverted through ports in the stage-cementing collar and out into the annular area, by-passing the lower part of the casing where the slurry has already been placed. Several isolated zones can also be cemented in one operation by placing the shut-off plugs and stage-cementing collars at the desired places.

CHAPTER 21

Drilling Economics

The drilling industry, like any other competitive industry, is continually seeking ways of reducing costs. As the search for oil is concentrating in the deeper horizons, the per foot cost of drilling is mounting steadily. Indeed, one of the major problems confronting the drilling industry today is the cost of deep drilling. Figures 21-1

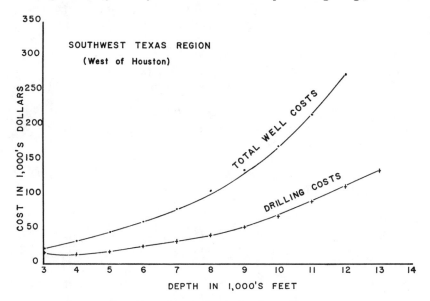

FIG. 21-1. Cost vs. depth, Southwest Texas.

through 21-3 are graphs showing average drilling costs in relation to total depth of the well. Of course, drilling costs vary from area to area, but in any area the general relationships hold true.

There are a number of reasons for increased per foot drilling cost as depth increases:

1. *Harder formations.* At deeper depths the overburden pressure on the rocks is greater, and as a result the rocks are more compact and more difficult to penetrate.
2. *Less percentage of time on bottom.* At these greater depths, bit life is shorter than at shallower depths, thus more round trips per unit depth drilled are required. Also, each of these round trips at

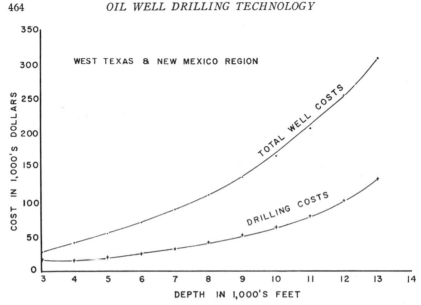

Fig. 21–2. Cost vs. depth, West Texas and New Mexico.

greater depths requires a longer period of time. Thus, the time-on-bottom drilling is a lower percentage of the total time at greater depths.

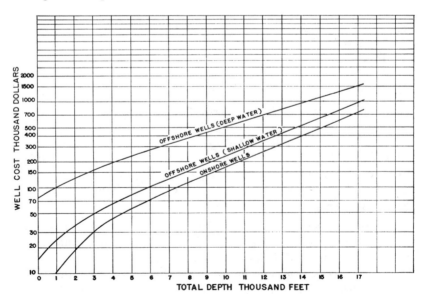

Fig. 21–3. Estimated average well cost, state of Louisiana.

3. *Increasing severity of problems.* Although it is not necessarily true in all cases, normally the severity of the problems encountered in drilling operations becomes more acute at greater depths. Abnormal pressure formations are encountered in many areas which require increased drilling-mud weights and closer control over the drilling mud. In turn, the heavier muds reduce the penetration rate. More time is required to control lost circulation in deeper formations than at shallower depths. Increased temperatures at the greater depths require the use of special drilling fluids. High-strength drill pipe is required for drilling below 15,000 feet. All of these factors increase unit drilling cost.
4. *Drilling-rig design.* The drilling rig must be designed for the maximum depth at which it will be operated. Therefore, for deeper drilling, more horsepower, greater capacity, and increased strength are required, even though much of the drilling will be through the shallower formations where a heavier rig is not necessary. This larger, heavier rig is another factor in the increased costs of deep drilling. Theoretically, the use of more than one rig to drill a deep well would be economically feasible—a small rig to drill the shallower depths and a larger rig to continue the deep-drilling operations—but usually such a plan would not be practical in fact, because of the extra moving, rigging-up, and rigging-down costs that would be incurred. Nevertheless, it is a consideration which should not be overlooked in the over-all problem of rig selection.
5. *Modifying factors.* Certain factors tend to modify or reduce the considerations which have previously been discussed. For example, in deep-drilling operations fewer rig moves are required in a given period of time, so that there is some economy in comparison with the rig operating at shallow depths where moves are relatively frequent. This factor of fewer jobs in a given period of time is attractive to all rig operators, as work-scheduling for the rig is reduced. But while such factors tend to offset some of the problems of deeper drilling, unfortunately, they do not compensate for all or even most of the factors causing increased per foot drilling costs and the greatly increased hazards associated with deep drilling.

Drilling-rig ownership can be classified into two broad groups, company-owned rigs and contractor-owned rigs. A great percentage of the rigs are owned by independent drilling contractors; in 1957, approximately 93 per cent of all wells drilled in the United States were drilled by independent contractors. This situation is due to the relatively temporary nature of drilling operations in any specific area. Most large oil companies and many smaller ones own rigs which are used only on company wells. A company-owned rig is often advantageous in conducting drilling research, in that drilling operations can be speeded up or slowed down during certain tests without the necessity of an elaborate contract between the driller and the producer. Thus a company rig increases the flexibility of the over-all testing program. In addition, new employees can be given training on a drilling rig with fewer legal complications than would be possible with contract rigs. There is also a slight advantage in that the drilling personnel are more familiar with company policies and operating procedures than are the contract drilling crews. On the other hand, there are certain disadvantages to company-owned rigs. Probably the principal disadvantage is the cost of operating the rig, for, almost without exception, a drilling contractor can operate his rigs more economically than an oil-producing company. Drilling is the sole, or at least the major, occupation of the drilling contractor. He is a specialist in his business and can be expected to operate more economically than a company whose principal activity is the production of oil and whose drilling activities are intermittent. Another important factor in differences in operating costs is that a drilling contractor, by working for a number of companies, can concentrate his equipment within a comparatively small area of operation, thereby reducing the cost of moving men and equipment long distances for relatively short periods of time. Conversely, in order for the company-owned rig to be kept busy, it will probably have to be moved from one area to another at more or less frequent intervals. In addition to the added moving costs which are incurred, it is also necessary to pay the drilling crews somewhat higher wages in order to overcome the basic dislike most people have of moving frequently from one place to another.

At the present time, most oil-producing companies that do operate some of their own drilling rigs have only enough rigs to do a fraction of their required drilling. This allows the company to maintain permanent drilling crews, for regardless of the amount of drilling activity, the company can always keep its own rigs busy. Thus, even those companies that own drilling equipment normally contract a large portion of their drilling. Another factor in a comparison of contract drilling costs and oil-company drilling costs is that the contractor can normally keep his rigs drilling at their approximate optimum-depth range, while the oil company, with a limited number of rigs and a wide depth range of drilling, cannot do so.

Equipment Costs

Rigs. In order to keep drilling costs to a minimum, proper selection of the drilling rig is important. Using a rig which is larger and heavier than actually required is inefficient and uneconomical, while, on the other hand, using a rig which is too small for the job not only is unsafe but is also inefficient and uneconomical. Therefore, the drilling rig should be carefully selected to provide a good balance of reserve power for emergencies, safe and efficient operation, and maximum over-all usage.

The trend in drilling-rig design is toward increased portability, which is another way to reduce the over-all drilling costs. The cost of moving, rigging up, and rigging down may be a sizable fraction of the entire drilling cost. Smaller drilling rigs can be designed for maximum portability; however, as rig size increases, some portability must be sacrificed. Various schemes have been devised to increase rig portability. The use of portable masts in place of conventional bolted derricks has resulted in substantial savings, although the tall masts have presented problems in moving from one location to another. To overcome some of the moving problems, telescoping masts have been used, as well as masts which can be laid down and then dismantled in sections before moving to the next location.

Increasing the portability of the draw works presents a difficult problem, as the draw works is a unit piece of equipment and very little can be done to transport it in sections. The depth rating of the rig depends largely upon the size and strength ratings of the derrick and draw works. Therefore, it again becomes obvious that the use of a draw works larger than necessary will result in increased drilling costs.

Bits. The bit is the focal point of the entire drilling operation. Where other conditions are approximately equal, the performance of the bit determines the cost per foot of drilling. Factors involved in bit performance include the bit itself, the manner in which it is used, and the resistance of the rock interval which is drilled. At any particular depth interval, the cost of operating the drilling rig and drill string can be expressed in terms of dollars per hour. Each time a bit is worn out, it must be replaced, and this requires both a new bit and a round trip hoisting the drill pipe. Therefore, the total feet drilled per bit becomes the drilling unit in determining cost per foot. This may be expressed as follows:[1]

$$C = \frac{(T+t)R + B}{F} \qquad (21\text{-}1)$$

where

C = drilling cost, dollars per ft.
T = hours on bottom with one bit
t = hours for round trip to replace bit
R = rig and drill-string cost, dollars per hr.
B = bit cost, dollars
F = feet drilled per bit

Where the drilling rate is fairly constant, the relation can be expressed as

$$C = \frac{(F/r + t)R + B}{F}$$

$$C = \frac{R}{r} + \frac{tR + B}{F} \qquad (21\text{-}2)$$

where

r = drilling rate, ft./hr.

Where the rig and drill string cost $40 per hour, the drilling rate is 6 feet per hour, five hours are required for a round trip, the bit cost is $250, and 40 feet are drilled per bit, the cost per foot would be

$$C = \frac{\$40}{6} + \frac{(5)(\$40) + \$250}{40}$$

$$= \$17.92 \text{ per foot}$$

In the above example, if the bit drilled 80 feet, the cost would be $12.40 per foot. If the bit drilled only 20 feet, the cost would be $29.27 per foot, where the same rate of penetration is assumed in all cases.

Both rig and drill-string costs and the required hoisting time for a round trip can be expressed as functions of drilling depth. Both of these quanities tend to increase linearly with depth.

Figure 21–4 is a graph of drilling-rig time versus depth which is used for estimating the cost of drilling a well. The preparation of such

---

[1] This equation was published by Henry B. Woods (*World Oil*, September, 1951). Equivalent equations have also been used by others.

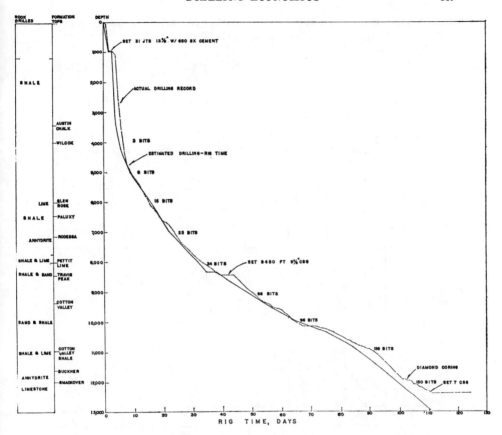

FIG. 21-4. Depth vs. time drilling curves, showing estimated and actual time.

a graph requires knowledge of the depth and thickness of the geological formations which will be enountered in the proposed well. It also requires a knowledge of the rate at which the formations can be drilled and the number of bits required to drill each formation. The estimated number of bits required is indicated on the depth-time chart. When such information is reliable, it is possible to make an accurate estimate of the cost of drilling a well. The time required for running and cementing casing and for routine maintenance must be included in the total.

A graphical method of analyzing drilling costs in terms of bit performance has been discussed by Yeatman and Woods.[2] Total feet

[2] Yeatman, "Practical Utilization of Rate of Penetration Drilling Data," *API Production Bulletin 225*, Vol. 21M [IV] (1940); H. B. Woods, "Analysis of Rock Bit Performance Data," *World Oil*, September, 1951.

drilled per bit is plotted versus total hours per bit. It is based essentially on Equation (21–1) arranged in the following form:

$$F = \frac{R}{C}\left(\frac{B}{R} + t + T\right) \qquad (21\text{–}3)$$

The cost of the bit is in this manner expressed as an equivalent number of rig hours and added to the number of hours per round trip. Further refinements consist of including in the actual round-trip time a proportional amount of rig-maintenance time and time required for

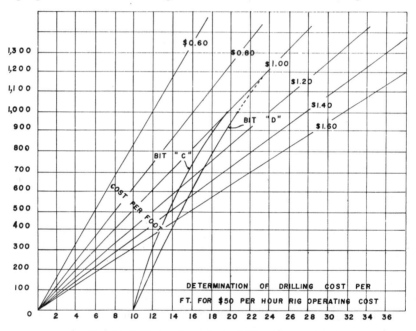

Fig. 21–5. Economic analysis of bit performance.

changing the rotary hoisting line. By this method, several hours of rig time are already charged against the bit before it actually begins to drill. Figure 21–5 illustrates a comparison of the performance of two bits. Assumed data are $50 per rig hour and ten hours for combined round-trip and bit cost. The slanting lines on the chart correspond to various drilling costs per foot (feet drilled at left; hours, bottom). The example provides a comparison of two different types of bits drilling in soft formations, where the teeth on bit C had become worn to the extent that its performance line was tangent to the $1.05 per foot line.

Power plants. For any drilling job there is an optimum size of power plant which should be used, as there is also a most desirable type of power plant. In determining the type of power plant, the rig operator must consider not one particular job, but the work which the rig will be doing throughout most of its life. The best power plant to use will depend upon the area of operations, availability and cost of fuel, and the frequency and length of moves that will be required. For example, where natural gas and water are not available, the use of steam would be inadvisable as a power source.

In general, for a given type of power plant, as the horsepower increases, the physical size of the plant increases. There are advantages in additional horsepower over the minimum required, as the rate of doing work is increased as a result of increased horsepower. Increased horsepower generally results in increased speed of hoisting, consistent with safe drilling practices, with less time consumed in round trips of the drill pipe. Particularly in deeper drilling the increased horsepower is advantageous in circulating more drilling fluid in order to obtain faster penetration rates.

Reducing the size and weight per horsepower is one method of reducing over-all drilling costs. The design of rig power plants is tending in this direction. The use of superchargers and pre-cooling of input air results in more air being brought into the power cylinders of internal combustion engines. With more air available, the fuel input can be correspondingly increased, with resulting greater horsepower output for the engine.

SLIM-HOLE DRILLING

Slim-hole drilling is frequently considered as a means of reducing drilling costs. The term *slim hole* is relative and generally refers to any size of hole that is smaller than the usual diameter of wells drilled in any particular area. For example, a $7\frac{7}{8}$-inch-diameter hole between 3,000 and 10,000 feet in West Texas is not considered a slim hole, but in this depth interval in areas of softer formations such a diameter might be considered a slim hole.

There are certain advantages in drilling smaller holes: Smaller and therefore less expensive bits are required. It should be possible to use smaller-diameter pipe and drill collars, which would effect a further reduction in cost. Since the drill strings will be lighter, smaller drilling rigs could be used for drilling to any given total depth. Since the volume of the hole is smaller, less drilling mud and less cement will be required, with another saving in cost.

The accompanying disadvantages of drilling smaller holes are as follows: Smaller holes generally require a better quality of drilling mud throughout because of the greater danger of sticking the drill pipe. The better-quality mud tends to give a slower rate of penetration. The smaller annular clearance tends to produce greater pressure drops when circulating and greater pressure surges when hoisting, so that the probabilities of mud losses into formations are increased. Rates of penetration are lower for the smaller bits because smaller rolling cutter bits are inherently weaker than the larger bits. Where penetration rates are slower, mud-maintenance costs per volume of mud are higher. Smaller-diameter drill collars do not have the stiffness of larger drill collars and would deviate more from the vertical for a given bit loading per inch of hole diameter.

While advantages may be gained in some instances by reducing hole diameter, unless conditions are favorable, the result may be lower penetration rates, mud-circulation losses, and crooked holes. The matter of choosing the optimum size of hole diameter for the major portion of the hole is certain to receive continued study in the future.

The relationships between drilling cost and the hole and drill-collar sizes were analyzed by Woods and Lubinski[3] with particular reference to performance of the drilling equipment in drilling a reasonably straight vertical hole. They concluded that the lowest drilling cost could be realized by adapting the size of hole—slim, conventional, or oversize—to the nature of the rock formations to be drilled. The factors which tend to increase the size of the most economical hole are formation crookedness, formation hardness, and depth. The conclusion was reached that in hard and severe crooked-hole formations the most economical size of drill collars would extend to $11\frac{1}{4}$, 14, or 16 inches in diameter.

Air-Gas Drilling

Where air or gas can be circulated rather than drilling mud, savings in drilling cost are reported. A comparison will be made for the following assumed conditions:

Rig and drill pipe cost	$1,000 per day
Drilling rate with mud	6 ft./hr.
Hours on bottom drilling with mud	12 hrs./bit
Feet drilled per bit, with mud	72 ft./bit.
Time required for round trip	6 hours
Cost of bit	$ 250 per bit

The cost per foot can be calculated as follows:

$$C_M = \frac{\$1,000}{(24)(6)} + \frac{(6)(\$1,000)}{(24)(72)} + \frac{\$250}{72}$$

[3] H. B. Woods and A. Lubinski, "Practical Charts for Solving Problems on Hole Deviation," *API Drilling and Production Practice* (1954), 56–71.

= $6.95 + $3.48 + $3.48

= $13.93 per foot of hole, drilled with mud

For drilling with air, it will be assumed that the cost of operating the compressors amounts to $500 per day more than the cost of operating the mud pump. Accordingly, the rig cost will total $1,500 per day. It is common practice in drilling with air or gas to reduce both

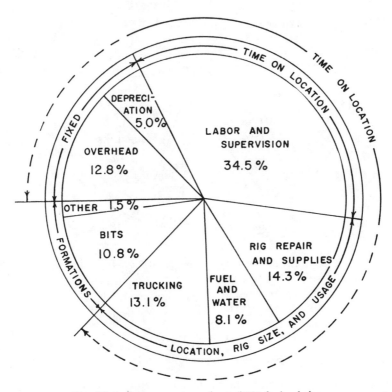

FIG. 21-6. Cost percentage chart, 5,000-ft. land rigs.

the weight on the bit and the rotary speed. Under these conditions the bit can be rotated on bottom for a considerably longer period of time. Under such conditions it will be assumed that the penetration rate is doubled and becomes 12 feet per hour. The time rotating on bottom would probably be about 30 hours. Thus the bit would drill a total of 360 feet. Probably the same type of bit could be used, but it will be assumed that a carbide-studded bit costing $1,000 is used for longer bit life. The cost per foot would be as follows:

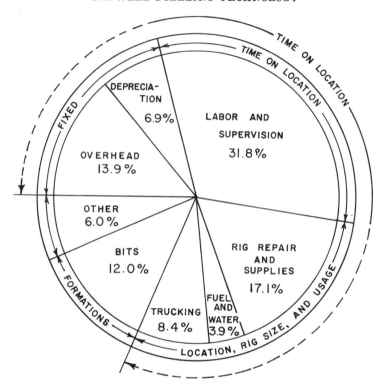

Fig. 21-7. Cost percentage chart, 5,000–7,500-ft. land rigs.

$$C_A = \frac{\$1,500}{(24)(12)} + \frac{(6)(\$1,500)}{(24)(360)} + \frac{\$1,000}{360}$$

$$= \$5.20 + \$1.04 + \$2.78$$

$$= \$9.02 \text{ per foot of hole drilled with air}$$

For the 360-foot interval, the indicated savings would be

$$(360 \text{ ft.})(\$13.93 - \$9.02) = \$1,770.00$$

In the above example, the cost per foot of rotating on bottom decreases from $6.95 for mud to $5.20 for air. The cost of round trips to replace bits decreases from $3.48 to $1.04 per foot, and the cost per foot for the bit decreases from $3.48 to $2.78. The greatest savings effected under the assumed conditions is in the cost of round trips to replace worn bits. If natural gas were available under pressure, the cost of the compressors would be eliminated and greater savings would be possible.

Methods of Reducing Drilling Costs

In searching for methods of reducing drilling costs, it is necessary to have reliable cost data on all parts of the drilling operation. With a knowledge of all applicable costs, methods of reducing them can be explored. Although attempts should be made to reduce all costs, regardless of how small the cost of the individual item may be, it is

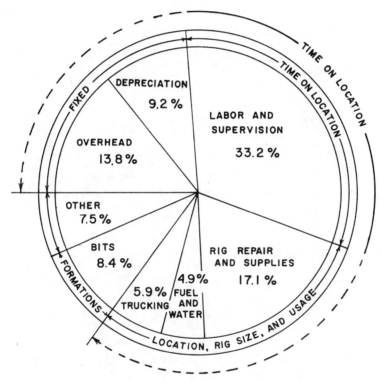

FIG. 21-8. Cost percentage chart, 7,500–15,000-ft. land rigs.

obvious that larger dollar savings can be effected by considering the larger cost items first. Figures 21–6 to 21–11 show the percentage cost distribution for drilling a typical well. Examination of the time breakdown shows that actual time in drilling is the largest item, which is as it should be, since hole is drilled only when the bit is on bottom. Round-trip time and down time for repairs or fishing should be kept to a minimum.

Increasing bit life has a two-fold advantage: bit costs are reduced, and round-trip time is reduced as a result of increased bit life. Drilling-bit performance could possibly be increased through several means: (1) determining optimum weight on the bit, (2) use of a different bit design, (3) increasing bit life by using a different type of drilling fluid, and (4) determining optimum speed of rotation.

The amount of rig depreciation chargeable to a particular drilling job depends upon the number of days required to drill the well, as total rig life is usually estimated on the basis of the number of days the rig will be used. Thus rig depreciation chargeable to a specific well

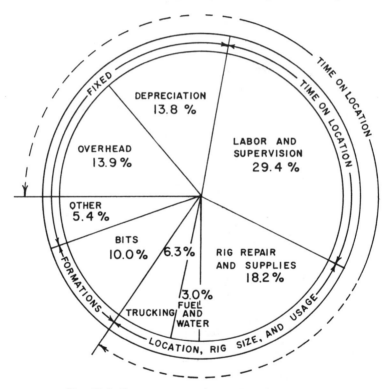

Fig. 21–9. Cost percentage chart, 15,000-ft. land rigs.

will be reduced if the number of days required to drill the well can be reduced. Daily rig depreciation can also be reduced by finding ways to reduce initial rig cost. In this connection, selecting the proper rig for the work to be performed is of the utmost importance. It is obvious that the use of a heavy rig, where daily rig-depreciation costs are high, would not be competitive for shallow or medium drilling operations.

Each of the other cost factors involved in drilling-rig operation should be studied carefully. This type of cost study is especially ad-

DRILLING ECONOMICS

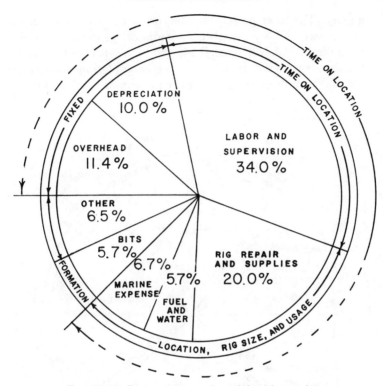

FIG. 21–10. Cost percentage chart, inland barge rigs.

vantageous when a drilling job can be compared to previous drilling operations in the same field or area.

The factors relating to drilling practices which generally reduce drilling costs are these:

1. The use of the maximum possible weight on the bit, depending on the formation, the bit, and the capacity of the drill collars and rate of circulation.
2. The use of slower rotary speeds in harder formations, which permit carrying heavier weights on the bit and give more total hours rotating on bottom and more footage drilled per bit (since drilling rate increases less than linearly with rotary speed).
3. The relationships between weight on the bit and total hours rotating and total footage drilled may require better definition in some instances. For example, the regular rock bits and carbide-studded bits used in air drilling generally reduce drilling costs through long runs at reduced weights.
4. The use of air or gas for circulating whenever possible saves

money. The next preferable method is the use of water or oil where possible. Next in order, economically, seems to be the use of the lightest low-viscosity, high filtration-loss muds which are consistent with safety and good well-completion practices. The use of oil-emulsion muds generally appears favorable. The use of jet bits with high circulation rates appears favorable in most softer formations.

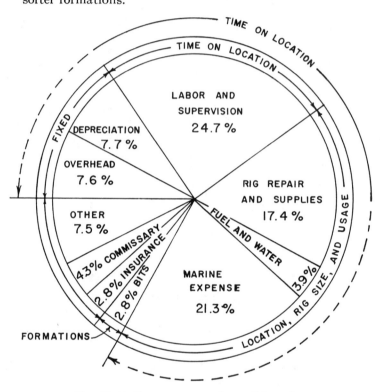

Fig. 21–11. Cost percentage chart, offshore rigs.

Payment of Drilling Charges

As more than 90 per cent of all drilling is performed by independent contractors, a formal written agreement or contract is necessary. The principal purpose of the contract is to set forth the duties and responsibilities of the drilling contractor and the owner of the well. Since the duties and responsibilities of each party have been more or less well established by custom, the item of greatest importance in the contract is the compensation the contractor is to receive for drilling the well.

There are as many different methods of paying drilling costs as

DRILLING ECONOMICS 479

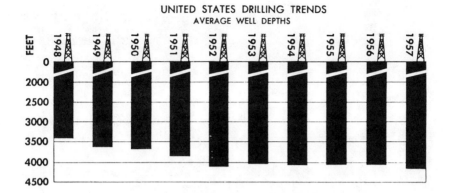

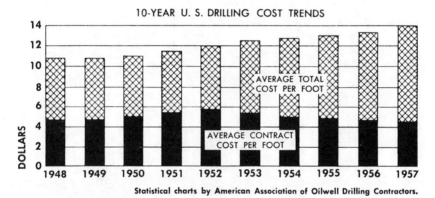

FIG. 21-12. Depth and drilling-cost trends (U. S.).

there are drilling contractors. However, over a period of years four different methods of payment have evolved and have been accepted as standard: (1) turn-key rates, (2) footage rates, (3) day rates, and (4) incentive rates (rates depending on penetration time).

In the *turn-key* type of drilling contract the contractor furnishes everything required to complete the well, including casing, cement, logs, formation testing, and drilling mud. The owner of the well assumes no liability whatsoever, the contractor being required to deliver to the owner a completed well or a properly plugged dry hole, as the case may be. The total cost of a well drilled on a turn-key basis is usually higher than on any other basis becuase the drilling contractor usually increases his normal charges to compensate for the additional risks he assumes. Although used infrequently, the turn-key contract has its utility. It has obvious advantages for an operator who has only

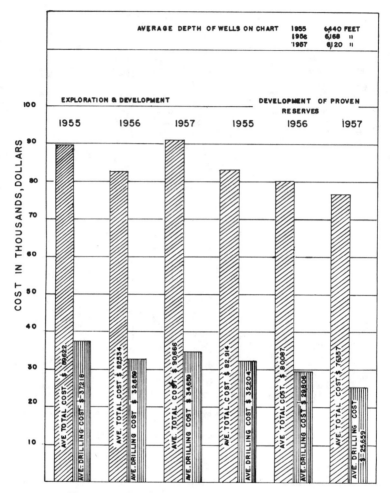

Fig. 21-13. Drilling and total costs of average wells.

a specified amount of money available for the well and is not in a financial position to assume the normal risks involved in drilling a well. In such a case the operator, in effect, pays the drilling contractor to assume all of the risks, in return for assurance that regardless of what happens during the course of drilling, the final cost will not exceed the original estimate.

The *footage-rate* contract has evolved as the standard drilling contract. In this type of contract the contractor agrees to drill the well to a certain depth for a specific amount per foot of hole drilled. Certain types of work, such as casing, logging, and testing are additional serv-

TABLE 21-1
Comparison of Drilling Costs

	1941	1945	1947	1948	1949	1950	1951	1952	1953	1954	1955	1956	1957
Labor costs	100%	140%	150%	170%	180%	190%	196%	197%	198%	200%	202%	204%	208%
Equipment costs	100%	110%	132%	145%	160%	175%	180%	183%	188%	190%	195%	200%	205%
Miscellaneous costs	100%	120%	160%	165%	170%	175%	181%	185%	187%	191%	194%	199%	204%
Total rotary-drilling costs	100%	122%	145%	160%	170%	180%	186%	189%	192%	194%	197%	201%	206%
Drilling prices* (footage basis)	100%	115%	110%	105%	101%	100%	103%	103%	102%	99%	96%	93%	90%
Average drilling time (days)	100%	95%	85%	80%	75%	65%	60%	59%	58%	56%	56%	54%	54%

* For drilling in same or similar fields to comparable depths.

Note: "Rate of Penetration" has been dropped because of its close relationship to drilling time in days.

ices, for which the owner is normally responsible, and he compensates the contractor who performs them on a different basis from the footage basis. The normal method of paying the contractor for his part in these extra services is the day-rate basis, which will be described in more detail below. Although the contractor is paid on a footage basis, payment is usually contingent upon the contractor's drilling the well to the desired depth. If something happens to prevent the contractor from reaching the contract depth—something not due to the fault of the owner—then the contractor is normally not entitled to any pay for the footage he has drilled. In this type of contract, there will be provisions specifying what each party to the contract will supply and be responsible for.

In the *day-rate* type of contract, the drilling contractor is paid at a specified rate for each day that he works on the well, regardless of the amount of hole he drills. The contractor furnishes only the rig and crew, while the owner furnishes all materials and services required. The day-rate contract is used where footage costs cannot be reasonably forecast. New drilling areas, marine drilling, deep drilling, drilling in areas

TABLE 21-2

UNITED STATES DRILLING-COST TRENDS

Year	Total New Wells Completed	Average Depth per Well	Average Contract Cost per Foot	Average Total Cost per Foot
1935	24,851	2,760	5.00	8.00
1936	28,962	2,797	4.75	7.75
1937	35,213	2,985	4.50	7.50
1938	29,127	3,110	4.25	7.50
1939	28,012	3,053	4.25	7.50
1940	31,149	3,088	4.00	7.50
1941	32,510	3,056	4.25	7.25
1942	21,990	3,088	4.00	7.50
1943	20,349	3,046	4.50	8.00
1944	25,786	3,272	4.75	8.25
1945	26,649	3,489	5.00	9.00
1946	30,230	3,345	5.00	9.50
1947	33,147	3,404	5.00	10.25
1948	39,477	3,463	4.75	10.75
1949	38,962	3,558	4.75	10.75
1950	43,204	3,689	5.00	11.00
1951	44,196	3,869	5.50	11.50
1952	45,879	4,060	5.85	12.00
1953	49,279	4,016	5.60	12.50
1954	53,650	4,025	5.10	12.75
1955	56,850	4,044	4.90	13.25
1956	58,271	4,022	4.75	13.50
1957*	53,668	4,120	4.60	14.00

* Estimated.

where drill conditions are extremely hazardous, and well-servicing operations are examples of instances in which day-rate contracts are used. The day-rate type of contract is being used more and more frequently as wells are drilled deeper and as marine drilling activity increases.

General Trends in Drilling Costs

The trends in drilling costs from 1941 to 1957 inclusive are shown in Table 21–1. It may be noted that during this period all costs, including labor, equipment, and miscellaneous costs, have more than doubled, in keeping with general trends in the economy. However, the drilling time has been greatly reduced during this same period, with the result that drilling costs on a footage basis have actually decreased.

The total new wells drilled from 1935 to 1957 is shown in Table 21–2. The average depth of the wells drilled has increased from 2,760 feet in 1935 to 4,120 feet in 1957. Table 21–3 shows that the percentage of wells drilled by independent drilling contractors has increased

TABLE 21–3

United States Contractor Drilling Trends

Year	Total New Wells Completed	Total Footage Drilled	Average Depth per Well	Indicated Work by Contractors		
				Per Cent	New Wells Completed	Footage Drilled
1935	24,851	67,844,939	2,760	68	16,899	46,134,559
1936	28,962	80,996,816	2,797	69	19,984	55,887,803
1937	35,213	105,099,189	2,985	71	25,001	74,620,424
1938	29,127	90,585,158	3,110	71	20,680	64,315,462
1939	28,012	85,523,094	3,053	70	19,608	59,866,166
1940	31,149	96,182,605	3,088	71	22,116	68,289,650
1941	32,510	99,347,714	3,056	72	23,447	71,503,354
1942	21,990	67,903,053	3,088	70	15,393	47,532,137
1943	20,349	61,991,857	3,046	75	15,262	46,493,893
1944	25,786	84,378,457	3,272	78	20,113	65,815,196
1945	26,649	92,982,113	3,489	80	21,319	74,385,690
1946	30,230	101,124,813	3,345	81	24,486	81,911,099
1947	33,147	112,816,124	3,404	81	26,849	91,381,060
1948	39,477	136,709,153	3,463	82	32,371	112,101,505
1949	38,962	138,616,941	3,558	85	33,119	117,824,400
1950	43,204	159,393,997	3,689	88	38,020	140,266,717
1951	44,196	173,315,000	3,869	90	39,776	155,984,000
1952	45,879	186,389,000	4,060	91	41,750	169,614,000
1953	49,279	197,920,000	4,016	91	44,844	180,107,200
1954	53,650	215,940,250	4,025	92	49,358	198,665,030
1955	56,850	229,901,400	4,044	92	52,302	211,509,288
1956	58,271	234,350,550	4,022	93	54,192	217,946,012
1957*	53,668	221,112,160	4,120	93	49,911	205,634,309

* Estimated.

TABLE 21-4

MONTHLY AVERAGE OF ROTARY RIGS ACTUALLY RUNNING IN UNITED STATES AND CANADA

Month	1942	1943	1944	1945	1946	1947	1948	1949	1950	1951	1952	1953	1954	1955	1956	1957
January	787	729	1,327	1,738	1,696	1,548	1,980	2,090	2,021	2,244	2,945	2,682	2,638	2,615	2,842	2,566
February	756	773	1,356	1,714	1,521	1,594	1,919	2,021	1,972	2,169	2,885	2,624	2,649	2,657	2,705	2,538
March	720	810	1,399	1,700	1,449	1,618	1,912	2,085	2,031	2,287	2,872	2,629	2,705	2,749	2,872	2,665
April	750	829	1,452	1,715	1,501	1,633	2,070	2,130	2,091	2,442	2,917	2,672	2,754	2,750	2,803	2,515
May	789	869	1,517	1,790	1,479	1,686	2,189	2,181	2,145	2,578	3,007	2,700	2,677	2,773	2,906	2,512
June	795	935	1,599	1,743	1,511	1,806	2,251	2,109	2,277	2,647	2,964	2,812	2,656	2,908	2,998	2,688
July	861	1,036	1,665	1,730	1,515	1,896	2,302	2,027	2,295	2,740	2,658	2,788	2,632	2,906	2,867	2,735
August	833	1,105	1,710	1,756	1,552	1,947	2,348	2,022	2,304	2,867	2,505	2,793	2,551	2,941	2,775	2,716
September	789	1,156	1,754	1,762	1,639	1,978	2,335	2,001	2,297	2,937	2,585	2,796	2,474	2,966	2,777	2,700
October	788	1,231	1,808	1,796	1,683	2,044	2,372	2,039	2,438	3,040	2,709	2,880	2,531	2,963	2,788	2,587
November	788	1,293	1,849	1,796	1,674	2,056	2,437	2,197	2,500	3,008	2,835	2,941	2,713	3,058	2,822	2,559
December	816	1,353	1,800	1,760	1,700	2,008	2,385	2,247	2,481	3,137	2,896	2,988	2,908	3,188	2,947	2,682
Average	788	1,009	1,603	1,744	1,576	1,817	2,208	2,096	2,238	2,683	2,816	2,777	2,657	2,873	2,842	2,622

TABLE 21-5
Survey of Drilling Prices

Location		Contract Depth, ft.	Drilling Price per foot	Rig Time Required, days	Date Drilled
County	State				
Martin	Texas	4,750	3.25	13	Feb. '58
Pecos	Texas	9,700	7.85	65	Jan. '58
Borden	Texas	8,500	5.25	36	Mar. '58
Dawson	Texas	5,000	3.75	22	Sept. '57
Wise	Texas	6,150	3.85	35	Aug. '57
Hutchinson	Texas	3,000	3.25	11	Aug. '57
Andrews	Texas	6,000	4.75	24	Aug. '57
Vintah	Utah	6,000	6.25	29	Feb. '58
San Juan	Utah	6,000	8.00	40	Dec. '57
Lea	New Mexico	10,850	6.85	52	Mar. '58
Lea	New Mexico	12,750	10.50	95	Sept. '57
Freemont	Wyoming	9,300	6.50	49	Mar. '58
Freemont	Wyoming	5,500	11.65	46	Dec. '57
Lincoln	Wyoming	12,000	11.40	100	Nov. '57
Washakie	Wyoming	9,000	8.50	54	Nov. '57
Washakie	Wyoming	10,150	8.75	64	Aug. '57
Rio Blanco	Colorado	6,400	5.40	30	Mar. '58
Dolores	Colorado	6,000	8.25	45	Dec. '57
McClain	Oklahoma	9,600	5.75	48	Mar. '58
McClain	Oklahoma	10,600	6.75	56	Oct. '57
Coal	Oklahoma	7,500	7.90	50	May '57
Carter	Oklahoma	7,000	4.50	27	Jan. '58
Cleveland	Oklahoma	6,850	4.00	27	Jan. '58
Cleveland	Oklahoma	8,150	4.25	33	Dec. '57

from 68 per cent in 1935 to 93 per cent in 1957. Table 21–4 gives the monthly average of rotary rigs actually running in the United States and Canada.

The average contract costs of drilling has remained fairly constant throughout the 1935–57 period, as indicated by Table 21–2. However, the total cost per well has steadily increased. In 1935 the contract price per foot amounted to about 62.5 per cent of the total cost. In 1957 the contract price per foot amounted to about 32.8 per cent of the total cost of drilling and completing the average well. The increases in total cost are generally attributed to increased prices in steel tubular goods, materials, and labor and the more extensive completion practices presently employed.

TABLE 21-6
Drilling and Completion Costs, 1956
West Texas–New Mexico Region Development Oil and Gas

Fields	No. Wells	Average Cost/Well	Average Depth	Average Cost/ft.
N. Russell	4	$158,754	10,988'	$14.43
Denton (Wolf)	3	160,141	9,447	16.95
Denton (Devon.)	2	317,376	12,587	25.21
Payton (Devon.)	1	148,592	6,955	21.36
Blk. 12 (Ellen.)	2	219,843	10,935	20.10
Shafter Lake	5	52,761	4,587	11.50
Sand Hills (Tubb)	2	68,068	4,385	15.52
McElroy (Gray.)	1	30,576	3,076	9.94
Langlie-Mattix	3	39,618	3,479	11.39
Field Avgs.	9	$132,859	7,383'	$16.27
Oklahoma-Kansas Region				
Sholem Alechem	5	$ 62,611	5,752'	$10.89
Ready	1	48,302	4,586	10.53
Fox-Graham	9	76,405	5,601	13.64
Grand Valley	4	101,434	6,966	14.56
Field Avgs.	4	$ 72,188	5,726'	$12.40
Gulf Coast Region				
E. Fulton (Singles)	3	$ 71,358	7,580'	$ 9.41
E. Fulton (Duals)	3	96,038	7,970	12.05
Mustang Is.	30	100,969	7,524	13.42
Redfish Bay	9	111,816	8,172	13.68
N. Cankton	1	93,616	9,205	10.17
Field Avgs.	5	$ 94,759	8,090'	$11.75
Dallas Eastern				
Ruston	1	$153,089	9,295'	$16.47
Ada	1	81,293	7,583	10.72
S. Hallsville	2	71,200	6,948	10.25
Field Avgs.	3	$10,1861	7,942'	$12.48
All Regions				
Field Avgs.	21	$107,803	7,316'	$13.91

About the Figures

Many of the figures used in this book were prepared and furnished specifically for it. The following companies and organizations deserve special mention of appreciation for making available such illustrative material.

    A-1 BIT AND TOOL COMPANY
    ACME FISHING TOOL COMPANY
    BAROID DIVISION NATIONAL LEAD COMPANY
    S. R. BOWEN COMPANY
    CONTINENTAL-EMSCO COMPANY
    EASTMAN OIL WELL SURVEY COMPANY
    HALLIBURTON OIL WELL CEMENTING COMPANY
    HOUSTON OIL FIELD MATERIAL COMPANY
    HUGHES TOOL COMPANY
    IDECO
    MAGNET COVE BARIUM CORPORATION
    MARTIN DECKER CORPORATION
    MID-CONTINENT SUPPLY COMPANY
    NATIONAL SUPPLY COMPANY
    OIL WELL SUPPLY COMPANY
    REED ROLLER BIT COMPANY
    SCHLUMBERGER WELL SURVEYING CORPORATION
    SHAFFER TOOL COMPANY
    SPANG AND COMPANY
    UNIVERSAL ENGINEERING COMPANY
    WILLIAMS BIT AND TOOL COMPANY
    AMERICAN ASSOCIATION OF OIL WELL DRILLING CONTRACTORS
    AMERICAN PETROLEUM INSTITUTE
    SOCIETY OF PETROLEUM ENGINEERS OF AIME

Index

Air-borne magnetometer: 36
Air-gas drilling: circulation requirements, 271–79; practices, 304–306
Archimedes' principle: 213

Balling, of shale particles: 296, 305
Barite in drilling fluids: 83
Bassinger rotary percussion tool: 352
Bentonite: 80 ff.; chemical reactions, 84–87; yield of, 116–18
Bernoulli's theorem: 242, 273
Bits: factors governing use of rotary, 172; drag, 173–74; disk, 174–75; rolling cutter (toothed-wheel), 175–83; button, 179; pegleg, 182; watercourses, 182; jet, 182–83; Zublin, 183; Zublin differential, 183; diamond, 184–88; selection of rotary, 280–82
Block-and-tackle systems: 190, 192–96
Blow-outs caused by pressure reductions: 384–85
Blow-out preventer: 72
Boiler: rating, 134; horsepower, 134; fuel requirements, 135; water requirements, 136
Brakes: draw-works, 232; auxiliary, 233
Breakdown pressure: see pressure
British thermal unit: 132
Buoyancy: 213–16

Cable-tool operations: 324–30; general drilling procedure, 325–28; sand line, 327; simple bailer, 327; sand pump, 327; drilling cables, 328; drill string, 328; cable-tool bit, 328; comparison of drilling rates, 328–29; drilling motion, 329–30
Casing: casing-bit-size program, 121, 123; conductor, 122; oil-string, 122; functions and requirements of various strings, 406–407; properties and requirements, 407–10; design safety factors, 410–13; bursting strength, 413; biaxial stresses, 414–16; mathematical relation for collapse resistance, 416–18; graphical solution of collapse change points, 418–20; examples, 420–36; analytical solution of collapse change points, 436–39; landing practices, 439–42
Catheads: 232

Cement: temperature effects of setting, 58–59; requirements for oil wells, 443; composition of oil well, 444; specifications, 446; water-cement ratios, 447; effects of pressure and temperature, 448; bentonitic, 450; pozzolanic, 452; perlite, 452; diatomaceous earth, 453; gypsum, 454; resin, 454; diesel oil, 454; contamination, 454; effects of fresh water, 454; effects of drilling mud, 454; effects of mud chemicals, 455; effects of saline water, 456
Cementing: operations, 443; squeeze, 456; methods, 456; equipment, 458; casing centralizers, 458; wall cleaners, 458; scratchers, 458; guide shoes, 460; float joint, 460; float collar, 460; stage, 462
Centrifuges, mud: use of, 241
Chemicals: used in drilling fluids, 84–87
Chemical tests: of drilling fluids, 102–103
Circulation: systems, 236–44; normal, 236; reverse, 236
Circulation requirements: drilling fluid, 241–42; air-gas, 271–79
Colloids: 80 ff.; secondary, 82
Company rigs: 465
Conductor (casing): 406
Contract rigs: 465
Contracts, drilling: 478; turn-key, 479; footage rate, 480; day rate, 482
Core analysis: 378
Core flushing: 378; effects of pressure differential, 378; effects of permeability, 378; effects of rate of penetration, 378; effects of mud-circulation rate, 378; effects of reservoir-fluid viscosities, 379; effects of core size, 379; effects of water loss of the mud, 379; effects of properties of the rock, 379; effects of original fluid content, 379
Cores, preservation of: 370
Coring: 359; rotary, 359; punch-type, 359; Poor Boy or Texas-type, 360; conventional, 361; bits, 361; diamond, 323–24, 366; core barrels, 364; wire-line, 364; wire-line core barrel, 365; reverse circulation, 367; side-wall, 368; side-wall punch-type, 368; side-wall rotary-type, 368; side-wall percussion-type, 368; cable-tool, 368, 381; chip,

370; preservation of original fluid content, 380
Corrosion: of drill pipe, 169
Costs: 463; rig, 467; bits, 467; power plants, 470; air-gas drilling, 472; reducing drilling, 475; effects of drilling practices on, 477; comparison of drilling, 481; trends in drilling, 482; in various areas, 483, 485, 486
Crude oil classification: 16; paraffin-base, 16; asphaltic-base, 16; mixed-base, 16

Dárcy's Law: 18
Density of drilling fluids: 88–89
Derricks: 198–204; loading, 201; efficiency factor, 202
Diesel engine rating: 140
Dip needle: 34
Directional drilling: concepts of directional progress, 306; causes of hole deviation, 307; uses of, 315, 318; surveying instruments, 319–20; deflecting tools, 320–23
Dog-legs, or sudden bends: 315–17
Drake, Col. Edwin L.: 3
Drake Well: 3, 324
Draw works: 230
Draw-works clutch: 234
Drift diameter (of casing): 408
Drillability, changes in: 285
Drill collars: functions, 170; operating practices, 170–71; effect on bit performance, 171; sticking in key seats, 171–72
Drill pipe: description and use, 166–70; tool joints, 167–68; upset, 168; failures, 168–70
Drill-stem testing: 371; procedure, 372; pressure recorders, 373
Drilling: purpose of, 343; trends in development, 357–58
Drilling conditions, factors involved in: 172
Drilling fluids: temperatures and temperature gradients, 56–58; pressures, 70–74; filtration loss from, 73; functions, 76; descriptions, 77 ff.; oil-base, 79; oil-emulsion, 79; water-base, 80–87; properties, 87–103; mixing and weighting calculations, 103–18; circulation requirements, 241–42; effects on drill string, 294–96; hydraulic effects on drilling rate, 296–301; composition of, effect on drilling rate, 299–304; effects on bit performance, 348–51; fluid-drilling systems, 350–51; pressure surges and anomalies, 384–405; mud losses caused by pipe movement, 385–89
Drilling industry, scope of: 3–4
Drilling time: 468
Drilling weight: 281–88, 292–94

Economics: 463
Efficiency: mechanical, 129; thermal, 132; boiler, 135
Electric power: 141–51; generator selection, 143; controls, 146
Engine performance: 131
Eötvös torsion balance: 28
Equivalent diameter: 245

Fanning friction factor: 247–50
Fatigue, failure of drill pipe by corrosion: 169
Fatigue failures, causes of: 288
Filtration loss: quality of muds, 98–101; effects on drilling rate, 302–304
Fishing, causes of: 331
Fishing tools: classification of, 331; inside tools, 332; spear, 332; taps, 333; outside tools, 334; die collar, 334; overshot, 334; hydraulic pulling tools, 335; jars, 336; hydraulic jars, 336; knuckle joints, 338; washover equipment, 338; mechanical cutters, 338; chemical cutters, 338; shaped charge, 339; back-off shot, 339; free-point indicator, 340; milling tools, 340; junk basket, 340; magnetic tools, 340; shaped-charge fragmentizer, 341; safety joints, 341; cable-tool devices, 342
Flow calculations: through porous media, 17–21; viscous, 250–52, 255, 259; Bingham plastic, 253–54, 256–57; Ostwald-type, 254–59; turbulent, 263–69; bit nozzles, 265–67; air-gas, 271–79
Flow history of drilling fluids: 91
Flow patterns: 243–45
Flow pressure losses: 242–79
Formation damage: 73–74; by drilling mud, 78
Formation evaluation: 371
Fracturing, formation: 65–66
Free-piston engines: 152
Friction factor: 247–50

Gas drilling practices: 304–306
Gel strength of drilling fluids: 96–98
Geological time: 6
Geophysical prospecting: definition, 22; surface methods, 23; subsurface meth-

ods, 23; gravity methods, 23; torsion balance, 25, 27; pendulum, 28; magnetic, 32; seismic, 37; electrical, 40; geochemical, 40; radioactive, 40
Gravimeter corrections: 26; tidal, 26; free-air, 26; Bouguer, 26; terrain, 27; latitude, 27; regional, 27; isostatic, 27
Gravimeter data, interpretation of: 31
Gravimeters: definition, 29; stable, 29; Hartley, 29; Gulf, 30; unstable, 30; Thyssen, 30; Humble, 30–31
Gravity: API conversion formula, 17; Baumé conversion formula, 17
Grief stem: 163–64; see also kelly

Heads (fluid pressures): see pressures
Heat, transfer of: by conduction, 53–55, 59–60; Fourier's Law, 59; thermal conductivities of rocks, 59; by convection, 59–60; by radiation, 60; Stefan's Law, 60
Hoisting system: component parts, 189–90; horsepower, 191
Horsepower: 126; brake, 127; indicated, 127; friction, 127; rating, 129; -hour, 132; cable-tool hoisting, 154; cable-tool drilling, 154; hydraulic and mud pump, 269
Hydrafrac: see fracturing, formation

Inert solids used in drilling fluids: 83–84
Internal combustion engines: 137 ff.; spark-ignition engines, 137; compression-ignition engines, 137; Otto cycle, 138; diesel cycle, 139

Kelly: 163–64; kelly saver sub, 164; kelly cocks, 164
Key seats: origin and sticking of drill collars, 171–72
Kick's Law: 347

Laminar flow: 244
Leasing: 41
Liner: 407
Logging-cable formation tester: 376
Logs: temperature, 55, 58; driller's, 381; cuttings, 381; mud, 382; electrical, 382; radioactivity, 382; micro-, 382; later-, 382; induction, 382; microliter-, 382; drilling-time, 383
Lost circulation caused by pipe movement: 385–89
Lost circulation material: 83
Lubricator: 72–73

Magnetic data: temperature corrections, 34; terrain corrections, 34; normal corrections, 35; interpretation, 35
Magnetic field strength: 33
Magnetostriction-vibration drill: 354
Masts: 204
Mechanical advantage: 195
Metallurgical influences on drilling tools: 351
Milling: 162
Mobility of drilling fluids: 92
Mud, drilling: see drilling fluids
Mud guns: 236
Mud pits and tanks: 236–37

Oil and gas lease: 43, 49
Oil saturation: 379
Ostwald plastic fluid: 254–59
Ownership of land: 41
Ownership of oil and gas: 41; ownership-in-place theory, 41; non-ownership theory, 42

Payment of drilling charges: 478
Pellet-impact drill: 353–54
Pennington drilling tool: 352
Percussion: theories of rock failure, 348; rotary percussion tools, 352–53; pellet-impact drill, 353–54; magnetostriction-vibration drill, 354; sonic drill, 354–55
Permeability: 18–21; Klinkenberg effect, 18
Petroleum reservoirs: classification of, 7; structural trap, 7; stratigraphic trap, 7; combination trap, 9
Pilot testing of drilling fluids: 102–103
Plastic flow: 244
Plug flow: 244
Poiseuille's Law: 250–52
Power plants: rotary, 125; design of, 126; basic units, 126; cable-tool, 153
Power transmission: 155–60; gears, 155; clutches, 157
Pressure: geological influence on, 61; blow-outs, 61; liquid hydrostatic, 61–62; mud and cement slurries, hydrostatic, 63; overburden (earth pressure), 63–67; formation rupture (or breakdown), 65, 66; formation, rock or reservoir, 67–70; abnormal, 69, 70; drilling-fluid (bore-hole pressures), 70–74
Pressure–gel strength relationships: 389–93
Pressure losses: see flow calculations
Pressure surges and anomalies (in drilling

fluids): 384–405; blow-outs caused by, 384–85; mud losses caused by, 385–89; pressure–gel strength relations, 389–93; viscous-flow effects, 393; acceleration effects, 393–95; plastic- and turbulent-flow effects, 395–404
Professional organizations: 4
Prony brake: 128
Property, description of: 45
Pumps, mud: 237–39

Reamers: 188; use of, 315
Reaming: 162
Reservoir fluids: 14
Reservoir rocks: classification of, 11; fragmental, 12; chemical, 13; miscellaneous, 13
Reynolds number: 245–47
Rigs: portability, 125; depreciation, 476; in operation, 484; *see also* company rigs *and* contract rigs
Rittenger's Law: 346
Rock failure, theories of: 346–48
Rock properties: porosity, 13; fluid saturation, 14; permeability, 14; effects on bit performance, 343; hardness, 343–44; fracture characteristics, 344; plasticity, 344–45; strength and porosity, 345
Rotary drilling practices: 280–324
Rotary hose: 239
Rotary operations: 162
Rotary speeds: effects on drilling rates, 288–92; practices, 292–94
Rotary table: 164–66

Sand content of drilling fluids: 101–102
Schmidt-type vertical magnetic-field balance: 34
Seismic data, interpretation of: 39
Seismic equipment: 39
Seismic exploration: 37; refraction, 38; reflection, 38
Settling velocity: of rock particles in mud, 270; of rock particles in gas, 271–72
Shale shaker: 241
Slim-hole drilling: 471
Sonic drill: 354–55
Source rock: 5

Starch used in drilling fluids: 82
Steam engine: 133
Steam power, improving efficiency of: 137
Straight-hole drilling: 306–15
Substructures, derrick: 205–208
Surge chamber, at mud pump: 239
Swivels: 162–63, 240

Temperature: effects of earth, 53; formation, 53; geothermal gradient, 53–55; logs, 55–56, 58–59; drilling-fluid circulation, 56–58; transient, 56–58; effects of setting of cement, 58–59
Thixotropy: 96
Tool joints: 167–69
Torque: 130, 194
Turbines: 151
Turbo drills: 355–57
Turbulent flow: 244

Velocity distribution in viscous flow: 251–53
Viscosity: of drilling fluids, 89–96; Newtonian fluids, 89–90; non-Newtonian fluids, 90–92; Bingham plastic fluids, 91–92; plastic, 92, 95; Stormer, 92–94; Fann, 94–96; apparent, 95; Marsh funnel, 96; turbulent-flow, 260–62; effects of, on drilling rates, 302–304
Viscous flow: 244

Water-loss quality of muds: 98–101
Weight indicators: 208–13
Weighting material: used in drilling fluids, 83; calculations, 103–18
Weight on bit: *see* drilling weight
Wells: function of, 119; planning, 120
Wetting characteristics of drilling fluids: 296
Wire line: 216–30; construction of, 216; description, 217; diameter, measurement of, 217–18; service, evaluation of, 219–30; work, ton miles, drilling, 220; round trips, 221–26; running casing, 226; coring, 227; calculating length on a drum, 229
Work: 126

Yield of clays: 116–18
Yield point, Bingham: 92, 95